^{13}C NMR SPECTROSCOPY

A WORKING MANUAL WITH EXERCISES

MMI PRESS POLYMER MONOGRAPH SERIES
Edited by Hans-Georg Elias

This series will present international accounts of research and developments in the specialized areas of macromolecular chemistry.

Volume 1 Thermodynamics of Polymer Solutions
By Michio Kurata
Translated from the Japanese by Hiroshi Fujita

Volume 2 Practical Macromolecular Organic Chemistry
By Dietrich Braun, Harald Cherdron, and Werner Kern
Translated from the German by Kenneth J. Ivin

Volume 3 ^{13}C NMR Spectroscopy: A Working Manual with Exercises
By Eberhard Breitmaier and Gerhard Bauer
Translated from the German by Bruce K. Cassels

In preparation:

Polymer Analytics
By Martin Hoffmann, Herbert Krömer, and Rainer Kuhn
Translated from the German by Horst G. Stahlberg

ISSN: 0275-7265

The publisher will accept continuation orders for this series, which may be cancelled at any time and which provide for automatic billing and shipping of each title in the series upon publication. Please write for details.

^{13}C NMR SPECTROSCOPY
A WORKING MANUAL WITH EXERCISES

Eberhard Breitmaier and Gerhard Bauer
University of Tübingen
Tübingen, Federal Republic of Germany

Translated from the German by

Bruce K. Cassels
Université de Paris XI
Paris-Sud
Orsay, France

harwood academic publishers
chur · london · paris · new york

Published under license by OPA Ltd. for
Harwood Academic Publishers GmbH.

Harwood Academic Publishers

Poststrasse 22
7000 Chur
Switzerland

P.O. Box 197
London WC2 4DL
England

58, rue Lhomond
75005 Paris
France

P.O. Box 786
Cooper Station
New York, NY 10276
United States of America

Originally published in German in 1977 as ^{13}C-NMR-Spektroskopie: Eine Arbeitsanleitung mit Übungen by Georg Thieme Verlag Stuttgart.

Library of Congress Cataloging in Publication Data

Breitmaier, E.
^{13}C NMR spectroscopy.

(MMI Press polymer monograph series; v. 3)
Translation of: ^{13}C NMR Spektroskopie.
Bibliography: p.
Includes index.
1. Nuclear magnetic resonance spectroscopy—Handbooks, manuals, etc.
2. Carbon—Isotopes—Spectra—Handbooks, manuals, etc. I. Bauer, Gerhard.
II. Title. III. Series.
QD96.N8B7313 1984 543′.0877 84-8942
ISBN 3-7186-0022-6

ISBN 3-7186-0022-6. ISSN 0275-7265. Printed in Great Britain by Bell and Bain Ltd., Glasgow.

Contents

Preface

In the last 15 years, carbon-13 resonance has proved to be a powerful method for the determination of structure, especially in organic chemistry. As a result, ^{13}C NMR spectroscopy is currently part of the curriculum in analytical organic chemistry at many important universities worldwide. This book will enable both professional chemists and advanced students to use ^{13}C NMR spectra in the solution of chemical structural problems.

The text consists of two parts. The first part (Chapters 1–9) is intended as a practical introduction to ^{13}C NMR. A short explanation of the essential basic concepts (Chapter 1) is followed by a description of the methods of measurement, insofar as they are valuable and routinely applicable to spectral assignments (Chapters 2 and 3). After a brief discussion of chemical aids for assignments (Chapter 4) and some references to solvents and standards (Chapter 5), the most important relationships between ^{13}C shifts and structural features are reviewed (Chapter 6).

Chapters 7 and 8 deal with the practical analysis of ^{13}C shifts and the coupling constants of basic classes of organic compounds. The first part ends with a discussion of the measurement and use of ^{13}C spin-lattice relaxation times (Chapter 9).

The second part (Sections A and B) is a collection of 76 exercises and solutions representing as many classes of organic compounds as possible. The corresponding spectra were prepared in such a way that they can be easily interpreted. These exercises offer the reader the opportunity to practice spectral interpretation using assignment techniques, to learn to recognize functional groups and structural features, and, finally, to go through some examples of structural elucidation.

We deliberately dispensed with any extensive discussion of the fundamentals and dynamic effects (temperature dependence) of NMR spectroscopy because enough books and reviews are available on these subjects. We have attempted to write a text that was as exact, easy to understand, meaningfully illustrated, and serviceable as possible. However, because nothing is free from errors and omissions, we shall appreciate suggestions for its improvement.

We are grateful to the following for making available some valuable compounds: Professor Drs. E. Graf, G. Jung, K. A. Kovar, R. Tschesche, F. Vogtle, and W. Voelter; Docent Dr. H. Schott; Drs. G. Hambrecht, I. Ruppert, and F. Ciper; and Messrs. H. Behr, W. Best, W. Bieber, R. Hanke, R. Kruse, C. Skotsch, and G. Wolff. We thank Ms. E. Ocherbeck and Ms. R. Scheffler for their careful typing of the manuscript and Dr. G. Zartner, Ms. G. V. Chamier, and Mr. Th. Dambacher.

E. Breitmaier
G. Bauer

PART I

CHAPTER 1

Basic Principles and Quantities of Nuclear Magnetic Resonance Spectroscopy

1.1 Chemical Shifts

Nuclear magnetic resonance spectroscopy investigates atomic nuclei possessing a mechanical spin, the nuclear spin p, such as proton ^{1}H or the stable isotope of carbon, ^{13}C. These nuclei have a magnetic moment μ because of their mechanical spin and nuclear charge. Its value is γ times the nuclear spin p, where γ is a nuclear constant called the gyromagnetic ratio:

$$\mu = \gamma p \tag{1.1}$$

When a spinning atomic nucleus is placed in a constant (i.e., static) magnetic field, its magnetic vector $\vec{\mu}$ precesses around the direction of the magnetic field vector H_0 (see Fig. 1.1) at a frequency that increases if the static magnetic field H_0 is adjusted to a higher value (Eq. 1.2):

$$\nu = \frac{\gamma H_0}{2\pi} \tag{1.2}$$

The precession or Larmor frequency ν_0 of an atomic nucleus lies in the radio frequency range. When $H_0 = 2.114$ Tesla (T), ν_0 is about 90 Megahertz (MHz) for protons and about 22.63 MHz for ^{13}C.

The importance of nuclear magnetic resonance spectroscopy as a tool for elucidating structures is based on the fact that the precession frequency of a spinning nucleus bonded in a molecule depends on its molecular environment. This is called the **chemical shift** of the Larmor frequencies. The shift arises from the shielding of an atomic nucleus by intramolecular magnetic fields. The amount of shielding is proportional to the external field H_0 and is de-

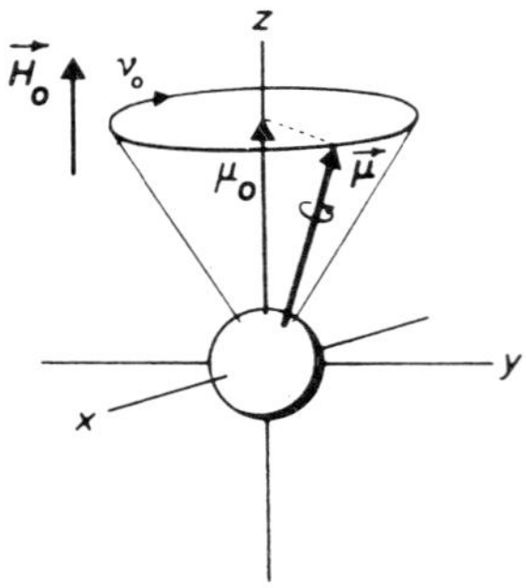

Fig. 1.1. Nuclear precession in the x, y, z system of coordinates. The z axis is defined as the direction of the magnetic field H_0.

scribed by the **magnetic shielding constant** σ_i, which has a specific value for each kind of nucleus. In most diamagnetic organic compounds, the shielding field $\sigma_i H_0$ counteracts the external field H_0. The nucleus i then experiences a somewhat weaker field ($H_0 - \sigma_i H_0$); therefore, Larmor's equation (1.2) must be corrected to Eq. (1.3):

$$\nu_{0_i} = \frac{\gamma}{2\pi} H_0 \quad (1 - \sigma_i) \qquad (1.3)$$

Because according to equations (1.2) and (1.3) the Larmor frequencies depend on the strength of the magnetic field H_0, there is no absolute scale of chemical shifts. Thus it is convenient to choose a reference substance with an extremely low Larmor frequency and to measure the **frequency difference** $\Delta\nu$ (in Hz) relative to this standard. The Larmor frequencies of the methyl protons and carbon atoms of tetramethylsilane [TMS, for short; formula $(CH_3)_4Si$] have proved to be useful as references in proton and ^{13}C nuclear magnetic resonance spectroscopy because most ^{1}H and ^{13}C nuclei in organic compounds precess faster than in TMS. According to Eq. (1.3), the chemical shift scale of frequencies obtained in this way expands with the strength of the magnetic field H_0. To obtain values that do not depend on the field of the particular instrument, the frequency differences $\Delta\nu$ (in Hz) are divided by the TMS Larmor frequency, which is proportional to the strength of the magnetic field H_0—the so-called operating frequency of the spectrometer. The δ scales thus arise for ^{1}H and ^{13}C chemical shifts:

$$\delta = \frac{\text{frequency difference } \Delta\nu \text{ relative to TMS [Hz]}}{\text{operating frequency of the spectrometer [MHz]}} \qquad (1.4)$$

As 1 Hz $= 10^{-6}$ MHz, the unit of the scale is 10^{-6} or one part per million (ppm). ^{1}H or ^{13}C Larmor frequencies larger than the precession frequencies for

TMS correspond to positive values, but according to Eq. (1.3), they also imply weaker shielding fields $\sigma_i H_0$. Therefore, *large ppm values* mean *weak shielding* ("downfield" resonances) and *vice versa*.

1.2 Coupling Constants

The spectrum of ^{1}H or ^{13}C Larmor frequencies of a compound is called its ^{1}H or ^{13}C nuclear magnetic resonance spectrum (NMR spectrum). The signals (resonances), however, are not usually simple (singlets) but rather **multiplets**. In a ^{13}C spectrum, the signals for CH, CH_2, and CH_3 carbon atoms, for instance, are split into *doublets* (intensity ratio 1:1), *triplets* (1:2:1), and *quartets* (1:3:3:1). With n neighboring protons, then, one sees an $(n + 1)$-fold splitting. Protons on neighboring carbon atoms can lead to further but finer *long-range splitting* (see Fig. 1.2).

The signal splittings are a consequence of indirect spin-spin coupling. A nucleus in a magnetic field H_0 is not only exposed to intramolecular shielding; it also experiences the smaller magnetic field of a neighboring spinning nucleus. According to the direction in which this neighboring nucleus precesses, the H_0 field at the observed nucleus can be weakened or strengthened somewhat. If the neighboring nucleus can precess in several orientations, the signal caused by the observed nucleus splits. The breadth of the splitting depends only on the additional field caused by the neighboring nucleus, not on the external field H_0 or the shielding field $\sigma_i H_0$. These H_0 field-independent separations of the lines of a multiplet are called **coupling constants** (J) and are measured in Hz.

One can infer how many protons (one, two, or three) are bonded to a ^{13}C nucleus from the multiplicity of its multiplet (e.g., doublet, triplet, quartet, in the ^{13}C spectrum). Above all, the coupling constant reflects the kind of bonds intervening between a proton and a ^{13}C nucleus (σ, π, substituents, and their interactions), because indirect spin-spin coupling is transmitted by the bonding electrons.

The determination of the chemical shifts, which (in the frequency scale) depend on the external field, and of the coupling constants, which do not, is best understood looking at spectra such as those recorded in Figure 1.2.

1.3 Spin-Lattice Relaxation Times

The magnetic vector $\vec{\mu}$ of a nucleus (see Fig. 1.1) can precess only at certain angles to the direction of the magnetic field vector H_0 (directional quantiza-

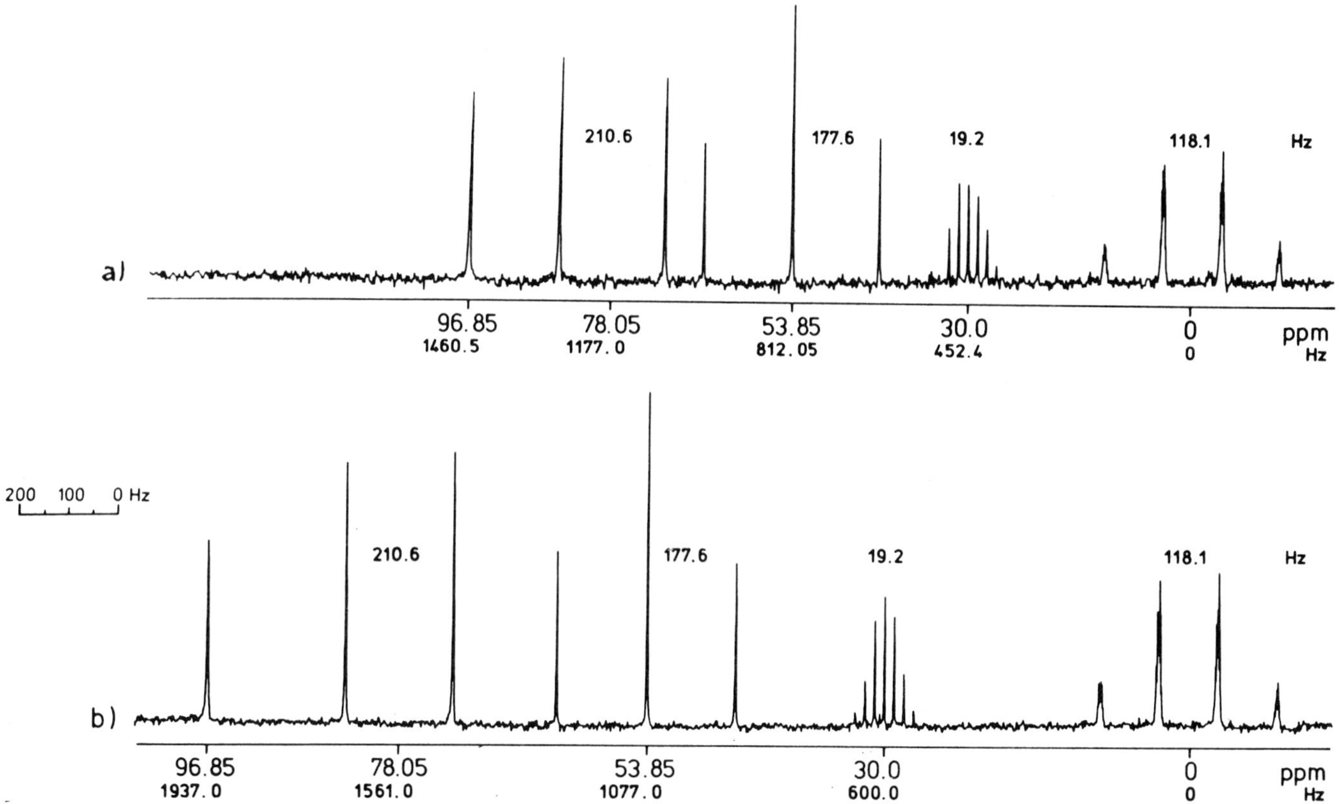

Fig. 1.2. ^{13}C NMR spectrum of a mixture of tetramethylsilane (0 ppm), hexadeuterioacetone (CD_3 signal at 30.0 ppm), dichloromethane (53.85 ppm), chloroform (78.05 ppm), and carbon tetrachloride (96.85 ppm). A: Magnetic field strength = 1.4 T; operating frequency = 15.08 MHz. B: Magnetic field strength = 1.87 T; operating frequency = 20.115 MHz. Both spectra were recorded on the same frequency scale. Although the shifts in spectrum B expand by a factor of 4/3, the carbon-proton and carbon-deuterium coupling constants remain the same.

tion). The number m of these orientations is given by the spin quantum number I, according to Eq. (1.5):

$$m = 2I + 1 \tag{1.5}$$

Therefore, for nuclei with $I = ½$—e.g., ^1H, ^{13}C, ^{15}N, ^{19}F, and ^{31}P—there are only two possible orientations for nuclear precession: around the direction of the magnetic field vector $\vec{H}_0$ or around the opposite direction (see Fig. 1.3A). Both orientations differ in energy and thus correspond to two nuclear energy states (see Fig. 1.3B). Precession around the vector $\vec{H}_0$ involves less energy and is preferred. Consequently, one must supply energy to a nucleus precessing around $\vec{H}_0$, and thus excite it, if it is to invert its precession orientation. As everyone knows, force—and therefore energy—is required to turn a compass needle away from the North-South direction.

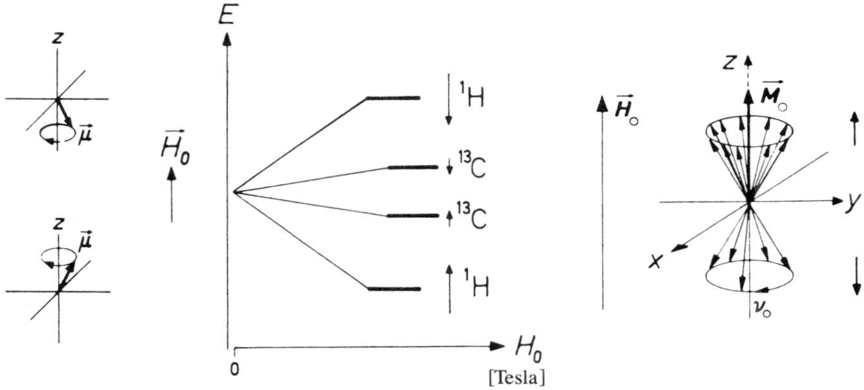

Fig. 1.3. a: For nuclei with $I = 1/2$, there are two precession orientations relative to the magnetic field $\vec{H}_0$ (nuclear energy states). b: The energy difference between the nuclear energy states increases with the strength H_0 of the magnetic field. It is larger for nuclei with greater magnetic moments (^1H). c: More nuclei occupy the lower energy level. At equilibrium, the excess magnetization $\vec{M}_0$ results.

The population of both nuclear energy levels in the magnetic field $\vec{H}_0$ follows a Boltzmann distribution: The larger the energy difference between both levels, the more crowded the low-energy state is. Thus more nuclei will precess in the direction of the field vector than against it, but with a different phase (see Fig. 1.3C). The equilibrium magnetization $\vec{M}_0$ along the z axis results from this excess and from the even distribution of the magnetic vectors on both precession cones. Yet, there is no magnetization in the x, y plane.

In the nuclear resonance experiment, transitions between both energy levels are induced. The equilibrium of the nuclei in the $\vec{H}_0$ field is disturbed, possibly until an equal distribution between both states is reached, so that no magnetization remains along the $\vec{H}_0$ field direction (z axis). The excited nuclei then transfer their excitation energy to the environment, the "lattice," to re-establish equilibrium. The Boltzmann distribution of the nuclear spins is set up again, and the equilibrium magnetization M_0 regenerates. This process is called **spin-lattice relaxation**. Assuming a constant increase of the magnetization M_z along the direction of the $\vec{H}_0$ field ("longitudinal magnetization") with a time constant T_1, the following differential equation can be written for the spin-lattice relaxation:

$$\frac{dM_z}{dt} = -\frac{M_z - M_0}{T_1} \qquad (1.6)$$

The time constant T_1 is the time after which a spin system is still M_0/e away from complete relaxation (see Fig. 1.4). It is called the **spin-lattice relaxation time.** Equilibrium magnetization also builds up in a non-magnetized sample via spin-lattice relaxation as soon as it is put in the magnetic field H_0 for a nuclear resonance measurement.

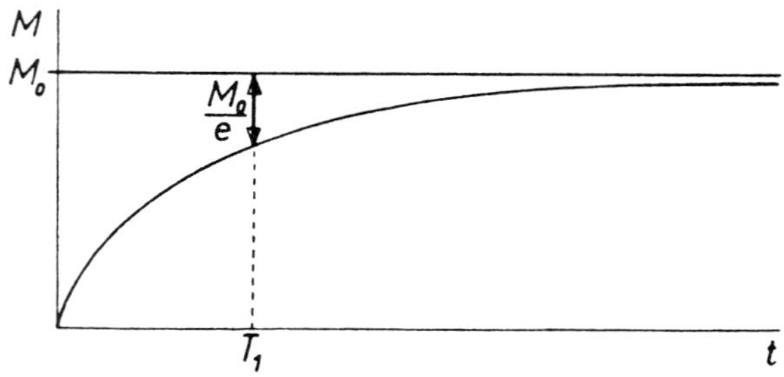

Fig. 1.4. Time course of spin-lattice relaxation.

Whereas chemical shifts and coupling constants lead mainly to conclusions concerning molecular constitution (molecular statics), the spin-lattice relaxation times T_1 reflect the mobility of molecules in solution (molecular dynamics). The knowledge of relaxation times in ^{13}C NMR spectroscopy is often not only a valuable aid for signal assignments but is also useful when optimizing the conditions under which the spectra are recorded.

CHAPTER 2

Measurement of ^{13}C-NMR Spectra

2.1 The Sensitivity Problem

The only carbon isotope with a nuclear spin, $^{13}_{6}C$, has a natural abundance of barely 1.1%. Its magnetic moment is merely one-fourth of the proton's moment [$\gamma_C = 1/4\ \gamma_H$; Eq. (1.1)], and the same is true for the excitation energy of ^{13}C nuclei. The population difference between the nuclear magnetic energy levels, however, also depends on this energy difference (see Fig. 1.3), and so does the intensity of the NMR signal. Both its low natural abundance and its small magnetic moment make carbon-13 considerably less sensitive (by a factor of 5,700) than the proton as an NMR probe. Therefore, more sensitive techniques had to be developed before ^{13}C resonance could become useful as a routine method of elucidating chemical structure. The most effective methods of improving the sensitivity of the ^{13}C NMR experiment are the pulse-Fourier transform technique and proton broad-band decoupling, as discussed in the following sections.

2.2 The Pulsed Nuclear Magnetic Resonance Experiment

The recording technique that is still often used in proton resonance—the continuous scanning of the chemical shift range with a radio frequency source and a receiver—is known as the **frequency sweep** or **continuous wave** mode [1,2,3]. This technique operates monochromatically: i.e., at any instant during the recording of the spectrum, excitation and signal reception occur at a single frequency **(single channel excitation).** The recording time is correspondingly long. For example, it takes 500 seconds to record a ^{1}H NMR spectrum spanning 10 ppm (1,000 Hz) at the usual scanning rate of 2 Hz/second.

The measurement time would be considerably shorter if all possible transitions of a nucleus were stimulated at the same time (polychromatically, **multiple channel excitation**). It can be shown [4] that a radio frequency pulse of duration t_p (a few microseconds) includes other frequencies in addition to the carrier frequency ν_1: the so-called Fourier components, which extend over the range $\nu_1 - 1/t_p$. It therefore behaves like a multiple-channel source.

When a radio frequency pulse acts on an assembly of nuclei at equilibrium with the static H_0 field (see Fig. 1.3 C), the precession double cone is rotated through a certain angle. This twist or flip angle can be adjusted by the duration of the pulse, or **pulse width** t_p. For instance, if the pulse width is adjusted so that the precession orientation of all the nuclear spins in Figure 1.3 C is rotated through 90°, one speaks of a *90° pulse*. After this pulse has ended, the excess magnetization M_0 lies at right angles to its normal position (see Fig. 2.1 B). The nuclear spins precess as usual around the direction of the H_0 field on the tilted double cone but *in phase* (Fig. 2.1 B). The transverse magnetization set up by the 90° pulse accordingly rotates in the x, y plane. It changes with time and thus induces an alternating voltage, the NMR signal, in the spectrometer's receiving coil, which is arranged, for example, along the y axis. This process of the nuclear resonance experiment is called **nuclear induction.**

Once the pulse is over, the nuclei lose their forced phase relationship via energy exchanges (collisions). They finally attain an even distribution on both precession cones, regardless of whether the Boltzmann distribution among both states is restored and, therefore, regardless of whether spin-lattice relaxation is complete. The divergence of the nuclear spins, however, means that the transverse magnetization, M_y, fans out into components in the x, y plane so that the resultant is finally zero (see Fig. 2.1 D).

This decay of M_y, which is assumed to be constant, is called **spin-spin relaxation.** The process is characterized by a time constant, T_2, the spin-spin relaxation time, and is described by the following differential equation:

$$\frac{dM_y}{dt} = -\frac{M_y}{T_2} \qquad (2.1)$$

After the end of the exciting pulse, spin-spin relaxation can be observed as an exponential, and therefore continuous decay of the nuclear induction current in the receiver coil of the spectrometer (see Fig. 2.1 E). This is known as **free induction decay** (FID). The spin-spin relaxation time, T_2, is the time elapsed until the induction current has decreased to 1/e of its initial amplitude.

Spin-spin relaxation of a nucleus can proceed faster than its spin-lattice relaxation ($T_1 > T_2$). The opposite ($T_2 > T_1$) is not possible: The equilibrium magnetization of a nucleus, M_0, is the vector sum of longitudinal and transverse magnetizations so that $M_o = M_z + M_y$. Once spin-lattice relaxation is complete, $M_z = M_0$, M_y must be zero by then at the latest; therefore, spin-spin relaxation also must be complete.

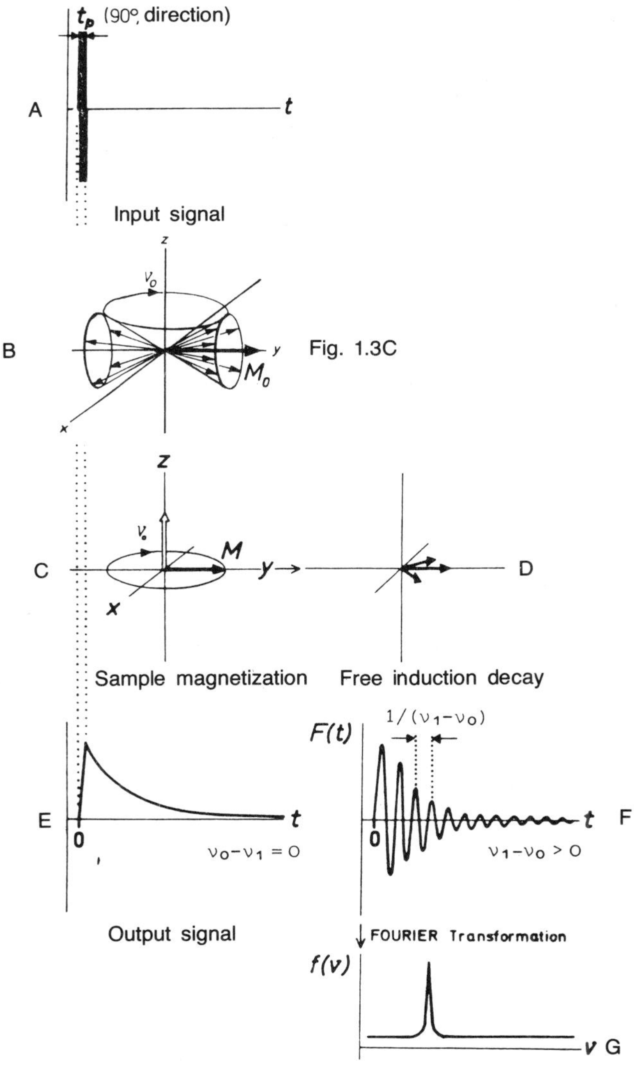

Fig. 2.1. Diagram of pulse-Fourier transform NMR spectroscopy: a radio frequency pulse of a certain width t_p with its magnetic field vector H_1 along the x axis (A) rotates all the nuclear spins through 90° (B). The magnetic moment vectors then precess in phase around the z axis, so that their resultant—the transverse magnetization set up by the pulse—rotates in the z, y plane (C) and, as a consequence of spin-spin relaxation, decays to zero (D, free induction decay). The rotating and collapsing transverse magnetization induces an exponentially decaying voltage in the receiver under resonance conditions ($\nu_1 - \nu_0 = 0$)(E). If the pulse's carrier frequency and the Larmor precession of the excited nucleus do not resonate ($\nu_1 - \nu_0 \neq 0$), an exponentially damped sinusoidal oscillation is received (F) with maxima separated by $1/(\nu_1 - \nu_0)$. The Fourier transform of this time function $F(t)$ is the frequency function $f(\nu)$, the NMR spectrum (G).

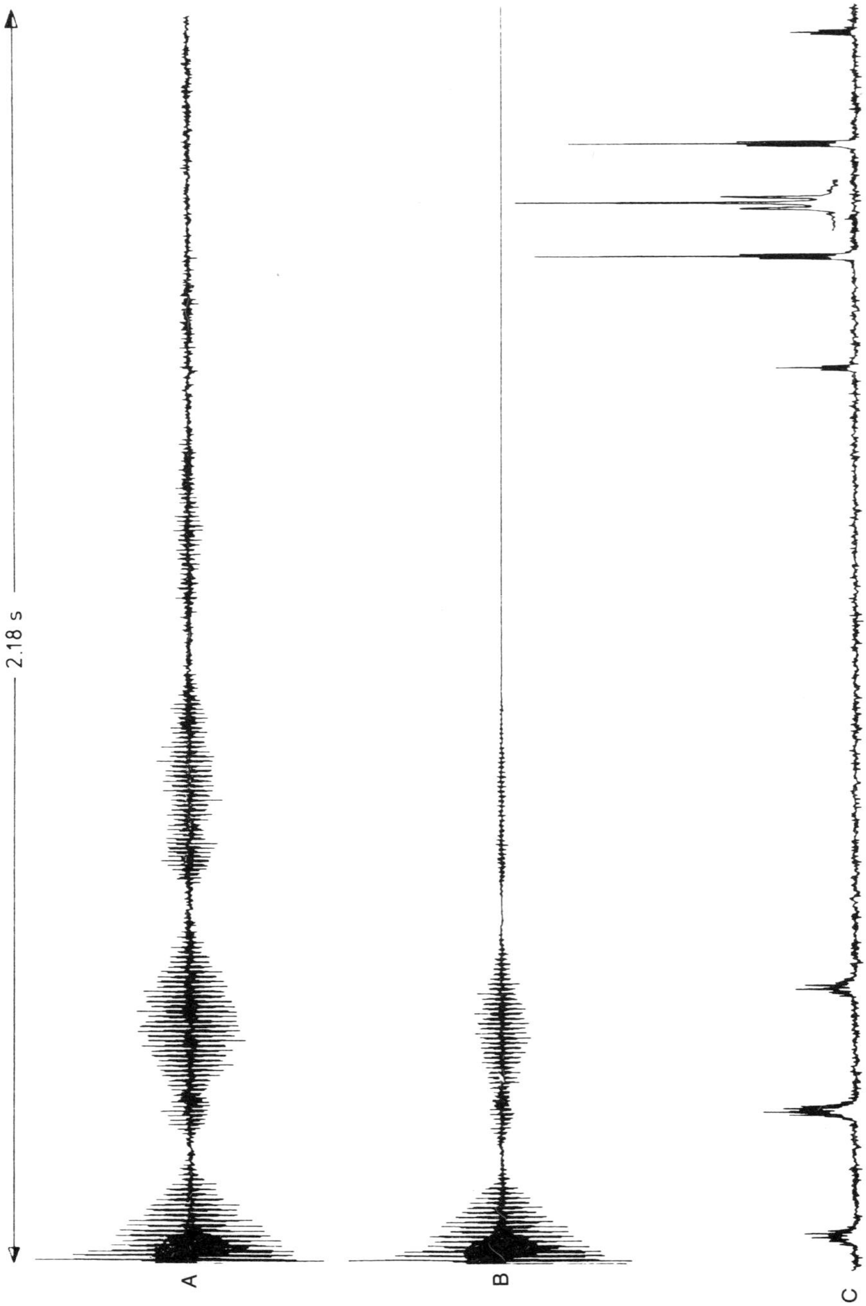
2.18 s
A
B
C

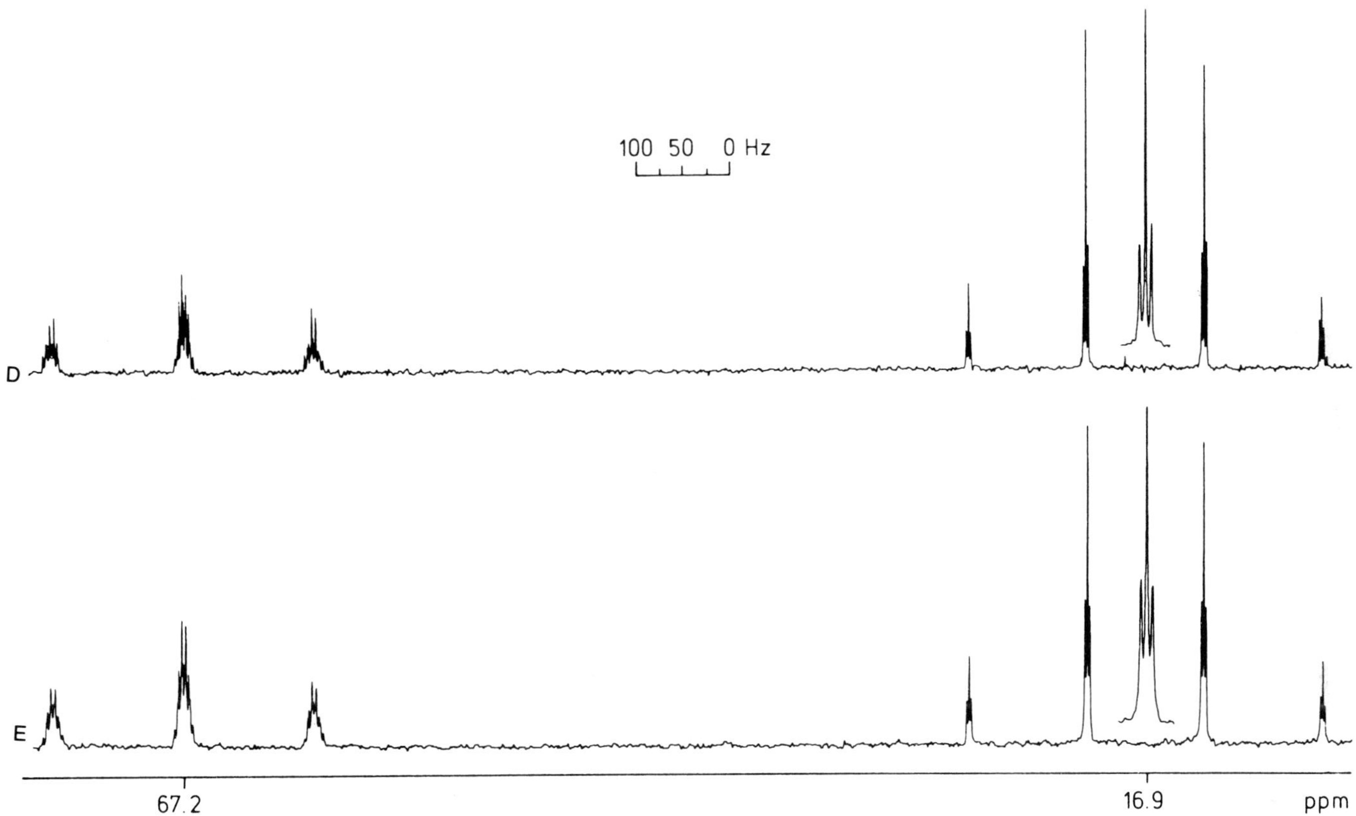

Fig. 2.2. Pulse interferogram and ^{13}C NMR spectra of diethyl ether, 95% in hexadeuterioacetone, 30° C, 75 accumulated scans. A: Interferogram without exponential multiplication. B: Interferogram with exponential multiplication for a line broadening of 0.5 Hz. C: ^{13}C NMR spectrum from interferogram A. D: ^{13}C NMR spectrum from interferogram A, but after noise smoothing. E: ^{13}C NMR spectrum from interferogram B without noise smoothing.

2.3 Pulse-Fourier Transform Method

If a radio frequency pulse (see Fig. 2.1 A) acts on a magnetized sample placed in the field, H_0, of an NMR spectrometer (see Fig. 1.3 C), a nuclear induction voltage builds up in the receiver coil (see Figs. 2.1 E and F). After the pulse has ended, the receiver records the decay of the induction voltage, or FID. If the carrier frequency, ν_1, of the pulse and the Larmor frequency, ν_0, of the excited nucleus are the same ($\nu_1 = \nu_0$, resonance situation), the FID signal is an exponential curve (Fig. 2.1 E). If the frequencies are different ($\nu_1 - \nu_0 = \Delta\nu$, "off-resonance" situation), the receiver records an exponentially decaying sinusoidal oscillation with maxima separated by $1/\Delta\nu$ (Fig. 2.1 F), and the frequency difference, $\Delta\nu$, can be determined. If several different nuclei of a kind are excited simultaneously, as is normally the case in polyatomic molecules, the decaying individual oscillations overlap, producing a **pulse interferogram** that is difficult to analyze (see Fig. 2.2 A) and is a function of time, *F(t)*. This interferogram can be converted by Fourier transformation into a function of frequency, $f(\nu)$, the NMR spectrum familiar to chemists (see Fig. 2.1 G) [4,5]. With the aid of a digital computer, usually built into the spectrometer, the Fourier transformation is achieved within a few seconds.

Although in the frequency sweep technique, the alternating electromagnetic field (the radio frequency) used to excite the nuclei is only a few milligauss strong—higher transmitter power levels lead to saturation effects so that the signals disappear—the radio frequency pulses are much stronger (several gauss). To conclude, the observed signals are considerably more intense. Furthermore, excitation and recording of the interferogram can be accomplished in less than a second. Hence, within the time required to record a conventional NMR spectrum, several hundred FID signals (scans) can be added coherently to each other (**interferogram accumulation**), averaging out the noise (statistical fluctuations). In doing so, the intensity of the NMR signals increases with the square root of the number of added individual measurements: for instance, 100 accumulated scans lead to a tenfold enhancement of sensitivity when compared with a single-scan experiment. Thus the advantages of the pulse-Fourier transform method (PFT-NMR) over the frequency-sweep technique are that the signals are stronger and the recording times are shorter. The net result is that for the same recording time, the PFT technique is about 100 times more sensitive than conventional NMR.

It is important to set the pulse carrier frequency, ν_1, outside the shift range to be measured, $\Delta = \nu_{0_{max}} - \nu_{0_{min}}$, so that $\nu_1 > \nu_{0_{max}}$ (see Fig. 2.3 A). If ν_1 lies within the shift range, Δ, all the NMR signals at frequencies $\nu_{0_i} > \nu_1$ are folded over on the spectrum by the amount $\nu_{0_i} - \nu_1 = \Delta\nu$ (see Fig. 2.3

B). Folded signals are recognizable in the absorption spectrum as phase distortions, which cannot be corrected (Fig. 2.3 B).

Of course, the pulse carrier frequency can also be set in the middle of the shift range to be measured. In this way, one half of the spectrum, Δ/2, is folded back on the other. This can be compensated for, however, by modifying the computer program for the Fourier transformation so that the spectral signals appear in the correct sequence. By measuring both halves of the spectrum, Δ/2, at the same time, one gains up to 40% ($> \sqrt{2}$) in sensitivity. **Quadrature detection** *(QD),* which can be used in addition to the PFT method, is based on this.

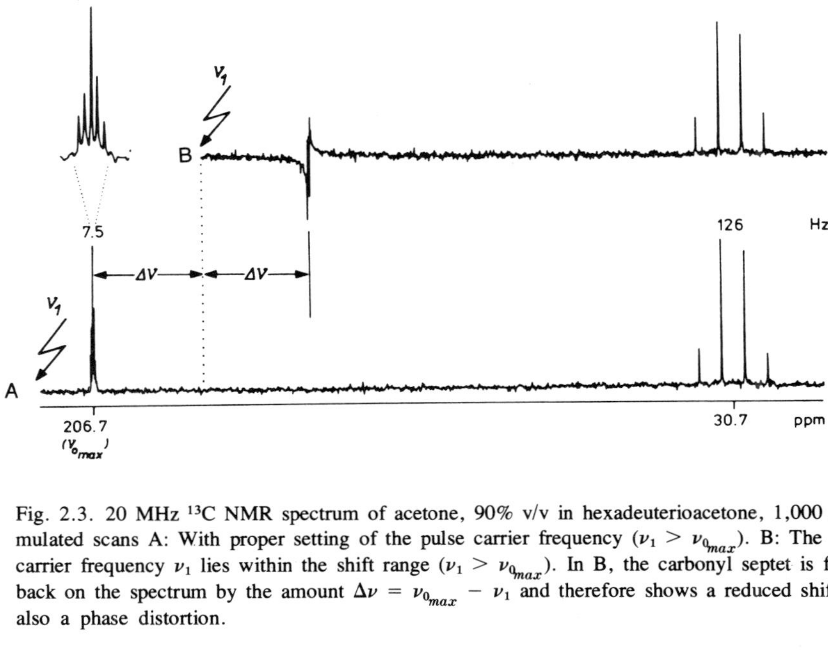

Fig. 2.3. 20 MHz ^{13}C NMR spectrum of acetone, 90% v/v in hexadeuterioacetone, 1,000 accumulated scans A: With proper setting of the pulse carrier frequency ($\nu_1 > \nu_{0_{max}}$). B: The pulse carrier frequency ν_1 lies within the shift range ($\nu_1 > \nu_{0_{max}}$). In B, the carbonyl septet is folded back on the spectrum by the amount $\Delta\nu = \nu_{0_{max}} - \nu_1$ and therefore shows a reduced shift and also a phase distortion.

2.4 Exponential Weighting and Smoothing

Another way to enhance the sensitivity (the **signal-noise** ratio) is to multiply the pulse interferogram by an exponential function, e^{-at}, before carrying out the Fourier transformation. The gain in sensitivity due to *exponential weighting,* though, is achieved at the expense of resolution (see Fig. 2.2 A, C and B, E).

Finally, the noise itself, i.e., the fluctuations of the spectrum's base line, can be smoothed. To do this, the first and the last of three neighboring extreme values of the noise are joined. With this *three-point smoothing,* there is no loss in resolution (see Fig. 2.2 C, D), but very weak signals may be averaged out.

References

1. H. Friebolin et al., *NMR-Spektroskopie, Eine Einführung mit Übungen,* Verlag Chemie GmbH, Weinheim, 1974, Taschentext Bd. 15.
2. H. Günther, *NMR-Spektroskopie, Eine Einführung,* Georg Thieme Verlag, Stuttgart, 1973.
3. H. Suhr, *Anwendungen der Kernmagnetischen Resonanz in der Organischen Chemie,* Springer Verlag, New York, 1965.
4. T. C. Farrar, E. S. Becker, *Pulse and Fourier Transform NMR,* Academic Press, New York, 1971.
5. J. T. Clerc, E. Pretsch, S. Sternhell, *^{13}C-Kernresonanzspektroskopie,* Akademische Verlagsgesellschaft, Frankfurt am Main, 1973.

CHAPTER 3

Proton Decoupling in ^{13}C NMR Spectroscopy

3.1 Purpose and Mode of Operation

Because of spin-spin coupling with bonded and with more distant nuclei, usually protons, the intensity of a carbon signal is distributed among several multiplet lines (compare Section 1.2.). Signal splitting, added to the unfavorable nuclear properties of carbon-13, thus further decreases the sensitivity of the ^{13}C experiment. Furthermore, the multiplets may overlap in molecules with several kinds of carbon, which complicates signal interpretation.

Signal splitting of the nucleus to be measured, A, is observed only if the coupled nucleus, X, maintains its orientation with respect to H_0 for a long enough period [1,2,3]. Its lifetime, τ, in a nuclear energy state (see Fig. 1.3 B) must be larger than the reciprocal, $1/J_{AX}$, of the coupling constant. If its lifetime is too short ($\tau < 1/J_{AX}$) and nucleus X therefore alters its precession orientation too rapidly, its time-averaged contribution to the field is zero. The splitting of the signal owing to nucleus A then breaks down, and the multiplet lines join together, giving one signal. **Spin decoupling** is achieved by irradiating the sample with a second frequency apart from the observation frequency, in resonance with the Larmor frequency of a bound proton: e.g., (^{13}C {^1H} **double resonance**). (The notation ^{13}C {^1H} NMR is used for all measurements when ^{13}C resonances are observed while decoupling the protons from them.) The proton then alters its precession orientation as in an NMR experiment, oscillating between its energy states so that its time-averaged contribution to the field becomes zero. The original ^{13}C-^1H multiplet related to the bonded carbon collapses to a singlet in which the intensities of all the multiplet lines are accumulated. This procedure, called **proton decoupling,** enhances the sensitivity of ^{13}C resonance and simplifies the spectrum.

The sensitivity enhancement of the proton-decoupled ^{13}C signal is not only a consequence of the multiplet's collapse, but also of the **Nuclear Overhauser Effect** (NOE) [4]. This effect results from the excitation of protons in the decoupling experiment, which raises the population of the energy-poorer ^{13}C levels over their Boltzmann distribution values. In this way, the sample in the H_0 field becomes more strongly magnetized (see Fig. 1.3 C), so that the ^{13}C signals gain intensity. This enhancement can attain a factor of two; thus a

CH doublet with the relative intensity ratio 1:1 becomes a singlet with intensity 4.

3.2 Broad-Band Proton Decoupling with High Transmitter Power

The most effective way to simplify a molecule's ^{13}C NMR spectrum is to decouple all its protons simultaneously. To do so, one cannot decouple with a single frequency but instead must use a frequency band comprising the entire proton shift range (see Fig. 3.1 A). The frequency band required is generated by modulating the decoupler frequency, which should lie as near as possible to the middle of the proton shift range, with low frequency noise (see Fig. 3.1 B). This decoupling procedure is known as **broad-band** or **noise decoupling.**

A broad-band proton-decoupled ^{13}C NMR spectrum obtained with a sufficiently strong decoupling power contains, at most, n singlets for the molecule's n kinds of carbon atom. At most, because some of these nuclei may be chemically equivalent (isochronous) because of molecular symmetry, among other things. Nine single signals are recognizable in the broad-band, proton-decoupled ^{13}C NMR spectrum of camphor for the ten chemically different carbon atoms in the molecule (see Fig. 3.2 A). The signal intensities, however, cannot be compared directly because the individual nuclear Overhauser intensity enhancement for each carbon signal may be different. If several pulse interferograms are accumulated to improve the signal-noise ratio, signal intensity is also affected by the spin-lattice relaxation time (T_1): Because slowly relaxing nuclei cannot follow an excessively fast sequence of pulses, a reduced stationary magnetization is set up, compared with the equilibrium value, M_0, and the signal is therefore suppressed. A quantitative interpretation of broadband decoupled ^{13}C NMR spectra—e.g., by using integration, as is

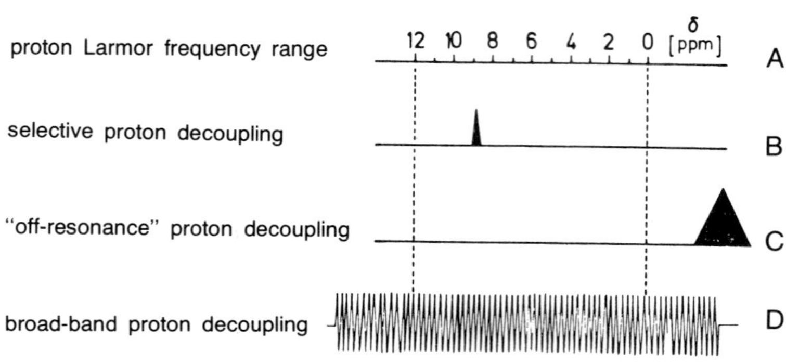

Fig. 3.1. Position and bandwidth of the proton decoupling frequency relative to the proton shift range (A) in broad-band (B), off-resonance (C), and selective decoupling (D). (Source: E. Breitmaier, G. Bauer, *Pharmazie in Unserer Zeit,* **5**, 97, 1976.)

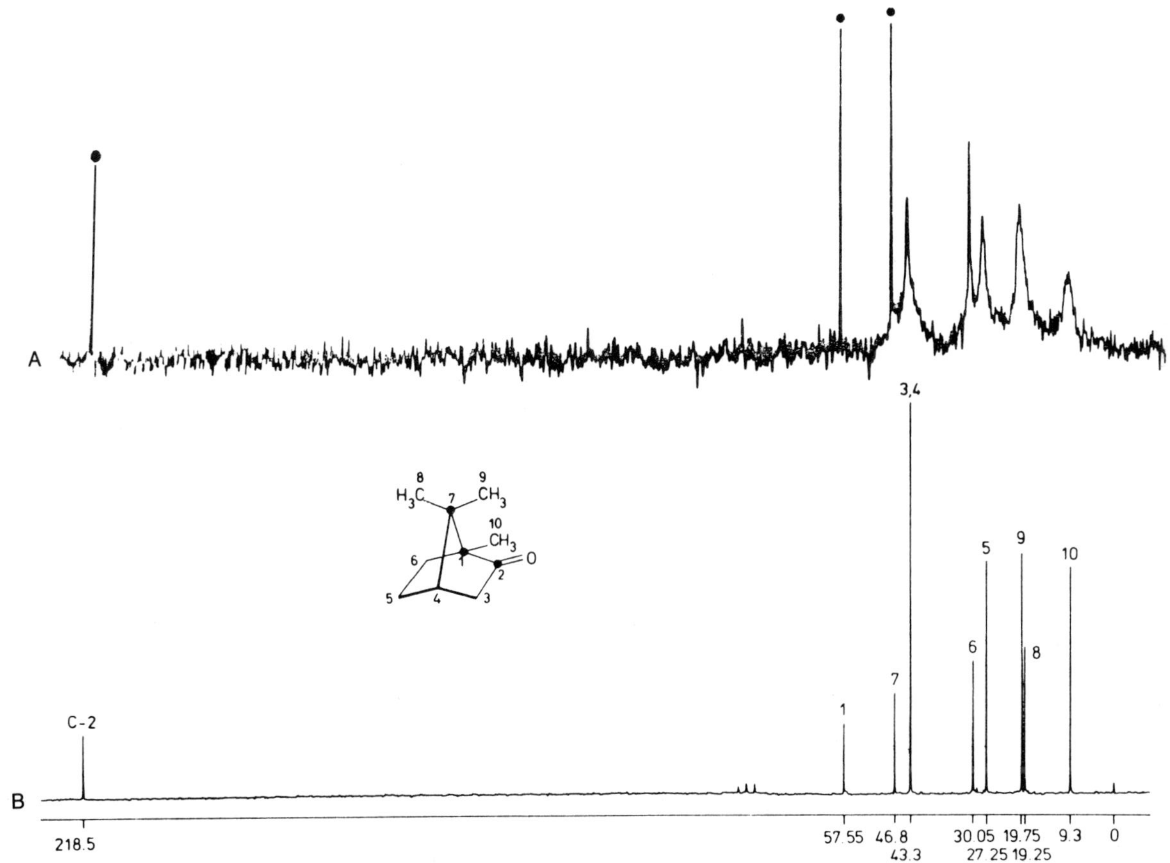

Fig. 3.2. Broad-band proton-decoupled ^{13}C NMR spectra of camphor, saturated solution in deuteriochloroform, 25°C, 20 MHz, 200 accumulated scans. A: High decoupler power (about 5 watts). B: Low decoupler power (about 0.05 watts); only the carbon atoms not bonded to hydrogen appear as sharp, intense singlets.

customary in PMR spectra—is impossible without additional experimental efforts (compare Section 3.7).

3.3 Low-Power Noise Decoupling

Broad-band proton decoupling only generates intense singlets for the carbon atoms bonded to hydrogen if the transmitter power is high enough (2 to 10 watts). If the transmitter power is too low, only quaternary carbons show up as sharp, intense singlets. The CH_n groups, on the other hand, produce broadened low-amplitude signals as a result of incomplete decoupling (see Fig. 3.2 B). Further discrimination between CH_n and CR_4 signals occurs if the decoupling frequency is adjusted slightly outside the proton shift range. Therefore, **low-power noise decoupling** allows quaternary carbons to be localized in the spectrum.

3.4 Off-Resonance Proton Decoupling

In broad-band proton-decoupled ^{13}C NMR spectra, it is no longer possible to distinguish between CH, CH_2, and CH_3 carbon atoms. The necessary spin-spin coupling information is lost as a result of complete decoupling.

Off-resonance proton decoupling permits only the coupling of protons bonded directly to the carbon atom to be seen, thus determining their number. To do this, the decoupling frequency, F_2 is adjusted about 100 to 500 Hz away from the proton resonance range (therefore, the term *off*-resonance), and no noise modulation is applied (see Fig. 3.1 C). If the transmitter power, $\gamma H_2/2\pi$, is high enough and is kept sufficiently constant, the residual splitting, J_R, for each carbon atom in a molecule is proportional to the offset, Δf_2, of the decoupling frequency F_2 from the respective carbon-bound proton resonance (see Eq. 3.1):

$$J_R = \frac{2\pi\ J_{CH}}{\gamma H_2}\ \Delta f_2 = \text{const.}\ \Delta f_2 \qquad (3.1)$$

The residual splittings for a particular kind of proton are finer the closer the decoupler frequency comes to this particular proton's resonance frequency. As the C-H multiplet signals move closer together, multiplet everlapping becomes rarer. In this way, it is often easier to distinguish quaternary, CH, CH_2, and CH_3 carbon atoms than in the coupled ^{13}C NMR spectrum, as can be seen by comparing Figures 3.3 B and C. Off-resonance spectra are not only clearer than coupled ^{13}C NMR spectra but fewer interferograms must be added to obtain them (see Fig. 3.3 B and C). This saving in acquisition time is once again a result of the signal enhancement owing to the NOE, which is also present in the off-resonance experiment, and of the breakdown of long-range couplings.

In off-resonance decoupling, multiplets are distorted in such a way that the lines facing away from the decoupling frequency are more intense than those facing toward it. For example, if one decouples at or near the TMS proton frequency, the multiplet lines facing the TMS signal are weaker (see Fig. 3.3 E).

A problem sometimes arises because off-resonance decoupling can also generate higher-order splittings. If the protons of a methylene group CH_AH_B are nonequivalent ($\delta_A \neq \delta_B$), a doublet of doublets (AB pattern) [3] rather than a triplet is observed in the off-resonance experiment owing to the circumstance that $\Delta f_A \neq \Delta f_B$ in Eq. (3.1). Similarly, if bonded carbon atoms have neighboring protons, several transitions may be coupled so that higher-order splittings occur in the $^{13}C\{^1H\}$ off-resonance spectrum. Generally speaking, higher-order off-resonance multiplets are to be expected if the C-H residual coupling, J_R, and the vicinal H-H coupling are of the same order of magnitude in systems such as $-CH_2-CH_2-$, $-CH-CH_2-$, or $-CH=CH-$ [5]. Such higher-order multiplets, however, are often useful for signal assignments since they permit identification of the signals caused by the bonding of the CH_n groups to each other.

3.5 Frequency-Dependent Off-Resonance Decoupling

The linear relationship between the residual coupling, J_R, and the decoupling frequency offset can be used as an assignment aid. To do this, several off-resonance decoupled spectra with different decoupling frequencies, F_2, are recorded, keeping the transmitter power constant. If the position of the ^{13}C residual multiplet lines is plotted versus F_2 (in Hz or ppm), n + 1 straight lines intersecting at a point arise for a CH_n group's (n + 1) multiplet (see Fig. 3.4). The coordinates of this point are the chemical shifts of both coupled nuclei. If the ^{13}C shift is known and has been assigned, the intersection produces the assignment of the decoupled bonded proton.

3.6 Selective Proton Decoupling

If the signals in a compound's ^1H NMR spectrum have been assigned unambiguously, individual proton types can be decoupled *selectively* when the ^{13}C NMR spectrum is recorded in order to assign the ^{13}C signals. To do so, the Larmor frequency of the proton to be decoupled must first be determined in the ^1H NMR spectrum. One then irradiates as sharply as possible at this frequency while recording the ^{13}C NMR spectrum. The carbon atom with the

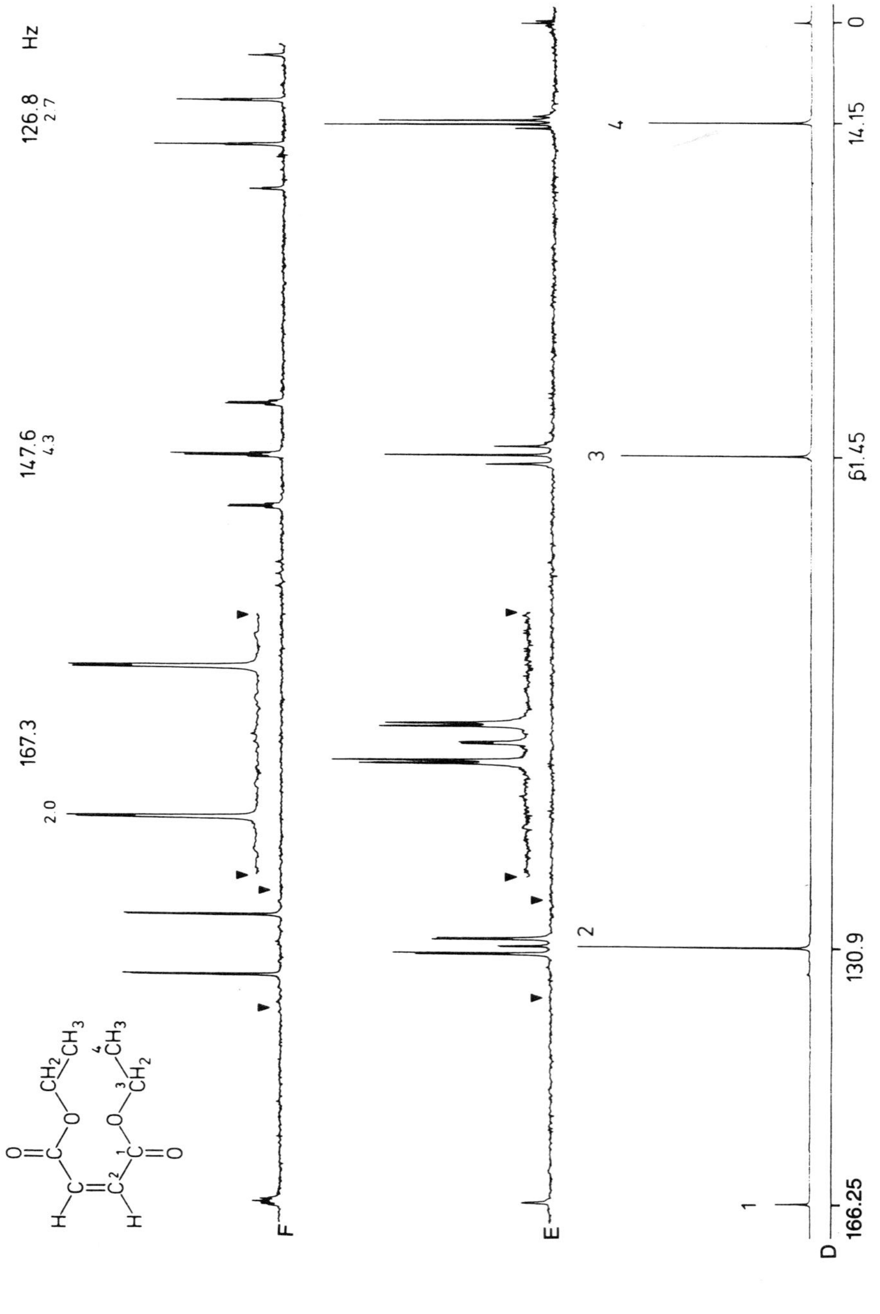
Hz
126.8
2.7
147.6
4.3
167.3
2.0
F
E
D
1
2
3
4
166.25
130.9
61.45
14.15
0

B

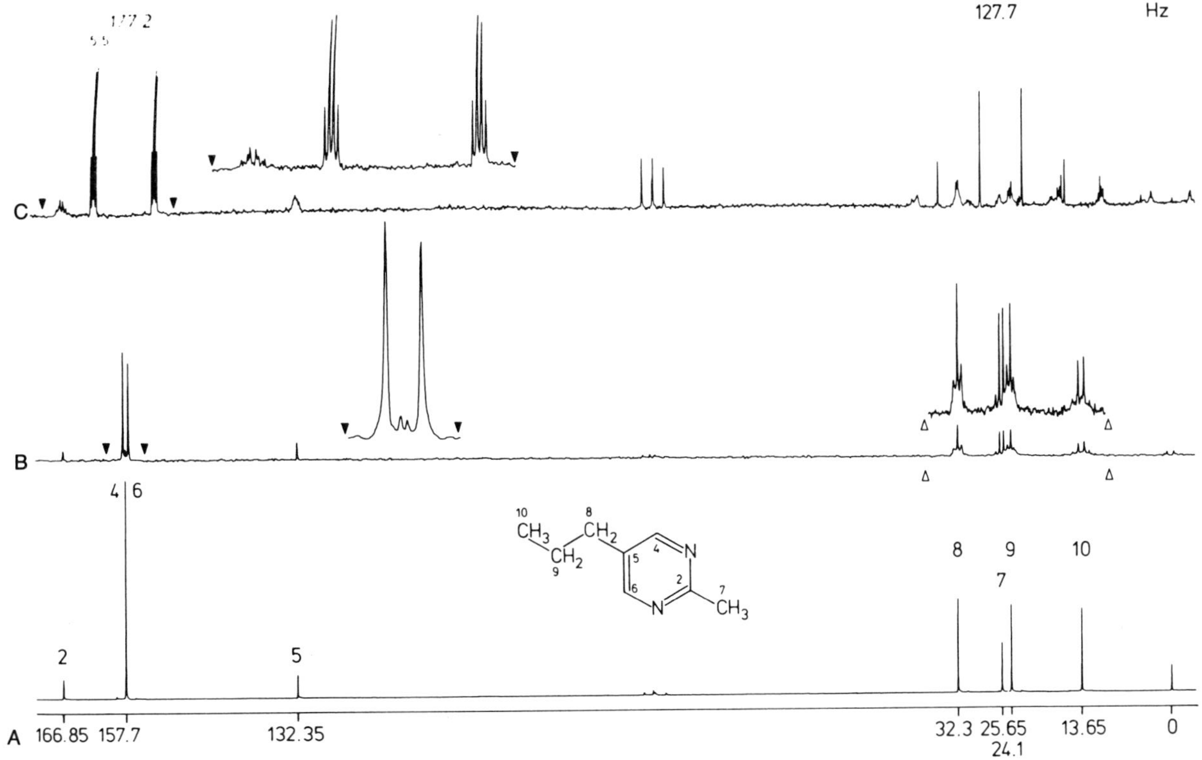

Fig. 3.3. 20 MHz ^{13}C NMR spectra of 2-methyl-5-propylpyrimidine (A-C) and maleic acid diethyl ester (D-F), both 50% v/v in deuteriochloroform at 25° C. A and D: Broad-band proton-decoupled, decoupling power 0.5 watts, noise-modulated, 50 accumulated scans. B and E: Off-resonanceδcoupled, decoupling power 0.5 watts, without noise modulation, 100 accumulated scans. C and F: Coupled, 1000 accumulated scans. Higher-order splittings can be seen in spectrum B for the signal at 157.7 ppm, and even better for the resonance at 130.9 ppm. in spectrum E. The ranges marked with ▼, with the higher-order splittings, are shown above the off-resonance spectra B and E.

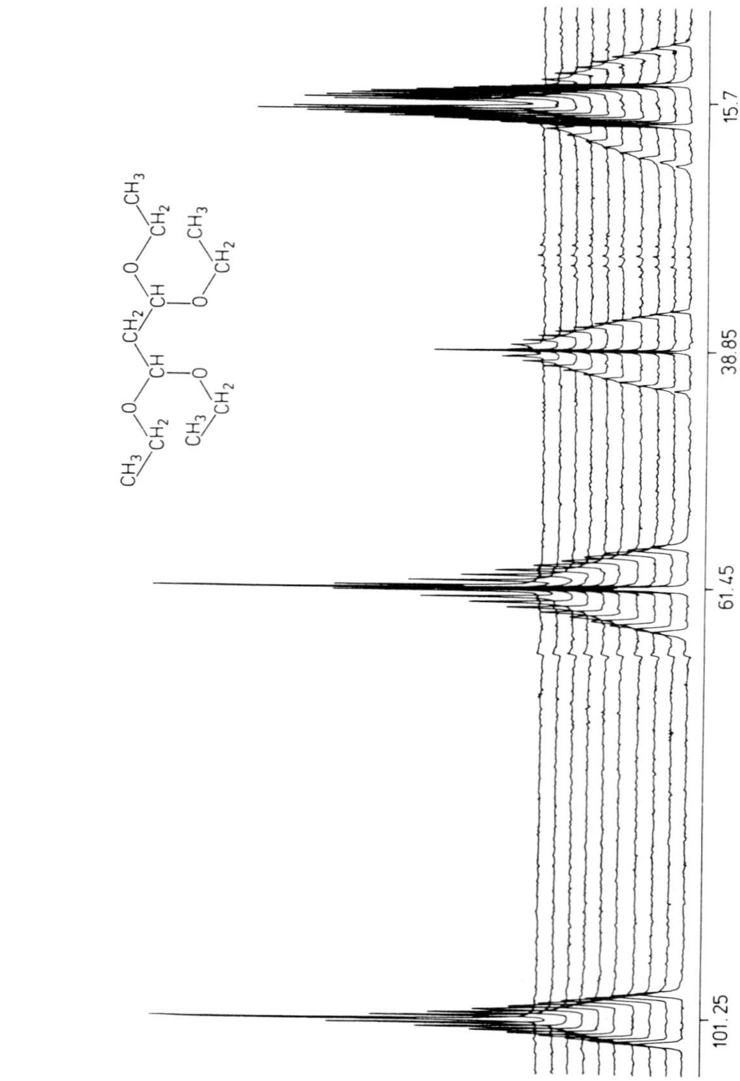

A

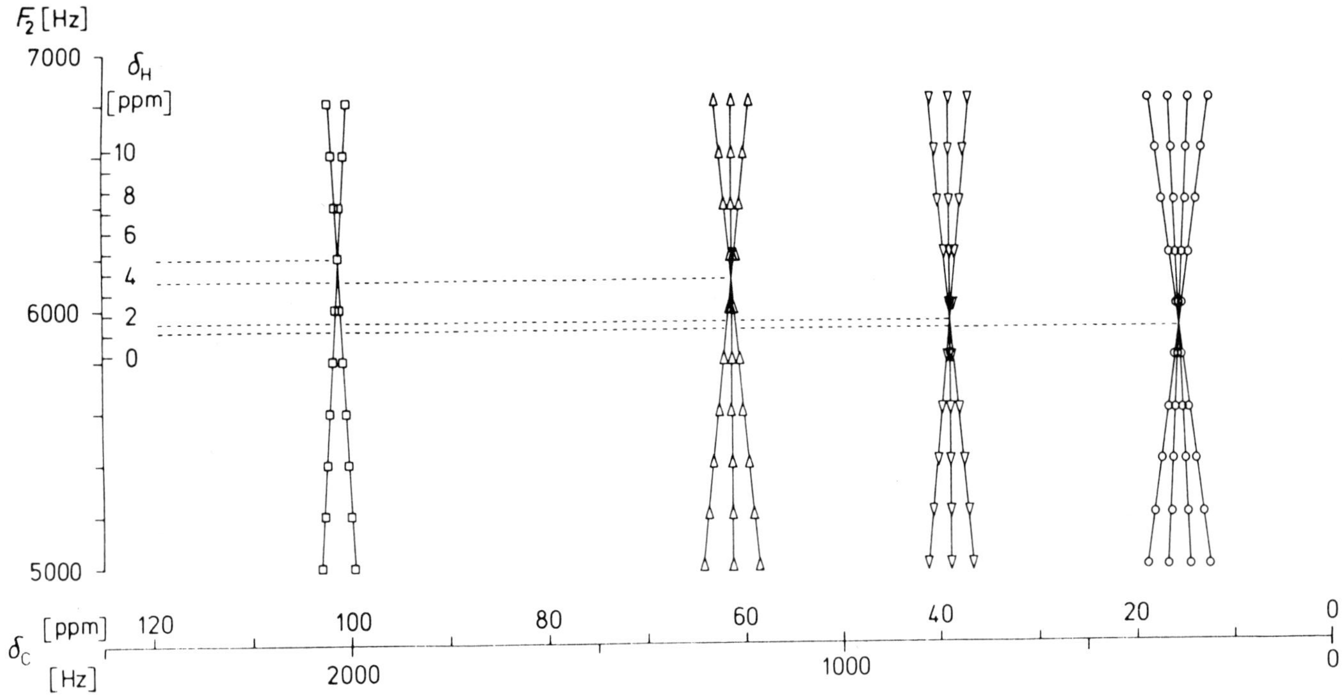

B

Fig. 3.4. Frequency-dependent off-resonance proton decoupling in the 10 MHz ^{13}C NMR spectrum of 1,1,3,3-tetraethoxy-propane (malonic dialdehyde tetraethyl acetal), 75% v/v in hexadeuterioacetone at 25° C. A: Computer-controlled series of measurements for frequency offsets Δf_2 2,200 to 400 Hz from TMS, 300 accumulated scans for each single experiment. B: Evaluation of a series of measurements for smaller decoupler frequency offsets from TMS (instrument-specific F_2 values from 5,000 to 6,800 Hz, corresponding to Δf_2 1,000 to −800 Hz). The curves for the outer multiplet signals that converge with decreasing frequency offset Δf_2 in series A can be approximated by straight lines at lower offset values, as in B. The intersection of these straight lines yields the ^{1}H and ^{13}C shifts of the coupled nuclei.

directly bonded hydrogens that are being decoupled then appears as an intense singlet (see Fig. 3.5 A-F). Off-resonance conditions prevail for all other carbons.

The practical applicability of selective decoupling often fails because the decoupler must affect both ^{13}C satellites in the ^{1}H NMR spectrum. If the decoupling frequency is set to coincide with the main signal, its intensity must be high enough to include the ^{13}C satellites in its frequency distribution. Because the C-H coupling constants lie between 120 and 300 Hz, the distribution curve of the decoupler frequency should have this width. If other proton signals appear within this range, however, selective decoupling can lead to ambiguous results.

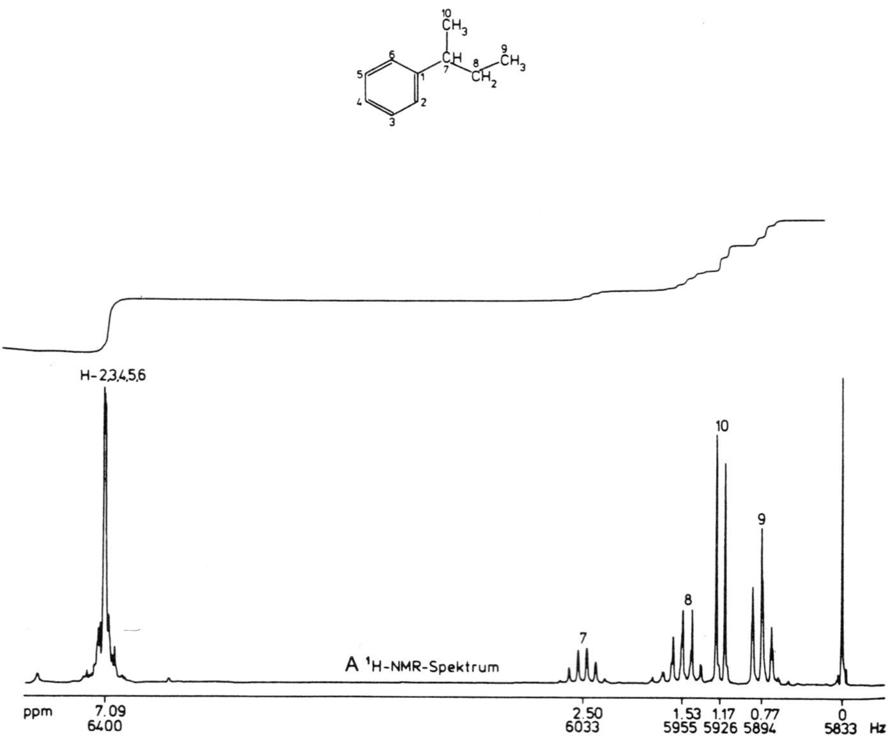

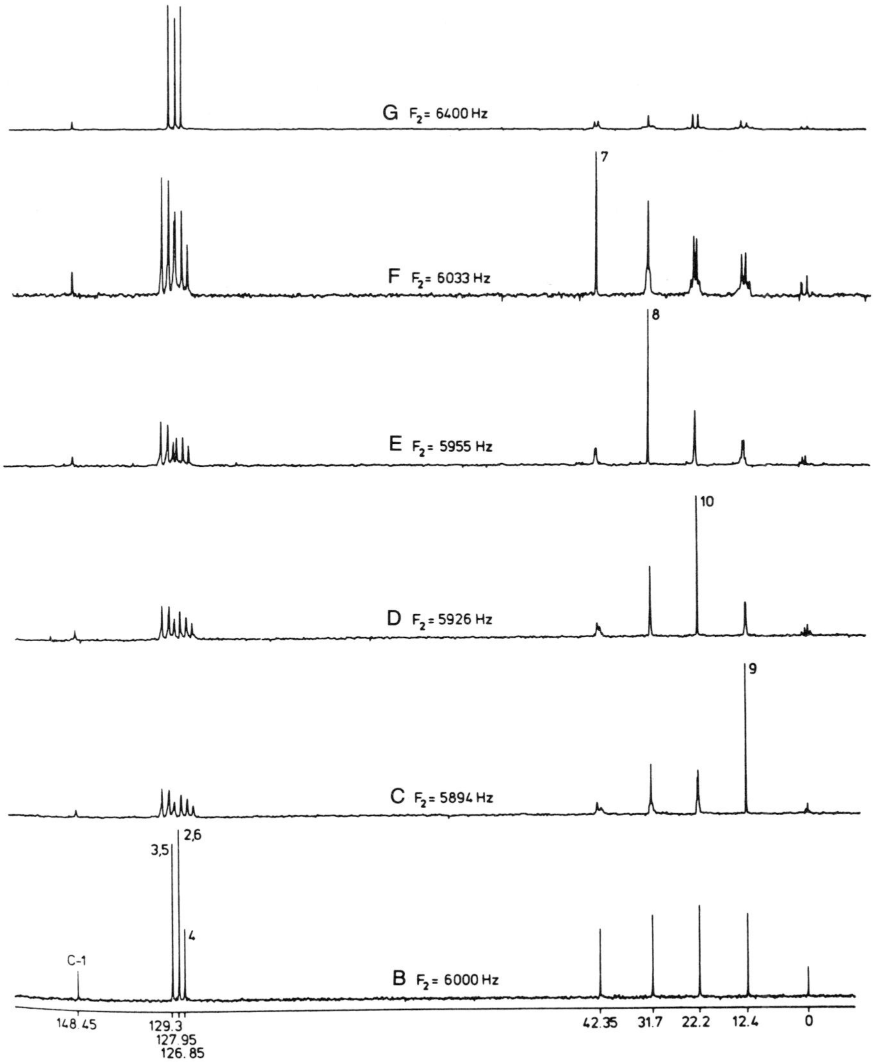

Fig. 3.5. Selective proton decoupling in *sec*-butylbenzene, 90% v/v in hexadeuterioacetone, 30° C. A: ^1H NMR spectrum (80 MHz) with the instrument-specific F_2 positions, single experiment. B: Broad-band proton-decoupled ^{13}C NMR spectrum (20 MHz), single scan experiment. C-G: Selective proton decoupling at 0.77 ppm (C), 1.17 ppm (D), 1.53 ppm (E), 2.50 ppm (F), and 7.09 ppm (G), 100 accumulated scans each time.

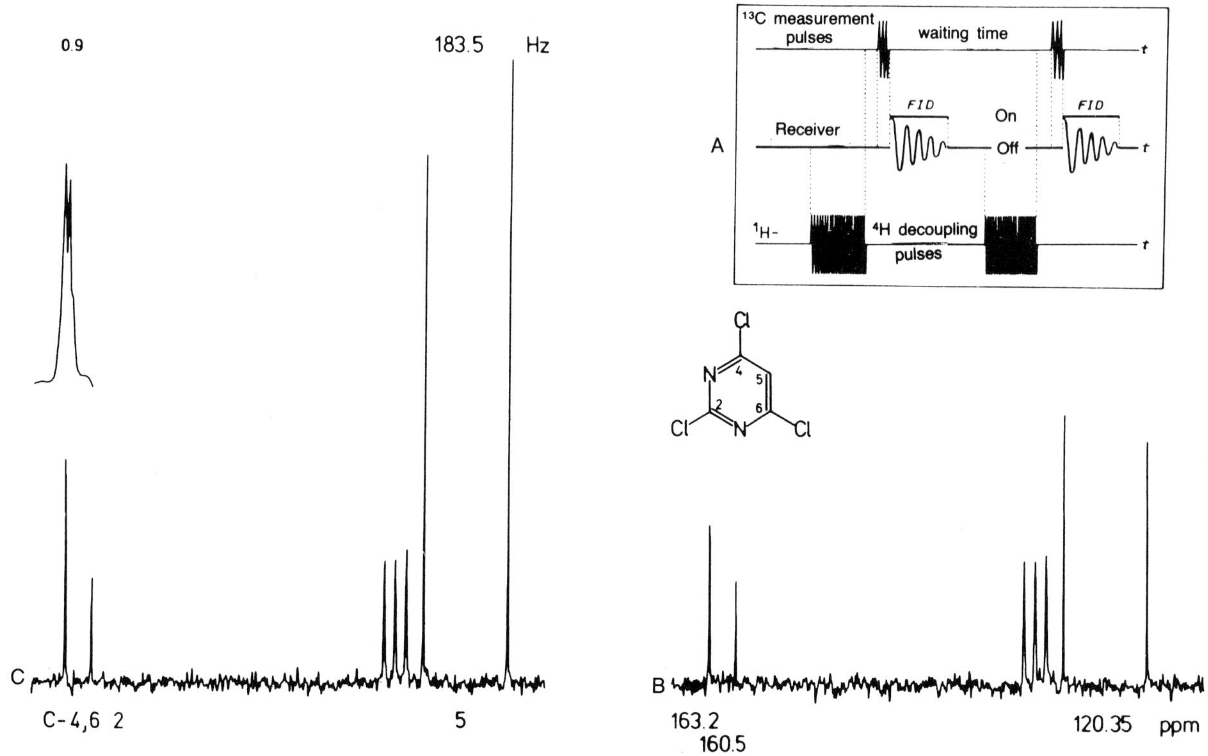

Fig. 3.6. Pulsed proton decoupling for the measurement of C-H coupling constants profiting from nuclear Overhauser enhancements; sample: 2,4,6-trichloropyrimidine, 75% v/v in hexadeuteriobenzene at 25° C. A: Alternating pulse sequence in measurement and decoupling channels. B: Undecoupled 20 MHz ^{13}C NMR spectrum, 50 accumulated scans. C: Pulsed proton-decoupled 20 MHz ^{13}C NMR spectrum, according to scheme A, 50 accumulated scans. For purposes of comparison, spectra B and C have been recorded using the same intervals between the measurement signals and have been standardized at the same noise level.

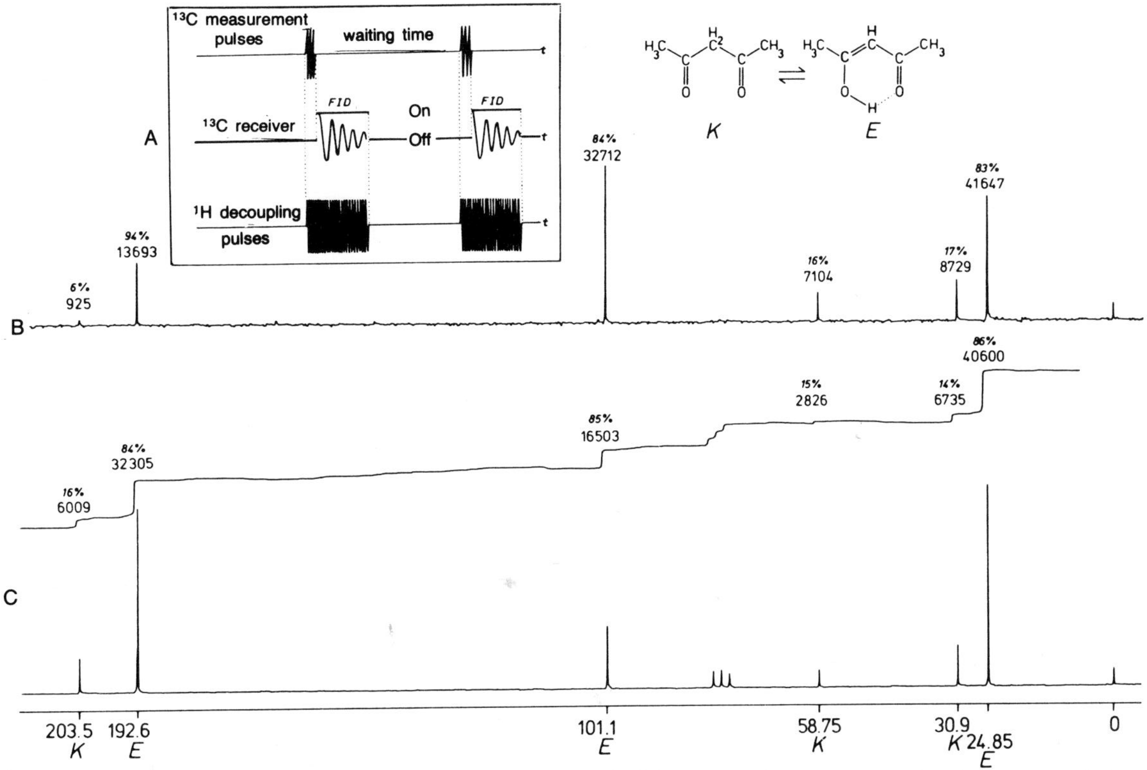

Fig. 3.7. Pulsed proton decoupling for the measurement of proton-decoupled ^{13}C NMR spectra with suppressed nuclear Overhauser effect; sample: acetylacetone, 50% v/v in deuteriochlorogorm at 25° C. A: Pulse sequence in measurement and decoupling channels. B: Broad-band proton-decoupled 20 MHz ^{13}C NMR spectrum, 200 accumulated scans. C: Pulsed proton-decoupled 20 MHz ^{13}C NMR spectrum, according to scheme A, 100 accumulated scans, waiting time between subsequent measurement pulses, 60 s. The integration of spectrum C shows 15 ± 1% keto and 85 ± 1% enol form, in agreement with the known values from proton resonance.

3.7 Pulsed Proton Decoupling

When the decoupler is switched on in a ^{13}C experiment, the protons are decoupled immediately. However, the change in the population of the ^{13}C energy levels, which is responsible for the NOE, is a spin-lattice relaxation process and proceeds much more slowly. Correspondingly, when the decoupler is switched off, proton decoupling is interrupted at once, and the NOE decays slowly. The two pulsed $^{13}C\{^{1}H\}$ double resonance variants discussed below are based on these facts.

3.7.1 Measurement of Nuclear Overhauser-Enhanced Coupled ^{13}C NMR Spectra

If the decoupler is shut off during acquisition of the interferogram but is on long enough between subsequent interferograms, a **coupled spectrum** is recorded. Owing to the slowly decaying NOE, signal intensities are enhanced by up to a factor of two. In this experiment, ^{1}H and ^{13}C nuclei are excited alternately by high frequency pulses; thus the method is referred to as **gated decoupling.** The alternating pulse sequences are shown schematically in Figure 3.6 A. Figure 3.6 C depicts the result in the case of 2,4,6-trichloropyrimidine: When the alternating pulse sequence is applied, the signals caused by carbon atoms coupled to hydrogen are nearly twice as intense as in the coupled spectrum, but the noise amplitude is the same (see Fig. 3.6 B).

3.7.2 Measurement of Broad-Band Proton-Decoupled Spectra with Suppressed Nuclear Overhauser Effect

If the decoupler is switched on only during acquisition of the interferograms and if sufficiently long waiting times are programmed between successive interferograms with the decoupler shut off, one obtains broad-band proton-decoupled ^{13}C NMR spectra without large intensity differences because of suppressed nuclear Overhauser effects. If the waiting time between each pulse of exciting and decoupling radio frequency is so long that even the most slowly relaxing carbon can follow, spin-lattice relaxation and NOE no longer affect the signal intensities. This mode of pulsed $^{13}C\{^{1}H\}$ double resonance, known as **inverse gated decoupling,** permits the recording of decoupled ^{13}C NMR spectra that can be evaluated quantitatively. Figure 3.7 shows the application of this technique to the tautomers of acetylacetone.

Long waiting times between pulses, which affect the recording time unfavorably, are a drawback in both pulsed double-resonance techniques.

References

1. H. Friebolin et al., *NMR-Spektroskopie, Eine Einführung mit Übungen,* Verlag Chemie GmbH, Weinheim, 1974, Taschentext Bd. 15.
2. H. Günther, *NMR-Spektroskopie, Eine Einführung,* Georg Thieme Verlag, Stuttgart, 1973.
3. H. Suhr, *Anwendungen der Kernmagnetischen Resonanz in der Organischen Chemie,* Springer Verlag, New York, 1965.
4. J. H. Noggle, R. E. Schirmer, *The Nuclear Overhauser Effect: Chemical Applications,* Academic Press, New York, 1971.
5. E. W. Hagaman, *Org. Magn. Res.* **8,** 389 (1976).

CHAPTER 4

Chemical Methods of Signal Assignment

4.1 Deuteriation

If the decoupling procedures described in Chapter 3 do not lead to the unambiguous assignment of the resonances in a ^{13}C NMR spectrum, one can resort to *labeling* of individual positions of the molecule. For instance, the specific deuteriation of certain carbon atoms is a fairly safe method for the spectral assignments in ^{13}C NMR spectra. This is possible in many cases by chemical means such as the catalytic deuteriation of multiple C-C bonds, electrophilic substitution of aryl protons by deuterium, carbonyl reductions with $LiAlD_4$, or syntheses starting from simple deuteriated compounds.

Because deuterium's nuclear spin I is unity, a deuteron has three possible nuclear spin orientations in a magnetic field (see Eq. 1.5). According to the multiplicity rule ($2nI + 1$), CD, CD_2, and CD_3 carbon atoms give triplets (1:1:1), quintets (1:3:5:3:1), and septets (1:3:5:7:5:3:1). These CD_n multiplets also appear in the proton-decoupled ^{13}C spectra. Owing to their larger gyromagnetic ratio (γ_H), protons precess about 6.5 times faster than deuterons in the H_0 field ($\gamma_H/\gamma_D = 6.5$). The C-D coupling constants for the labeled bonds are also smaller than the C-H coupling constants by the same ratio. For example, Figure 4.1 shows the proton-decoupled and the undecoupled ^{13}C NMR spectra of a mixture of hexadeuterioethanol and ethanol. In these spectra, one can recognize the different multiplicities and coupling constants of the CH_n and CD_n carbon atoms. In addition, the ^{13}C shifts show isotope effects: Bound deuterons shield the carbon atom more and more as their number increases, as Figure 4.1 illustrates.

In large molecules, C-D couplings are usually lost in the noise of the proton-decoupled ^{13}C NMR spectrum. The importance of deuterium labeling as an assignment aid is based on this fact, as Figure 4.2 shows in the case of D-ribose and 2-deuterio-D-ribose. In the CMR spectrum of this sugar (Fig. 4.2 A), no clear distinction can be made between C-2, C-3, and C-4 using off-

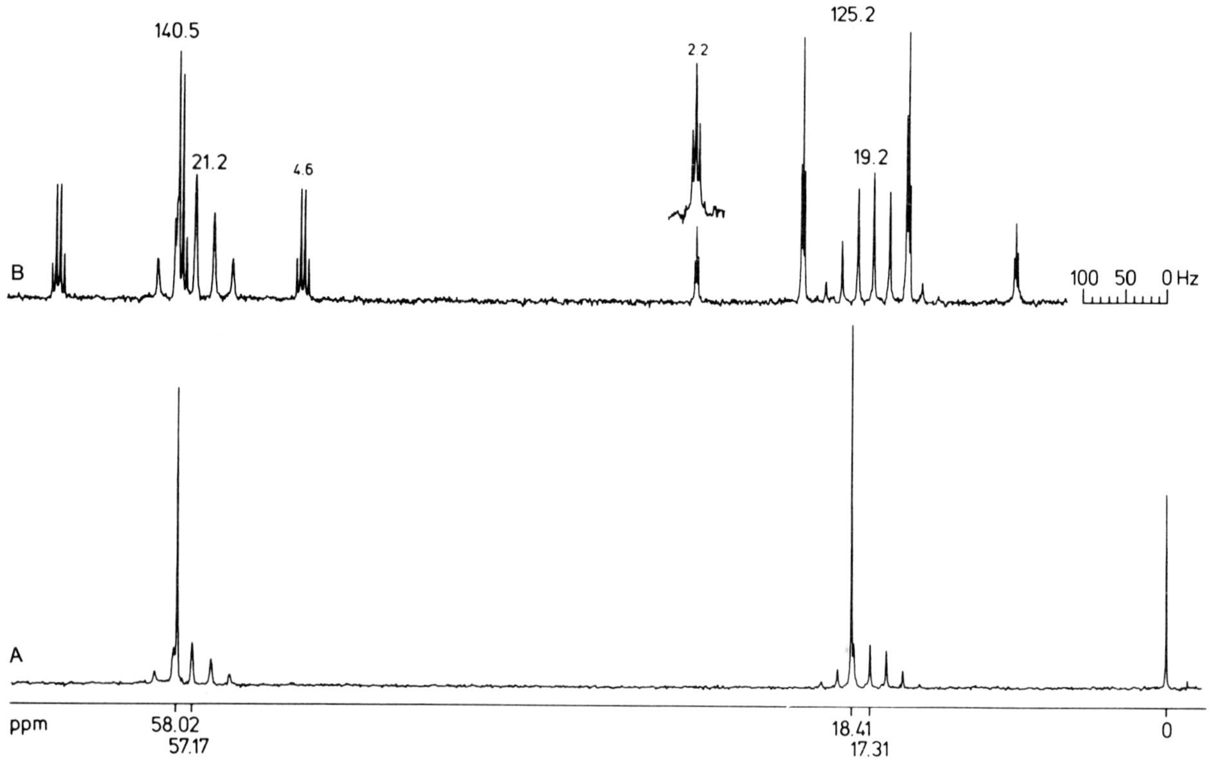

Fig. 4.1. 20 MHz ^{13}C NMR spectra of a mixture of ethanol and hexadeuterioethanol, 25:75 v/v, 25° C. A: Wide-band proton-decoupled, 200 accumulated scans. B: Coupled, 1,000 accumulated scans. The isotope effect on the ^{13}C shift is 1.1 ppm for the methyl group and 0.85 ppm for the methylene group. The ratio of the coupling constants, J_{CH}:J_{CD} is 6.5 for the methyl group and 6.6 for the methylene group.

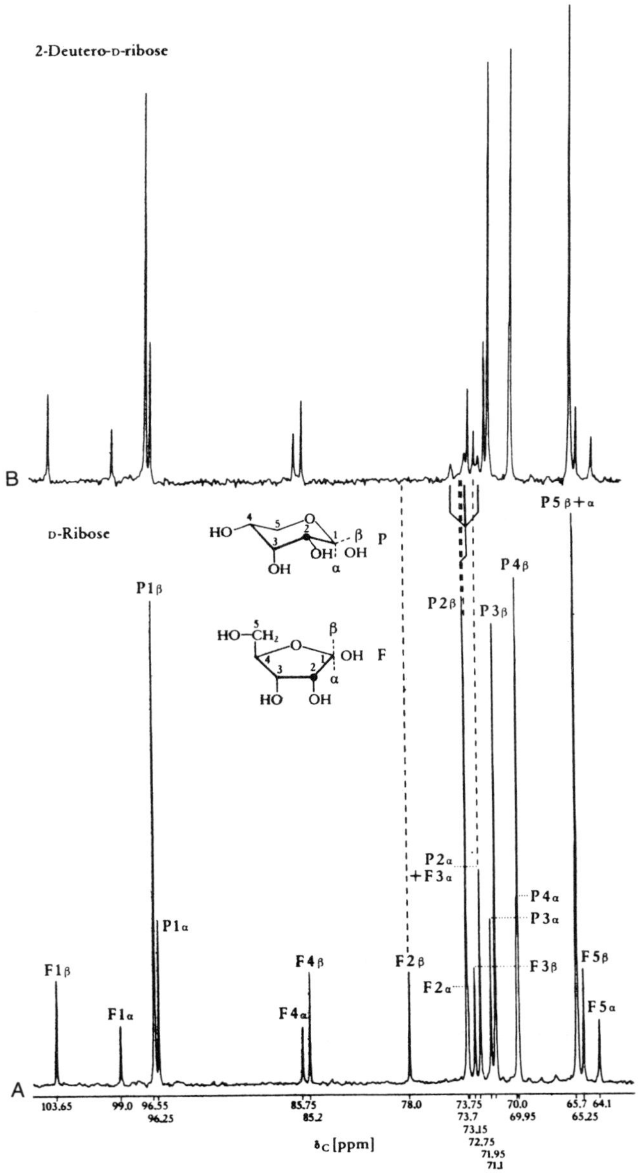

Fig. 4.2. 22.63 MHz ^{13}C NMR spectra of D-ribose (A) and 2-deuterio-D-ribose (B), mutarotated samples, 1 M in D_2O at 30° C, wide-band proton-decoupled. The signals for C-2 can be assigned by comparison of spectra A and B: the C-2 signal of β-2-deuterio-D-ribopyranose (P2 β) appears as a 1:1:1-triplet because of C-D coupling. The C-2 C-D triplets of the other forms of ribose are too weak, and are lost in the noise. (Source: E. Breitmaier, G. Bauer, *Pharmazie in Unserer Zeit*, **5**, 97, 1976.)

resonance decoupling, for example. The resonance assignment is further complicated by the fact that D-ribose exists in the aa-and β-furanose forms in aqueous solution. C-2 can be deuteriated, however. In the spectrum of 2-deuterio-D-ribose (see Fig. 4.2 B), the C-2 signals of the furanoses and the α-pyranose disappear in the noise. Only the C-2 C-D triplet of the β-pyranose form can still be discerned. Comparison of spectra 4.2 A and B leads to the unequivocal assignment of the C-2 signals in all four forms of ribose.

4.2 ^{13}C and ^{12}C Enrichment

The labeling of particular carbon atoms in the molecule with ^{13}C is sometimes useful for signal assignment because the ^{13}C-enriched positions (concentration of ^{13}C larger than the natural abundance, 1.1%) produce stronger signals in the CMR spectrum. Conversely, one can also label with ^{12}C-enriched carbon (concentration of ^{13}C less than 1.1%). In the ^{13}C spectrum of such ^{12}C-enriched compounds, the signals caused by the labeled carbon atoms are weaker and may be lost in the noise. The advantage of ^{12}C labeling over ^{13}C labeling is that there are no ^{13}C-^{13}C couplings (see Section 8.3) to make interpretation of the spectrum more complicated.

^{12}C and ^{13}C labeling, like deuteriation, are not merely signal assignment aids. One can also use accessible ^{13}C-, ^{12}C-, or deuterium-labeled starting compounds to elucidate reaction mechanisms by studying the ^{13}C NMR spectra of the isolated intermediate compounds or end products. The biosynthetic origin of carbon atoms in natural producs can also be determined by ^{13}C NMR. For example, the biosynthesis of sepedonin was carried out with ^{13}C-enriched formic acid (◆), $^{13}CH_3$-acetic acid (●), and finally with ^{13}COOH-acetic acid (▲) in the culture medium (see below) [1]. The origin of each sepedonin carbon atom can be deduced from the ^{13}C NMR spectra of the three isolates [1]:

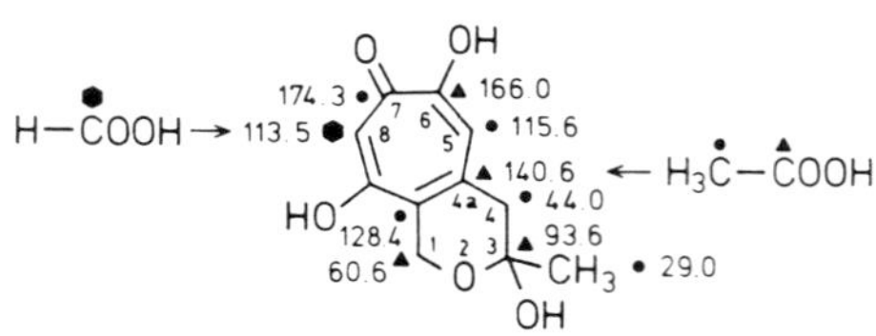

4.3 Paramagnetic Shift Reagents

Some paramagnetic lanthanide 1,3-diketone chelates known as **shift reagents** cause signal spreading in a substance's NMR spectrum when they are added

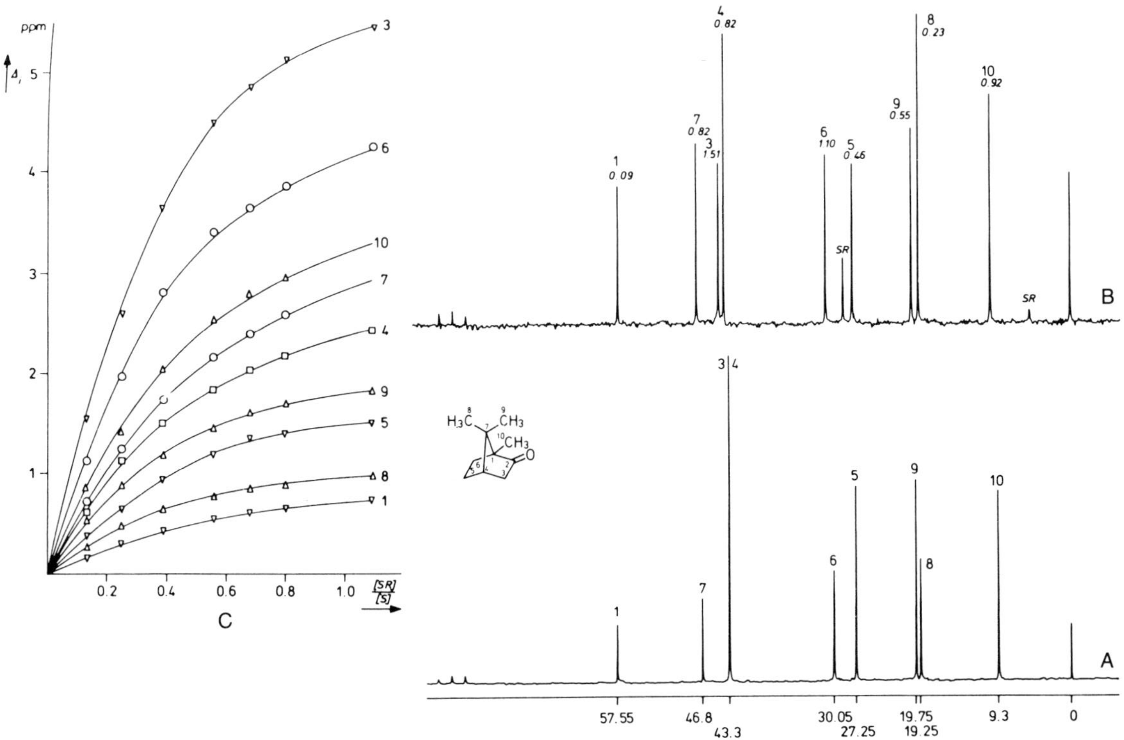

Fig. 4.3 Wide-band proton-decoupled ^{13}C NMR spectra of camphor, saturated solution in deuteriochloroform, 25° C, 20 MHz, 200 accumulated scans. A: Without shift reagent. B: With 100 mg of tris(1,1,1,2,2,3,3-heptafluoro-7,7-dimethyl-4,6-octanedionato) europium (III); the measured pseudocontact shifts, Δ_i, appear in italics below the number of each respective carbon atom. C: Pseudocontact shifts, Δ_i, of the camphor atoms as a function of the molar ratio of shift reagent to sample ([*SR*]/[*S*]). Tris-(4,4,4-trifluoro-(2-thienyl)-1,3-butanedionato)europium(III) was used (see E. Breitmaier, W. Voelter, ^{13}C *NMR Spectroscopy: Methods and Applications*, "Monographs in Modern Chemistry," Vol. 5, H. E. Ebel, ed., Verlag Chemie GmbH, Weinheim, 2nd ed., 1978).

to the sample solution in high enough concentrations [2]. It is a prerequisite that the compound being studied must not decompose the chelate (e.g., carboxylic acids and phenols) but, on the other hand, that it must be capable of coordinating with the lanthanide ion (e.g., alcohols, ethers, thiols, thioethers, ketones, esters, and amines). Europium (III) and praseodymium (III) chelates of the 1,3-diketones dipivaloylmethane (*I*), 1,1,1,2,2,3,3-hepta-fluoro-7,7-dimethyl-4,6-octanedione (*II*), and 1-(2-thienoyl)-3,3,3-trifluoro-acetone (*III*) are useful shift reagents because they dissolve well in the usual deuteriated solvents used in NMR spectroscopy and, in contrast to other paramagnetic compounds, produce only slight broadening of the sample's signals:

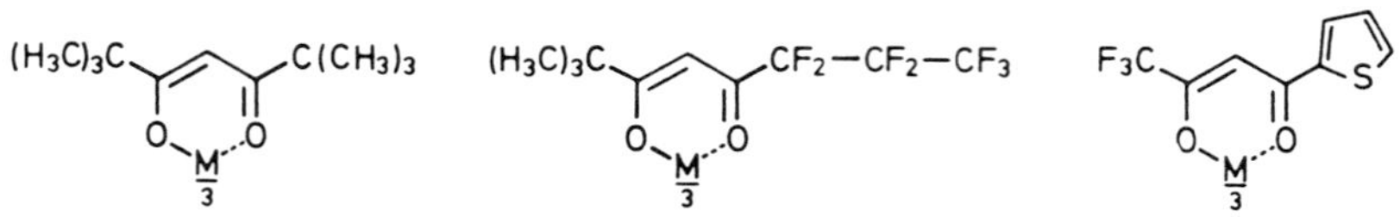

I II III

M = Eu, Pr

The **pseudocontact shift** of a compound's NMR signals by europium or praseodymium chelates depends on the *concentration* of these shift reagents [2]. The magnitude of the shift for a particular kind of nucleus increases as its *distance* from the group capable of coordinating with it *decreases*. (For more exact relationships, see Reference 2.) This is the basis for using shift reagents as assignment aids, especially in proton resonance. Figure 4.3 is an example in ^{13}C NMR. It is impossible to assign camphor's three methyl resonances by off-resonance decoupling, and the same is true for the methylene signals. If the pseudocontact shifts, Δ_i, of all the signals are measured at several shift reagent-substrate molar ratios, $[SR]/[S]$, one obtains the series of curves shown in Figure 4.3 C. As the distance between the carbonyl group and the methyl or methylene groups decreases in the order C-10, C-9, C-8 and C-3, C-6, C-5, respectively—as can be seen in the structural formula—the slope of the curves decreases too. Therefore, the methyl resonance with the largest slope corresponds to C-10, and the one with the smallest slope corresponds to C-8. The same is valid for the assignment of the methylene resonances.

References

1. A. G. McInnes, D. G. Smith, L. C. Vinning, L. Johnson, *Chem Commun*. 1971, 325.
2. R. v. Ammon, R. D. Fischer, *Angew. Chem.* **84**, 737 (1972).

CHAPTER 5

Sample Concentration: Solvents and Standards

When studying chemical structural problems by means of high-resolution NMR, solutions are used. The recommended volume for ^{13}C measurements is about 1 ml, and the amount of substance necessary to record a wide-band proton-decoupled ^{13}C NMR spectrum with a good signal:noise ratio, for compounds with 10 to 20 carbon atoms, is about 50 mg. Depending on the strength of the magnetic field, between five and ten thousand interferograms must be added together, which demands an acquisition time of one or two hours. Doubling the signal:noise ratio (*S:N*), or halving the amount of substance, requires that the measurement time be quadrupled, according to Eq. (5.1) (*n* is the number of accumulated scans):

$$(S{:}N)_n = (S{:}N)_1 \sqrt{n} \tag{5.1}$$

Even longer acquisition times or larger samples are needed for off-resonance experiments, especially for coupled ^{13}C NMR spectra, because the signals break down into multiplets. If less than 20 mg are available, only wide-band proton-decoupled ^{13}C NMR spectra are practical.

Deuteriated compounds are most commonly used as solvents for the following reasons. First, in the usual NMR spectrometers the field-frequency ratio is stabilized (locked) with the aid of the deuterium resonance, using one of the sample's deuterium signals. Second, the deuteriated solvent's signals appear as multiplets (see Sec. 4.1), which are unaffected by proton decoupling and are therefore weak. As a consequence, the solvent signals, in practice, seldom interfere with the evaluation of a spectrum and are easily recognizable by their triplet, quintet, or septet splittings. The ^{13}C shifts and coupling constants of some frequently used deuteriated solvents are shown in Table 5.1. By comparing these data with the data for the corresponding ^{1}H compounds, one can discern the isotope effect of about 1 ppm mentioned earlier. With extremely

TABLE 5.1. ^{13}C NMR DATA OF SOME COMMON DEUTERIATED SOLVENTS AND THEIR PROTONATED COUNTERPARTS

Compound	C-D Signal Multiplicity	C-D Coupling Constant [Hz]	^{13}C Shift [ppm]	Corresponding ^{1}H Compound ^{13}C Shift [ppm]	Corresponding ^{1}H Compound C-H Coupling Constant [Hz]
Dideuteriomethylene chloride	Quintet	27	53.1	53.8	177.5
Deuterio-chloroform	Triplet	32	77.0 ±0.5	78.0 ±0.5	210.5
Deuterio-bromoform	Triplet	31.5	10.2	10.3	205
Trideuterio-nitromethane	Septet	23.5	60.5	61.1	146.5
Trideuterio-acetonitrile	Septet	21	1.3	1.7	117.4
	Long-range multiplet	1	118.2	118.2	<5
Tetradeuterio-methanol	Septet	21.5	49.0	49.9	141
Hexadeuterio-ethanol	Septet	19.5	15.8	16.9	127
	Quintet	22	55.4	56.3	140.5
Decadeuterio-ethyl ether	Septet	19	13.4	14.5	127
	Quintet	21	64.3	65.3	142

Octadeuterio-1,4-dioxan	Quintet	22	66.5	67.6	146
Octadeuterio-tetrahydrofuran	Quintet	20.5	25.2	26.2	132
	Quintet	22	67.4	68.2	137
Hexadeuterio-acetone	Septet	20	29.8	30.7	126
	Long-range multiplet	<1	206.5	206.7	7.5
Hexadeuterio-dimethylsulfoxide	Septet	21	39.7	40.9	137
Tetradeuterio-acetic acid	Septet	20	20.0	20.9	129
	Long-range multiplet	<1	178.4	178.8	7.5
Heptadeuterio-dimethylformamide	Septet	21	30.1	30.9	138
	Septet	21	35.2	36.0	138
	Triplet	30	167.7	167.9	188
Octadecadeuterio-hexamethylphosphoramide	Septet	21	35.8	36.9	129.5
Dodecadeuterio-cyclohexane	Quintet	19	26.4	27.8	126
Hexadeuterio-benzene	Triplet	24	128.0 ±0.5	128.5 ±0.5	161
Pentadeuterio-pyridine	Triplet (3,5)	25	123.5	124.2	166
	Triplet (4)	24.5	135.5	136.2	163
	Triplet (2,6)	27.5	149.2	149.7	177

small samples (1 to 10 mg of substance), it can happen that the solvent's $^{13}CD_n$ signal fills up the computer memory before the sample's resonances reach a sufficiently high *S:N* ratio. For these cases, ^{12}C-enriched deuteriated solvents are available [$^{12}CDCl_3$, $^{12}CD_3OD$, $(^{12}CD_3)_2SO$], as are decoupling and computational procedures which may be used to depress the intense solvent signals.

The $^{13}CD_n$ multiplets listed in Table 5.1 can also be used as internal reference signals by recalculating the shifts with the aid of the values tabulated using tetramethylsilane as a standard. If D_2O is the solvent, tetramethylsilane can be put in a coaxially welded capillary tube as an external standard. In these cases, however, corrections accounting for the differing volume magnetic susceptibilities of D_2O and TMS must be made [1,2,3]. One can also add an aqueous solution of sodium 3-trimethylsilylpropanesulfonate as a internal reference because this compound's trimethylsilyl group has approximately the same ^{13}C shift as TMS ("water-soluble TMS").

$$\begin{array}{l} \qquad\qquad 57.25 \quad 21.15 \quad 1.7 \qquad 0 \text{ ppm} \\ Na^+ \; {}^-O_3S{-}CH_2{-}CH_2{-}CH_2{-}Si(CH_3)_3 \end{array}$$

References

1. H. Friebolin et al., *NMR-Spektroskopie, Eine Einführung mit Übungen*, Verlag Chemie GmbH, Weinheim, 1974, Taschentext Bd. 15.
2. H. Günther, *NMR-Spektroskopie, Eine Einführung*, Georg Thieme Verlag, Stuttgart, 1973.
3. H. Suhr, *Anwendungen der Kernmagnetischen Resonanz in der Organischen Chemie*, Springer Verlag, New York, 1965.

CHAPTER 6

^{13}C Chemical Shifts: A Survey

6.1 Comparison of ^{1}H and ^{13}C Shifts

The ^{13}C chemical shifts of the carbon atoms in organic compounds extend over about 240 ppm, if extreme cases such as tetraiodomethane (292.5 ppm) and the *t*-butyl cation (330 ppm) are disregarded. On the contrary, proton resonances cover a range of little more than 10 ppm. Because $\gamma_H = 4\gamma_C$ (see Sec. 1.1), proton resonance frequencies are four times larger than ^{13}C resonance frequencies; therefore, one finds that ^{13}C shifts cover a frequency range that is six times larger than that of proton shifts. Since the line widths of ^{1}H and ^{13}C signals are of the same order of magnitude, ^{13}C NMR shows up variations in the chemical environment much more clearly than ^{1}H NMR does. This can be seen in the ^{1}H and ^{13}C NMR spectra of *sec*-butylbenzene shown in Figure 6.1. Both spectra were recorded for the same solution, using the same magnetic field strength, and are drawn on the same frequency scale.

The characteristic ^{13}C shifts of functional carbon atoms in different classes of organic compounds are shown in Figure 6.2. Some of these groups, like the cyanide of isocyanate functions, do not appear directly in the ^{1}H NMR spectrum. The following section sketches some general relationships between ^{13}C shifts and chemical features of the corresponding carbon atom.

6.2 Carbon Binding State and ^{13}C Shift

The chemical shift of a nucleus depends on the atom's binding state in the molecule and on how neighboring groups influence this state. A magnitude that describes all these influences is the magnetic shielding constant σ_i, mentioned in Section 1.1, which can be taken as the sum of three components, σ_{dia}, σ_{para}, and σ', as shown in Eq. (6.1):

$$\sigma_i = \sigma_{dia} + \sigma_{para} + \sigma' \qquad (6.1)$$

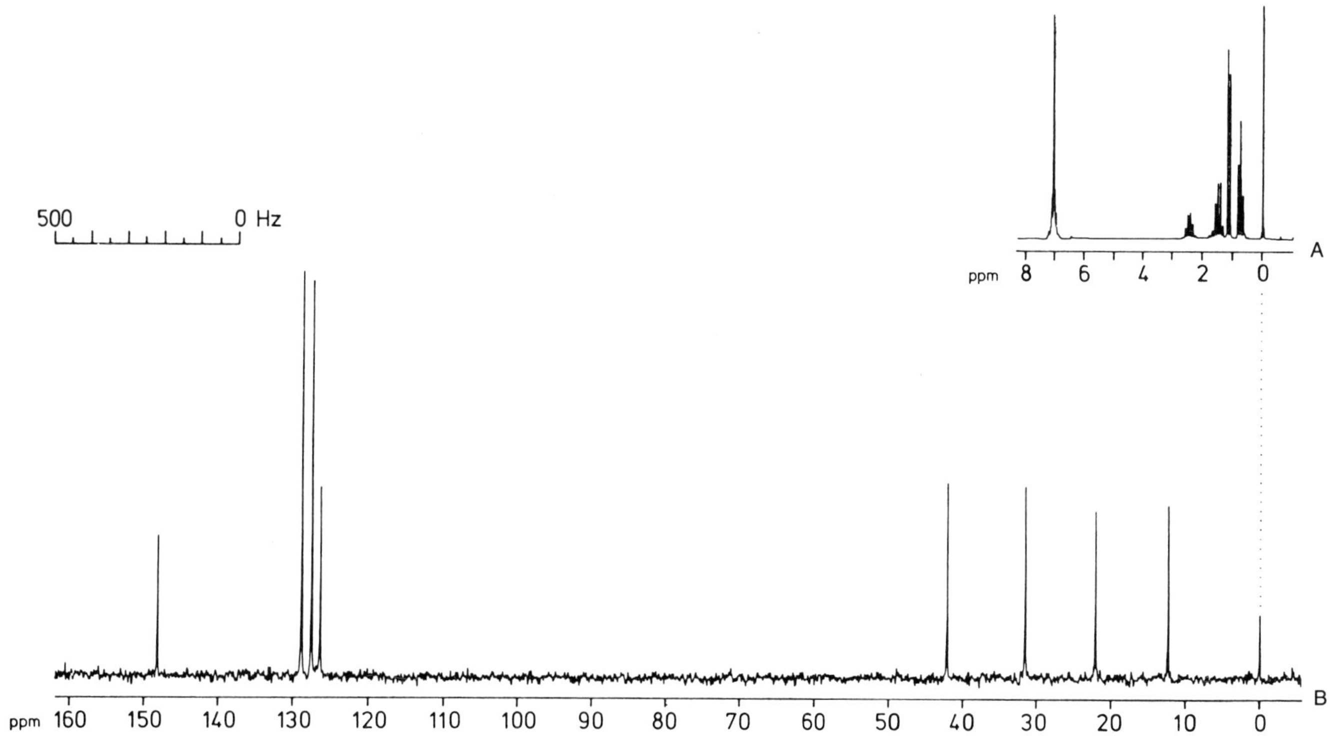

Fig. 6.1. Comparison of 80 MHz ^{1}H NMR spectrum (A) and proton-decoupled 20 MHz ^{13}C NMR spectrum (B) of *sec*-butylbenzene, 90% v/v in hexadeuterioacetone at 25° C. Both spectra were recorded at the same magnetic field strength and on the same frequency scale. The assignments were shown in Fig. 3.5.

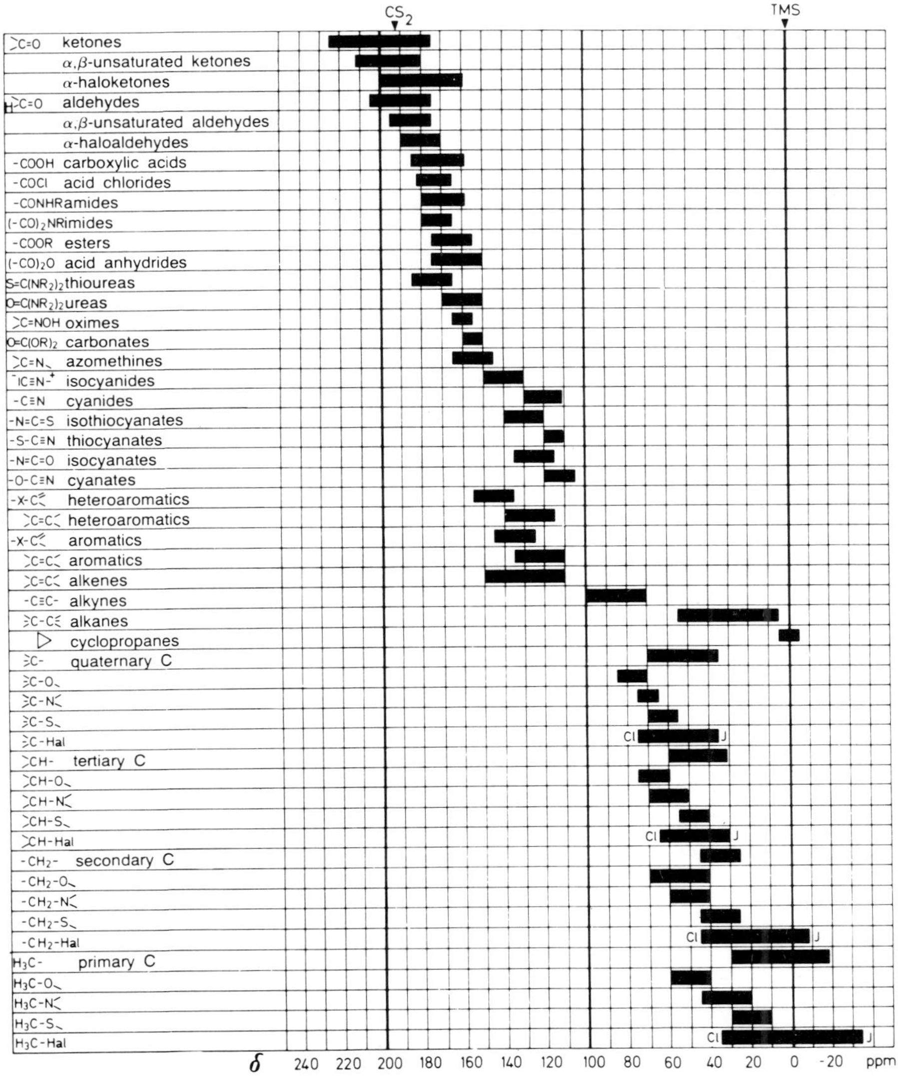

Fig. 6.2. ^{13}C shifts of functional carbon atoms in some classes of organic compounds, with TMS as a standard. Carbon disulfide was once commonly used as a standard with $\delta = 192.5$ ppm. (Source: E. Breitmaier, G. Bauer, *Pharmazie in Unserer Zeit*, **5**, 97, 1976.)

Fig. 6.3. Relationship between π electron density and ^{13}C chemical shifts of ring carbon atoms: A, in benzene and non-benzenoid aromatic ions; B, in benzene and pyridine.

The diamagnetic component σ_{dia} is a consequence of currents induced by the external magnetic field in the electron shells surrounding a nucleus. Intramolecular magnetic fields are thus produced, which according to Lenz's rule counteract the applied field $\vec{H}_0$ at the nucleus. The nucleus is *shielded*, and its resonance appears at a lower frequency (see Section 1.1). The smaller the distance r between the nucleus and the electrons, the larger the shielding field is at the nucleus ($\sigma_{dia} \sim r^{-1}$). For this reason, s electrons have a stronger shielding effect than do the electrons in the more extended p orbitals.

Unlike s electrons, p electrons can be excited in a magnetic field. This is the basis of the σ_{para} paramagnetic term. The latter opposes diamagnetic shielding and thus has a negative sign, and it increases the resonance frequency and therefore the chemical shift.

The intramolecular fields caused by neighboring groups (substituents) can weaken or strengthen the magnetic field at a nucleus. Such influences are described by the neighboring group term, σ'. Some groups ($-NR_2$, $-OH$, $-COOH$) can also interact intermolecularly. These interactions are subject to medium effects (solvent, pH) and may be taken into consideration by adding a medium effect term, σ_{med}, to Eq. (6.1).

As in hydrogen the electrons only occupy s orbitals, practically speaking, proton shifts depend on σ_{dia} and σ' only. In all nuclei such as ^{13}C, where electrons occupy higher (p) orbitals, the paramagnetic term σ_{para} predominates. Since this term depends on the electronic excitation, its value is inversely proportional to the electron excitation energy, ΔE. For example, this energy decreases from alkanes through alkenes to carbonyl compounds, as can be seen from the increasing wavelength at which these compounds absorb light in this series. The ^{13}C shifts increase in the same sequence (alkanes, 10-50 ppm; alkenes and aromatics, 100-150 ppm; carbonyl compounds, 170-220 ppm; see Fig. 6.2). σ_{para} also depends on whether the carbon atom under observation is singly, doubly, or triply bonded—the usual expression in MO theory is bond order—and how far the bonding orbitals extend in space.

The ^{13}C shift of nonbenzenoid aromatic ions correlates linearly with the relative π electron density at individual carbon atoms, taking the value for benzene as unity (see Fig. 6.3 A). With the negative ions, the higher π charge density at carbon leads to electron repulsion. This makes the carbon bonding orbitals extend further out so that the value of the paramagnetic component falls. In this way, the ^{13}C nuclei in the cyclopentadienyl anion are more strongly shielded than in benzene. The opposite is true for the carbon atoms in the tropylium cation, where six π electrons are shared by seven carbon atoms and where the shielding of the ^{13}C nuclei is less than in benzene (see Fig. 6.3 A).

In pyridine the carbon atoms at the α and γ positions are π electron-poorer than the carbon atoms in the benzene ring. This can be described using the following resonance formulas:

6A 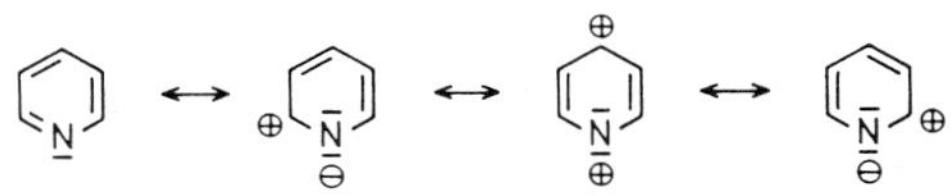

But the effect can also be seen in the rise of the ^{13}C shift of the pyridine carbon atoms with decreasing π electron density, $q\pi$, calculated using MO theory (see Fig. 6.3 B).

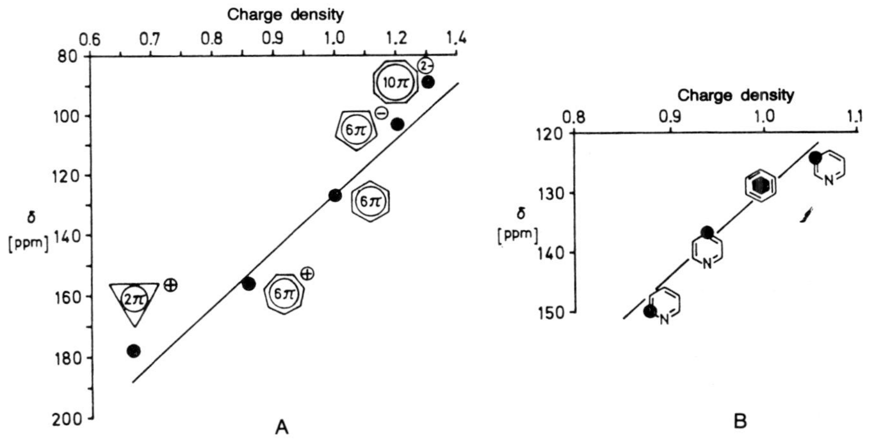

Fig. 6.3. Relationship between π electron density and ^{13}C chemical shifts of ring carbon atoms: A, in benzene and non-benzenoid aromatic ions; B, in benzene and pyridine.

Therefore, the resonance formulas familiar to chemists can be a key to the assignment of ^{13}C shifts. For example, the methoxy group in anisole raises the electron density at the *ortho* and *para* positions of the phenyl ring, leaving the *meta* carbon almost unaffected (see below). The consequence is that the *o* and *p* carbons are more shielded than the *m* carbons, which, considering their ^{13}C shift, are more like the benzene carbon atoms (128.5 ppm). An assignment of the *o* and *p* shifts is possible in this case by comparing the intensities since the NOEs are roughly the same. The *o* signal (113.5 ppm) is twice as strong as the *p* signal (120.5 ppm) because both *o* carbon atoms are equivalent.

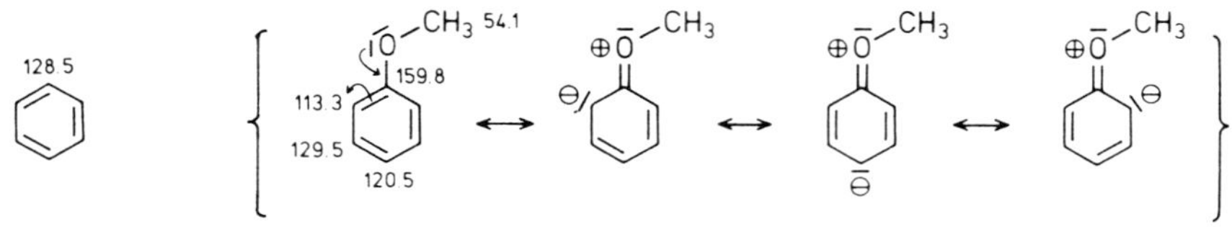

Benzene Anisole 6B

Conversely, one can evaluate the contribution of polar structures to a molecule's valence state from its ^{13}C shifts. Thus the CH_2 group's ^{13}C shift in ketene suggests considerable carbanionic character, and the chemical shift of C-3 in (*E*)-2-butenal (*trans*-crotonaldehyde) indicates that it carries a positive charge (vinylogy), as can be seen from the dipolar resonance formulas below:

Ketene:

(*E*)-2-Butenal:

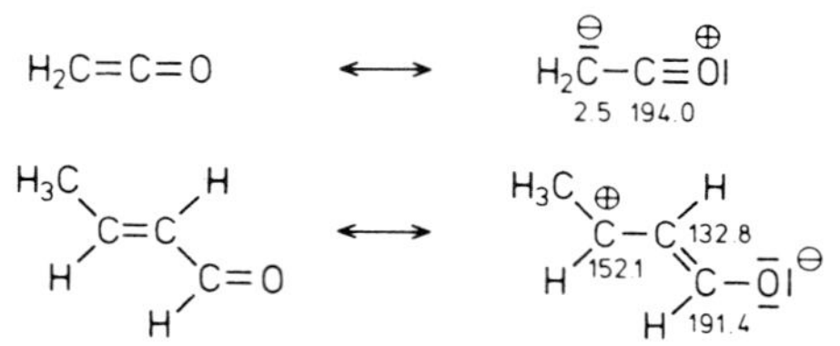

6C

One can see in all these examples that a carbanionic carbon atom is strongly shielded, whereas a positively charged carbon atom is weakly shielded. Extremely weak shieldings, and therefore large ppm values, are found in carbocations with a localized positive charge [1]. The *t*-butyl cation is an example of this. Delocalization of the positive charge leads to stronger shielding of the carbocationic carbon atom, as can be seen by comparing the ^{13}C chemical shifts of the *t*-butyl cation and the dimethylphenylcarbenium ion (see below):

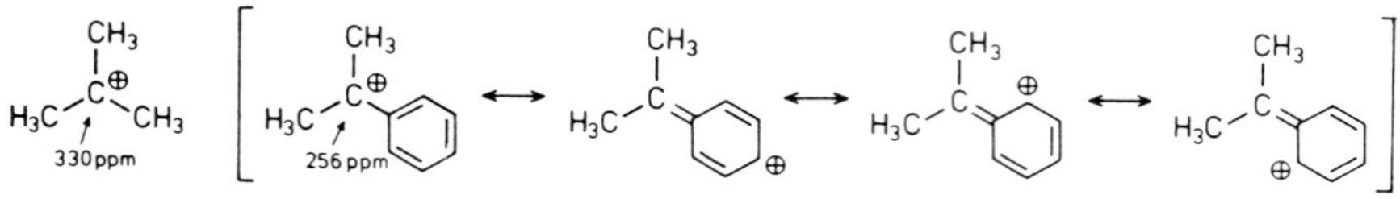

t-butyl-cation Dimethlphenylcarbeniumion 6D

^{13}C shifts can often be assigned by resorting to formalisms and concepts originating from organic chemistry. Additionally, strong analogies between ^{1}H and ^{13}C shifts exist: e.g., the increase of the ppm values for the alkane-alkene-alkyne series and within each kind of carbon with increasing electronegativity of a bonded substituent. Likewise, the ^{13}C resonances of cyclopropane carbon atoms and the cyclopropane proton resonances lie close to the corresponding TMS resonances. Yet, there are also enlightening differences. For instance, the resonances of aromatic and olefinic carbon atoms appear in the same range (100 to 160 ppm), whereas the aromatic ring protons differ from olefinic protons by large ppm values (**aromaticity criterion**) [1]. The reason for this difference is that the magnetic field caused by the ring current only affects nuclei over and around the aromatic ring but not the ring carbon atoms themselves.

Theoretical procedures exist for calculating ^{13}C chemical shifts: e.g., for hydrocarbons [3]. Empirical relationships between the type, number, distance, and configuration of substituents and a carbon atom's chemical shift have also proved to be useful for the practical application of ^{13}C shifts to chemical structural analysis. These relationships will be discussed in Chapter 7.

References

1. G. A. Oláh, *Angew. Chem.* **85**, 183 (1973).
2. H. Friebolin et al., *NMR-Spektroskopie, Eine Einführung mit Übungen*, Verlag Chemie GmbH, Weinheim, 1974, Taschentext Bd. 15.
3. H. Günther, *NMR-Spektroskopie, Eine Einführung*, Georg Thieme Verlag, Stuttgart, 1973.
4. H. Suhr, *Anwendungen der Kernmagnetischen Resonanz in der Organischen Chemie*, Springer Verlag, New York, 1965.
5. J. B. Stothers, *Carbon-13 NMR Spectroscopy*, Academic Press, New York, 1972.

CHAPTER 7

Interpretation of ^{13}C Chemical Shifts

7.1 Alkyl Group Chemical Shift Increments in Aliphatics

The analysis of many alkane ^{13}C NMR spectra has yielded several additivity relationships with which ^{13}C chemical shifts of alkane carbon atoms can be estimated by adding shift increments. According to the widely used Grant-Paul relation (Eq. 7.1) [1,2], the ^{13}C chemical shift δ_k of carbon atom k in the hydrocarbon chain

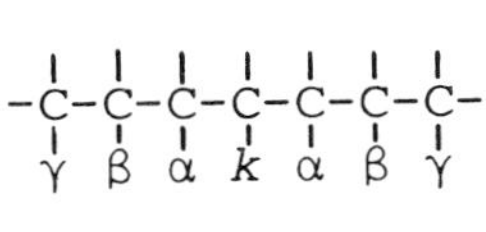

can be calculated from the chemical shift of methane, B (−2,3 ppm), and the shift increments, A_l, of the carbon atoms at positions l:

$$\delta_k = B + \Sigma A_l n_l + \Sigma S_{kl}$$

In this equation, n_l is the number of carbon atoms at the l position. The S_{kl} values are steric corrections, which must be applied to carbon atoms only at branching points (tertiary $-CHR_2$ and quaternary $-CR_3$). The chemical shift increments A_l and the corrections S_{kl} are listed in Table 7.1.

The use of the Grant-Paul relation (Eq. 7.1) is exemplified for carbon 3 in 2-methylhexane. According to the equation, two α increments (for C-2 and C-4), three β increments (for C-1, C-1′, and C-5), and one γ increment (for

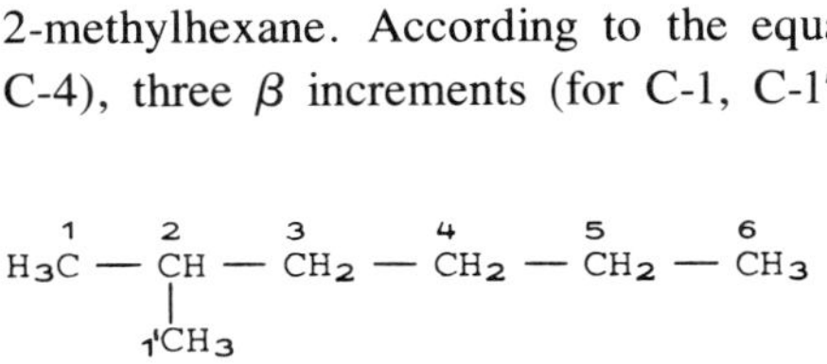

2-Methylhexane

	B	=	− 2.3 ppm
$\Sigma n_l A_l$	$2\ A_\alpha$	=	18.2 ppm
	$3\ A_\beta$	=	28.2 ppm
	A_γ	=	− 2.5 ppm
ΣS_{kl}	$S_{2^\circ 3^\circ}$	=	− 2.5 ppm

TABLE 7.1. ^{13}C CHEMICAL SHIFT INCREMENTS A_l (A) AND STERIC CORRECTIONS S_{kl} (B) FOR THE GRANT-PAUL RELATION (EQ. 7.1).
(The α-, β-, or γ-position with regard to the observed carbon atom k is designated position l. For primary, secondary, tertiary, and quarternary carbon atoms, CH_3, CH_2, CH, and CR_4, the abbreviations 1°, 2°, 3°, and 4° are used.)

A. A_l increments (±0.1 ppm):

A_α	9.1
A_β	9.4
A_γ	−2.5
A_δ	0.3
A_ϵ	0.2

B. S_{kl} corrections (±0.1 ppm):

	Neighboring Carbon Atom *l*			
Observed Carbon Atom k	1°	2°	3°	4°
1°	0	0	−1.1	−3.4
2°	0	0	−2.5	−7.5
3°	0	−3.7	−9.5	—
4°	−1.5	−8.4	—	—

C-6) must be added to the chemical shift of methane (B = −2.3 ppm). An $S_{2°3°}$ correction is also required because C-3 is secondary and is bonded to the tertiary carbon atom C-2.

The shifts of the other carbon nuclei can be determined in a similar fashion. The calculated values that appear in italics below agree well with the ^{13}C chemical shifts measured in deuteriochloroform. The signal caused by the equivalent methyl groups (C-1 and 1′) can be recognized by its double intensity.

```
        28.5   39.1   29.7   23.4   14.4    ppm (calculated)
        28.55  39.45  30.3   23.5   14.3    ppm (CDCl3, 50 Vol.%, 30° C)
H3C --- CH --- CH2 -- CH2 -- CH2 -- CH3
22.95   |
21.1    CH3
```

Using increment calculations, carbon-atom chemical shifts in long-chain alkyl groups can also be estimated. For example, the signal assignment was achieved for the C-17 side chain carbon atoms in this way in some cholestane derivatives [3,4]. Increment calculations are the only possible assignment aid when wide-band proton-decoupled ^{13}C spectra alone can be recorded owing to lack of substance or when the multiplets in off-resonance or coupled spectra overlap too much.

7.2 Substituent Increments in Aliphatics

If one replaces a hydrogen atom in an alkane chain by group X, the ^{13}C chemical shifts of the neighboring carbon atoms change. Comparing the measured ^{13}C chemical shifts of alkanes, R-H, and alkyl compounds, R-X, shift increments can be determined for the carbon atom directly bonded to X (α-position) and the more distant carbon atoms (at the β-, γ-, δ-, or ϵ-positions):

$$Z_l = \delta_{RX_l} - \delta_{RH_l} \tag{7.2}$$

The substituent increments of some groups are listed in Table 7.2. Note that the great majority of these substituents raise the chemical shifts of α- and β-carbon atoms ($Z_l > 0$, deshielding). A carbon atom at the γ position, on the contrary, is always shielded ($Z_l < 0$). The effect of a δ substituent on the ^{13}C shift of a carbon atom is usually 0 ppm and seldom exceeds 1 ppm (e.g., for $-\overset{+}{N}R_3$). The α effect is predominantly *inductive*. As Table 7.2 shows, its value depends on the electronegativity of the substituent X: i.e., on how strongly it polarizes (deshields) the C-X bond. In this way, the increment Z_α for halogens decreases in the order $-$I, $-$Br, $-$Cl, $-$F, as shown in the table.

Carbon atoms at the β position experience the substituent's intramolecular fields as well as its inductive effect. Since most of these fields are unknown, no convincing general explanation for the β effect exists as yet.

The sign of the γ effect, which is opposed to the α and β effects, is a consequence of the **van der Waals repulsion** between substituent X and the γ proton. For X = CH_3, the CH_3 and γ protons come so close to each other that the σ electrons of the C-H bonds are repelled. As a result, the CH_3 and γ carbon atoms are more strongly shielded.

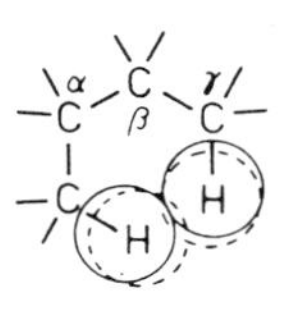

Shielding of γ carbon as a result of intramolecular van der Waals proton repulsion.

Substituent increments are useful aids for signal assignments. For example, using the $-NH_2$, $-\overset{+}{N}H_3$, $-$COOH and $-COO^-$ increments in Table 7.2 and the known ^{13}C chemical shifts of pentane, it is possible to predict the sequence in which the ^{13}C NMR signals appear for the amino acid L-lysine (see Tables 7.3 A and B). The α carbon atom of this amino acid is tertiary. Therefore, one must add the steric corrections $S_{2°3°}$ for C-α and C-β according to Table 7.1 B. Off-resonance decoupling does not allow any differentiation between the β, γ, δ, and ϵ carbon atoms (CH_2 groups).

The example of lysine also shows that the calculated values deviate consid-

TABLE 7.2. EMPIRICAL INCREMENTS [ppm] OF SUBSTITUENTS X REPLACING H IN ALKANES

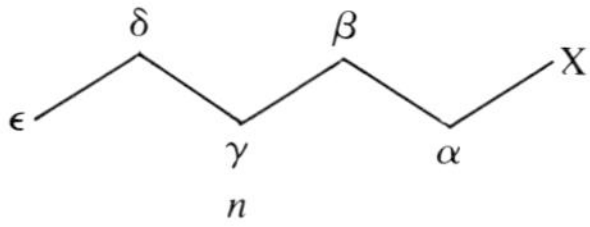

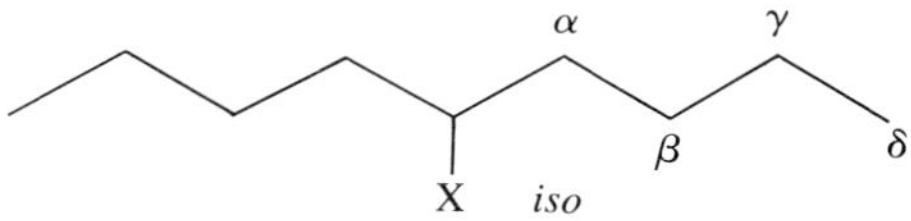

—X	Z_α		Z_β		Z_γ	Z_δ	Z_ϵ
	n	*iso*	*n*	*iso*			
—F	70	63	8	6	−7	0	0
—Cl	31	32	10	10	−5	−0.5	0
—Br	20	26	10	10	−4	−0.5	0
—I	−7	4	11	12	−1.5	−1	0
—O—	57	51	7	5	−5	−0.5	0
—$OCOCH_3$	52	45	6.5	5	−4	0	0
—OH	49	41	10	8	−6	0	0
—SCH_3	20.5	—	6.5	—	−2.5	0	0
—S—	10.5	—	11.5	—	−3.5	−0.5	0
—SH	10.5	11	11.5	11	−3.5	0	0
—NH_2	28.5	24	11.5	10	−5	0	0
—NHR	36.5	30	8	7	−4.5	−0.5	−0.5
—NR_2	40.5	—	5	—	−4.5	−0.5	0
—$\overset{+}{N}H_3$	26	24	7.5	6	−4.5	0	0
—$\overset{+}{N}R_3$	30.5	—	5.5	—	−7	−0.5	−0.5
—NO_2	61.5	57	3	4	−4.5	−1	−0.5
—NC	27.5	—	6.5	—	−4.5	0	0
—CN	3	1	2.5	3	−3	0.5	0
C=NOH *syn*	11.5	—	0.5	—	−2	0	0
C=NOH *anti*	16	—	4.5	—	−1.5	0	0
—CHO	30	—	−0.5	—	−2.5	0	0
CO	23	—	3	—	3	0	0
—$COCH_3$	29	23	3	1	−3.5	0	0
—COCl	33	28	2	2	−3.5	0	0
—COO−	24.5	20	3.5	3	−2.5	0	0
—$COOCH_3$ (and C_2H_5)	22.5	17	2.5	2	−3	0	0
—$CONH_2$	22	—	2.5	—	−3	−0.5	0
—COOH	20	16	2	2	−3	0	0
—Phenyl	23	17	9	7	−2	0	0
—$CH=CH_2$	20	—	6	—	−0.5	0	0
—C≡CH	4.5	—	5.5	—	−3.5	0.5	0

TABLE 7.3. DETERMINATION OF THE ^{13}C SIGNAL SEQUENCE OF L-LYSINE AND HEXANOL, USING THE SUBSTITUENT INCREMENTS IN TABLE 7.2 AND THE ^{13}C SHIFTS OF PENTANE AND HEXANE.

A. Lysine in the Dipolar Ion Form.

	H_2N —	CH_2 —	CH_2 —	CH_2 —	CH_2 —	CH (NH_3^+) —	COO^-
Pentane		13.7	22.6	34.6	22.6	13.7	
Z_{COO^-}		—	0.0	−2.5	3.5	24.5	
Z_{NH3+}		—	0.0	−4.6	7.5	26.0	
Z_{NH2}		28.3	11.3	−5.1	0.0	—	
S_{3o2o}		—	—	—	—	−3.7	
S_{2o3o}		—	—	—	−2.5	—	
δ calculated		*42.0*	*33.9*	*22.4*	*31.10*	*60.5*	*ppm*
δ measured[a]		40.0	30.7	22.4	27.2	55.3	ppm (in D_2O)

B. Lysine as an α-amino Carboxylic Acid.

	H_2N —	CH_2 —	CH_2 —	CH_2 —	CH_2 —	CH (NH_2) —	$COOH$
Pentane[b]		13.7	22.6	34.6	22.6	13.7	
Z_{COOH}		—	0.0	−2.8	2.0	20.1	
Z_{NH_2}		—	0.0	−5.1	11.3	28.3	
Z_{NH_2}		28.3	11.3	−5.1	0.0	—	
S_{3o2o}		—	—	—	—	−3.7	
S_{2o3o}		—	—	—	−2.5	—	
δ calculated		*42.0*	*33.9*	*21.6*	*33.4*	*58.4*	*ppm*
δ measured[b]		40.0	30.7	22.4	27.2	55.3	ppm (in D_2O)

C. Hexanol.

	H_3C —	CH_2 —	CH_2 —	CH_2 —	CH_2 —	CH_2 —	OH
Hexane	13.7	22.9	32.0	32.0	22.9	13.7	
Z_{OH}	—	—	*0.0*	*−6.2*	*10.1*	*49.0*	
δ calculated	*13.7*	*22.8*	*32.0*	*25.8*	*33.0*	*62.7*	*ppm*
δ measured	14.5	23.1	32.3	26.1	33.1	62.2	ppm

[a]E. Breitmaier, W. Voelter, *^{13}C NMR Spectroscopy: Methods and Applications,* "Monographs in Modern Chemistry," Vol. 5, H. E. Ebel, ed., Verlag Chemie GmbH, Weinheim, 2nd ed., 1978.

[b]J. B. Stothers, *Carbon-13 NMR Spectroscopy,* Academic Press, New York-London, 1972.

erably from the measured ones for the carbon atoms most strongly affected by interactions between the substituents (C-α and C-β). If the molecule contains only one polar substituent (hexanol, for example; see Table 7.3 C), the agreement is much better.

7.3 Substituent and Configurational Influences in Cyclohexane Derivatives

The α, β, and γ effects of substituents in monosubstituted cycloaliphatics [3-6] are qualitatively analogous to those in the aliphatics, as Table 7.4 shows for the example of monosubstituted cyclohexanes. The reference value here is the ^{13}C chemical shift of cyclohexane (27.1 $\pm$ 0.5 ppm, usually 27.6 ppm). Because of the molecular geometry, the van der Waals repulsion between substituent X and the protons is more marked in cyclohexanes than in open-chain aliphatics; therefore, C-4 is more strongly shielded by X ($Z_4 < 0$; see Table 7.4).

If a polysubstituted cyclohexane ring lies in a particular conformation at a sufficiently low temperature, the van der Waals repulsion between methyl and cyclohexane protons, for example, makes itself felt clearly, as Table 7.5 A shows for the methylcyclohexanes. It follows that axial (*a*) substituents shield more strongly than equatorial (*e*) substituents, particularly when occupying the

TABLE 7.4. EMPIRICAL INCREMENTS OF EQUATORIAL AND AXIAL SUBSTITUENTS IN CYCLOHEXANE. REFERENCE COMPOUND: CYCLOHEXANE WITH δ=27.6 ($\pm$0.5) PPM AND CORRECTION TERMS *S* FOR 1,1- AND 1,2-DIMETHYLCYCLOHEXANES[a]

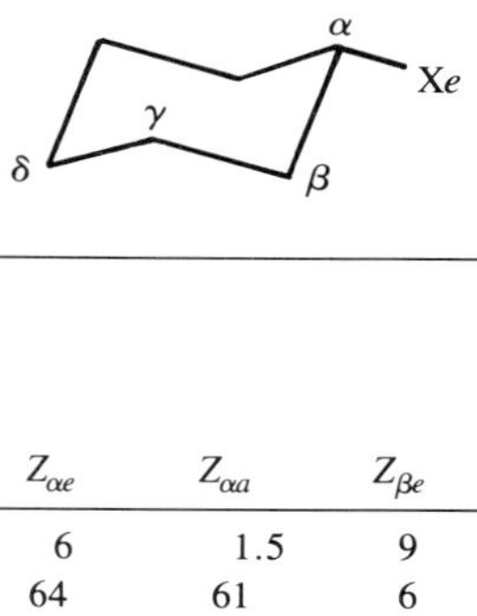

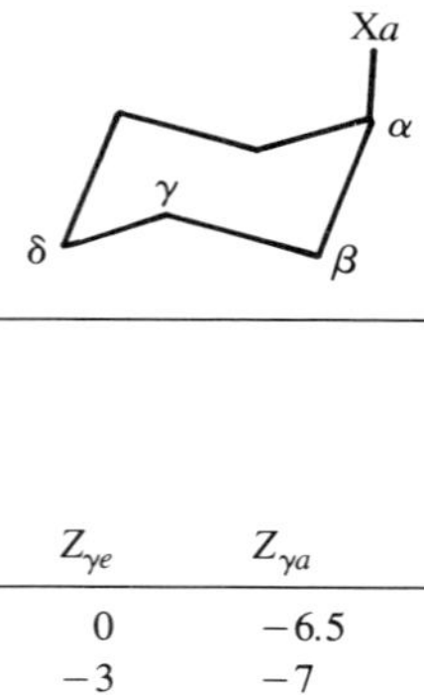

X	$Z_{\alpha e}$	$Z_{\alpha a}$	$Z_{\beta e}$	$Z_{\beta a}$	$Z_{\gamma e}$	$Z_{\gamma a}$	$Z_{\delta e}$	$Z_{\delta a}$
CH_3	6	1.5	9	5.5	0	−6.5	−0.3	0
F	64	61	6	3	−3	−7	−3	−2
Cl	33	33	11	7	0	−6	−2	−1
Br	25	28	12	8	1	−6	−1	−1
I	3	11	13	9	2	−4	−2	−1
CN	1	0	3	−1	−2	−5	−2	−1
NC	25	23	7	4	−3	−7	−2	−2
OH	43	39	8	5	−3	−7	−2	−1
OCH_3	52	47	4	2	−3	−7	−2	−1
$OCOCH_3$	46	42	5	3	−2	−6	−2	0
NH_2	24		10		−2		−1	

[a]Correction terms [ppm] for 1,1-Dimethylcyclohexane: $Z_{\alpha_a\alpha_e} = -3.8$, $Z_{\beta_a\beta_e} = -1.3$, and $Z_{\gamma_a\gamma_e} = 2.0$. Correction terms [ppm] for 1,2-Dimethylcyclohexane: $Z_{\alpha_e\beta_a} = -2.9$ (*cis*), $Z_{\alpha_e\beta_e} = -2.9$ (*trans*), $Z_{\alpha_a\beta_e} = -3.4$ (*cis*), $Z_{\beta_e\gamma_a} = -0.8$ (*cis*), and $Z_{\beta_a\gamma_e} = 1.6$ (*cis*).

TABLE 7.5. CONFIGURATION INCREMENTS IN METHYLCYCLOHEXANES (A) AND INCREMENT CALCULATIONS FOR THE RING CARBON ATOMS OF 1-*TRANS*-2-*CIS*-3-TRIMETHYLCYCLOHEXANE.

A. Configuration increments in methylcyclohexanes (from low temperature measurements).[a]

Configuration	Position of the Substituents	Z_α	Z_β	Z_γ	Z_δ
	*e*	6.0	9.0	0.0	−0.2
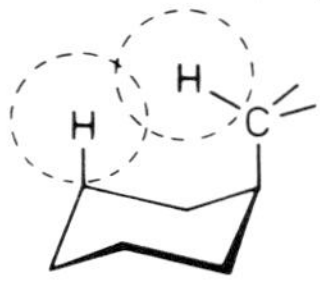	*a*	1.4	5.4	−6.4	−0.1
	$Z_{i_e} - Z_{i_a}$	4.6	3.6	6.4	0.1
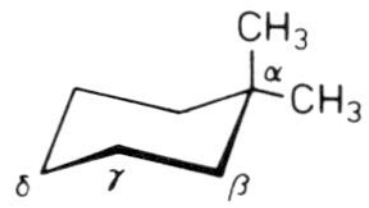	1,1-di- (*geminal*)	−3.8	−1.3	2.0	0.0
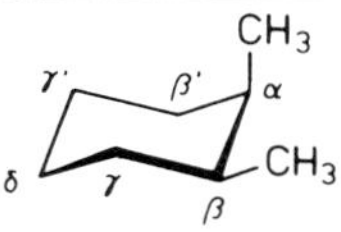	*1,2-cis*	−3.4 (α_a, β_e)	−2.9 (β_a, α_e) 1.6 (β_a, γ_e)	−0.8 (γ_a, β_e)	
	*1,2-trans*	−2.5 (α_e, β_e)			

B. Increment calculation for the ring carbon atoms of 1-*trans*-2-*cis*-3-trimethylcyclo-hexane.

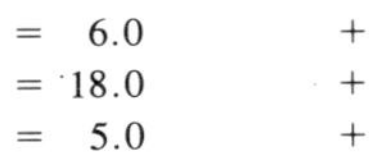

	C-1,3	C-2	C-4,6	C-5
	27.6	27.6	27.6	27.6
	$+ Z_{\alpha e}$ = 6.0			
	$+ Z_{\beta e}$ = 9.0	$+ Z_{\alpha e}$ = 6.0	$+ Z_{\beta e}$ = 9.0	$+ 2Z_{\gamma e}$ 0.0
	$+ Z_{\gamma e}$ = 0.0	$+ 2Z_{\beta e}$ = 18.0	$+ Z_{\gamma e}$ = 0.0	$+ Z_{\delta e}$ −0.2
	$+ Z_{\alpha e \beta e}$ = −2.5	$+ 2Z_{\alpha e \beta e}$ = 5.0	$+ Z_{\delta e}$ = −0.2	
δcalculated	*40.1*	*46.6*	*36.4*	*27.4 ppm*
δmeasured[b]	39.15	46.25	36.45	26.45 ppm

[a]D. K. Dalling, D. M. Grant, J. Am. Chem. Soc. **89,** 6612 (1967); and H. J. Schneider, R. Price, T. Keller, Angew. Chem. **83,** 759 (1971).

[b]E. Breitmaier, W. Voelter, *¹³C NMR Spectroscopy: Methods and Applications,* "Monographs in Modern Chemistry," Vol. 5, H. E. Ebel, ed., Verlag Chemie GmbH, Weinheim, 2nd ed., 1978.

γ position (see Table 7.4 B). In polysubstituted compounds, one must consider the steric interaction between the substituents themselves (see Table 7.5 A). The practical application of these increments to predict the ^{13}C shifts of 1-*trans*-2-*cis*-3-trimethylcyclohexane is illustrated in Table 7.5 B.

Other substituents shield the cyclohexane carbon atoms more strongly when they are axial in a fixed conformation, as Table 7.6 A shows for the meth-

TABLE 7.6. (A) ^{13}C CHEMICAL SHIFTS (δ [ppm]) OF ISOMERIC METHYLCYCLOHEXANOLS (B), ANOMERIC PYRANOSES (C), AND 2-*ENDO*-AND *EXO*-SUBSTITUTED NORBORNANES[a]

A. ^{13}C Chemical Shifts (δ [ppm]) of Isomeric Methylcyclohexanols.

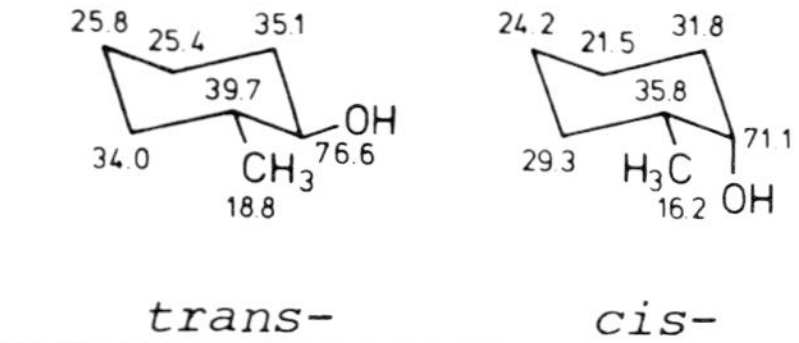

2-Methylcyclohexanol

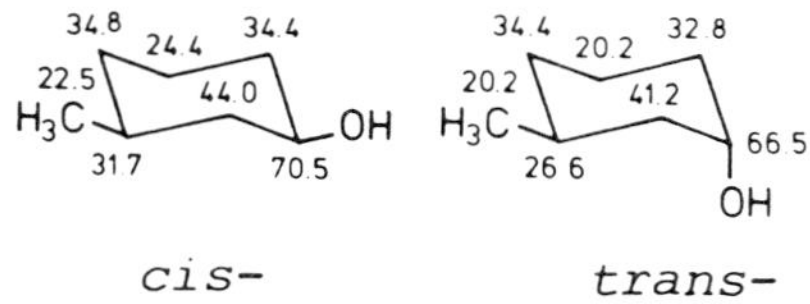

3-Methylcyclohexanol

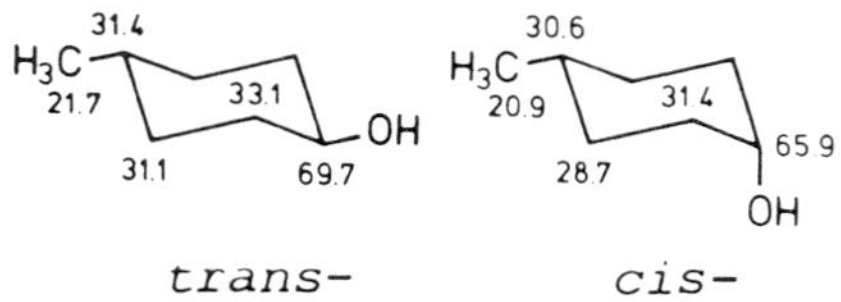

4-Methylcyclohexanol

B. Anomeric Pyranoses.

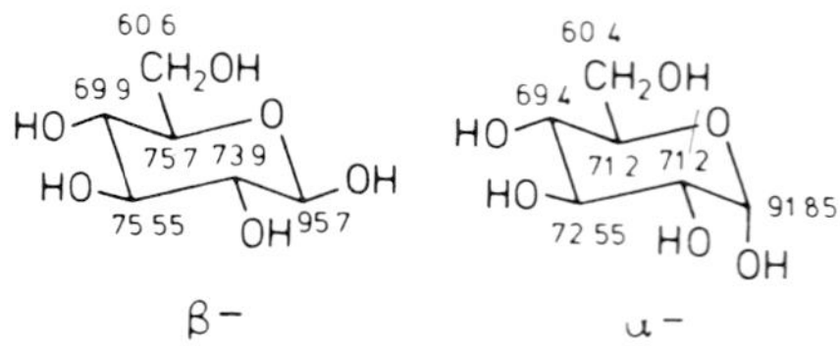

D-Glucopyranose (C1 conformation)

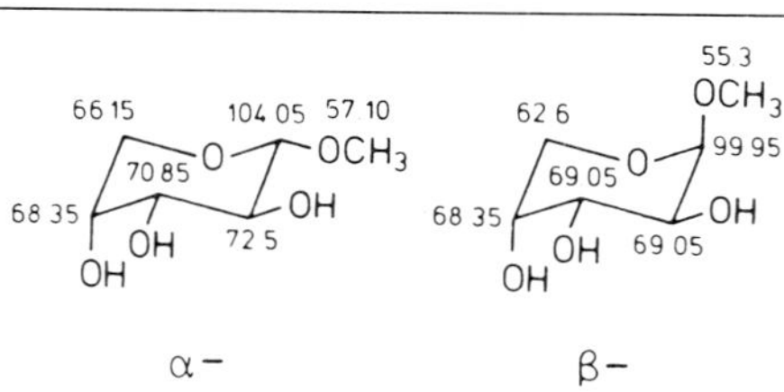

Methyl D-arabinopyranoside (1C conformation)

C. 2-*endo*-and *exo*-substituted Norbornanes

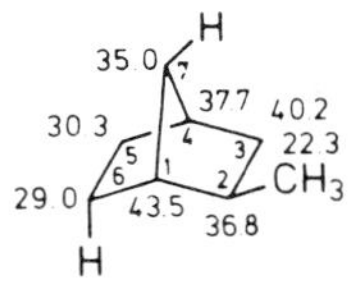

exo-

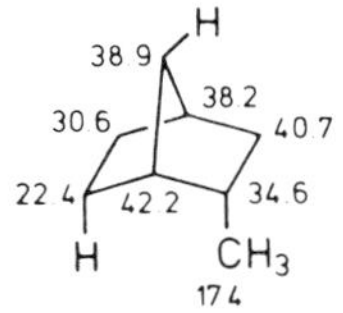

endo-

2-Methylnorbornane

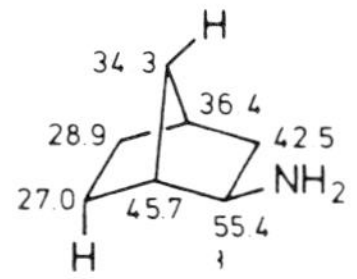

exo-

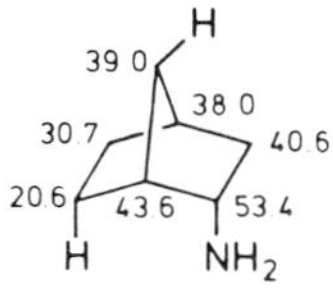

endo-

2-Aminonorbornane

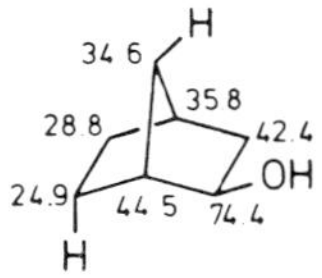

exo-

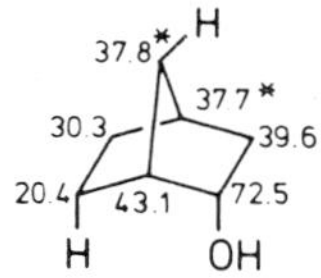

endo-

2-Norbornanol[b]

---

[a]J. B. Stothers, *Carbon-13 NMR Spectroscopy*, Academic Press, New York-London, 1972; G. C. Levy, G. L. Nelson, *Carbon-13 Nuclear Magnetic Resonance for Organic Chemists*, Wiley-Interscience, New York, 1972; and E. Breitmaier, W. Voelter, *^{13}C NMR Spectroscopy: Methods and Applications*, "Monographs in Modern Chemistry," Vol. 5, H. E. Ebel, ed., Verlag Chemie GmbH, Weinheim, 2nd ed., 1978.

[b]Assignments are interchangeable.

ylcyclohexanols. The same is true for the heteroanalogs of cyclohexane (α- and β-pyranose derivatives; see Table 7.6 B). Finally, the γ effect is also responsible for the fact that **endo** substituents on the norbornane ring shield C-6 strongly, whereas **exo** substituents exert this effect on C-7 (see Table 7.6 C).

7.4 Substituent and Configurational Influences in Alkenes

The ^{13}C chemical shifts of alkene carbon atoms can be determined from the increments A_l, the steric corrections S, and the chemical shift of ethylene (123.3 ppm), according to the empirical additivity relationship (Eq. 7.2) [7]. For example, the chemical shifts of the olefinic carbon atoms of *cis*-2-hexene are calculated and compared with the experimental values in Table 7.7 C, using the increments A_l and the steric corrections from Tables 7.7 A and B.

$$\begin{array}{cccccccc} |\gamma' & |\beta' & |\alpha' & k & \alpha| & \beta| & \gamma| & \\ -C- & C- & C- & C= & C- & C- & C- & C- \\ | & | & | & | & | & | & | & | \end{array}$$

$$\delta_k = 123.3 + \Sigma A_l + \Sigma A_l' + \Sigma S$$

Steric interactions, in the sense of the γ effect discussed for alkanes and cycloalkanes, are the cause of the greater shielding of *cis* alkyl carbon atoms in alkenes (see Table 7.8). The relationship found for olefinic carbon atoms in alkenes, $(\delta_{trans} - \delta_{cis}) > 0$, can reverse when polar substituents such as halogens are attached to the double bond.

TABLE 7.7. INCREMENTS FOR THE OLEFINIC CARBON ATOMS IN ALKENES (A), STERIC CORRECTIONS *S* FOR POLYSUBSTITUTION (B), AND ESTIMATION OF THE ^{13}C CHEMICAL SHIFTS FOR C-2 AND C-3 OF *CIS*-2-HEXENE (C) USING EQUATION (7.2).

A. A_l increments:

A_α	=	10.6 ppm
A_β	=	4.9 ppm
A_γ	=	−1.5 ppm
$A_{\alpha'}$	=	−7.9 ppm
$A_{\beta'}$	=	−1.8 ppm
$A_{\gamma'}$	=	1.5 ppm

B. Steric corrections for

each pair of *cis*-α,α' carbon atoms:	$S_{cis\,\alpha,\alpha'}$	=	−1.1 ppm
each pair of *geminal* α,α carbon atoms:	$S_{gem\,\alpha,\alpha}$	=	−4.8 ppm
each pair of *geminal* α,α' carbon atoms:	$S_{gem\,\alpha,\alpha'}$	=	2.5 ppm
one or more β substituents:	S_β	=	2.3 ppm

C. Estimated ^{13}C chemical shifts.[a]

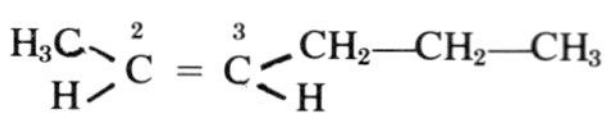

	C-2		C-3
	123.3 ppm		123.3 ppm
$+ A_\alpha$	10.6 ppm	$+ A_\alpha$	10.6 ppm
$+ A_{\alpha'}$	−7.9 ppm	$+ A_\beta$	4.9 ppm
$+ A_{\beta'}$	−1.8 ppm	$+ A_\gamma$	−1.5 ppm
$+ A_{\gamma'}$	1.5 ppm	$+ A_{\alpha'}$	−7.9 ppm
$+ S_{cis\alpha,\alpha'}$	−1.7 ppm	$+ S_{cis\alpha,\alpha'}$	−1.7 ppm

[a]Measured values: δ = 124.0 ppm   127.7 ppm
δ = 122.7 ppm   129.7 ppm

E. Breitmaier, W. Voelter, *^{13}C NMR Spectroscopy: Methods and Applications,* "Monographs in Modern Chemistry," Vol. 5, H. E. Ebel, ed., Verlag Chemie GmbH, Weinheim, 2nd ed., 1978.

As compared with olefinic carbon atoms (sp^2), those with linear bonding geometry (sp) are more strongly shielded in alkynes and, on the contrary, are much more weakly shielded in allenes (see Table 7.9). The large chemical shift ($\delta > 200$ ppm) of the central allenic carbon atom is ascribed to the paramagnetic component of the shielding constants (Eq. 6.1) and its causes.

TABLE 7.8. ^{13}C CHEMICAL SHIFTS OF *Z-E* ISOMERIC ALKENES.

Z		*E*	$\delta_Z - \delta_E$ [ppm] Olefin C-Atom	α-C-Atom
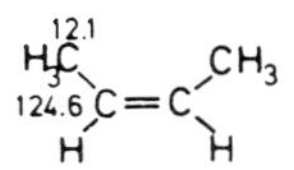	2-Butene	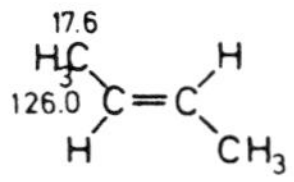	1.4	5.5
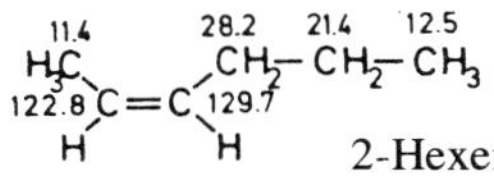	2-Hexene	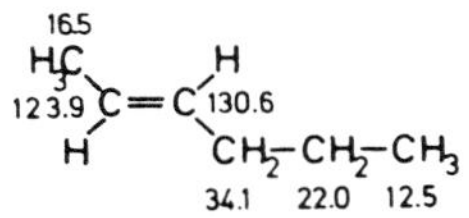	1.1 0.9	5.1 5.9
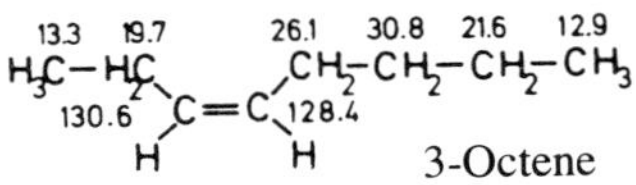	3-Octene	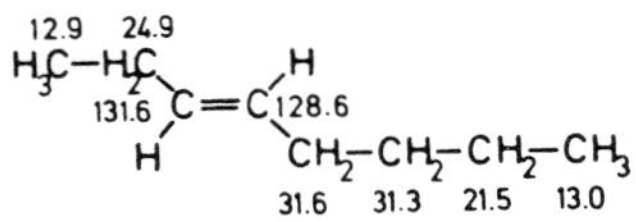	1.0 0.4	5.2 5.5
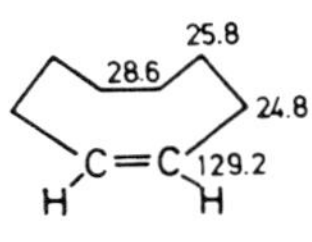	Cyclooctene	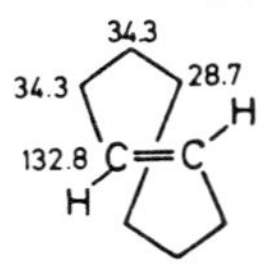	3.6	9.5
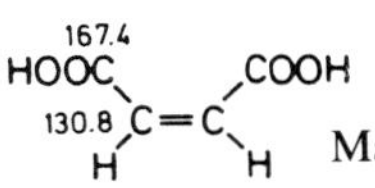	Maleic acid / Fumaric acid	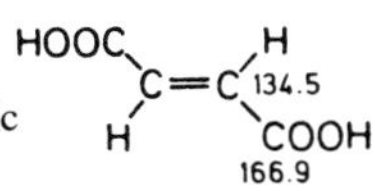	3.7	−0.5
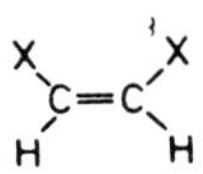	X=	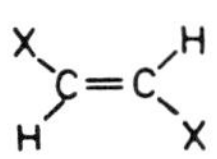		
120.8	CN	120.2	−0.6	
121.1	Cl	119.2	−1.9	
116.2	Br	109.2	−7.0	
96.3	I	79.2	−17.1	

TABLE 7.9. ^{13}C CHEMICAL SHIFTS OF SELECTED ALKYNES AND ALLENES.

Compound	Structure	^{13}C chemical shifts (ppm)
1-Hexyne	H—C≡C—CH$_2$—CH$_2$—CH$_2$—CH$_3$	C≡C: 67.4, 82.8; CH$_2$: 17.4, 29.9, 21.2; CH$_3$: 12.9
2-Hexyne	H$_3$C—C≡C—CH$_2$—CH$_2$—CH$_3$	H$_3$C: 1.7; C≡C: 73.7, 76.9; CH$_2$: 19.6, 21.6; CH$_3$: 12.1
3-Hexyne	H$_3$C—CH$_2$—C≡C—CH$_2$—CH$_3$	H$_3$C: 14.4; CH$_2$: 12.0; C≡C: 79.9
Allene	H$_2$C=C=CH$_2$	=C=: 213.5; CH$_2$: 74.8
3-Methyl-1,2-butadiene	H$_2$C=C=C(CH$_3$)$_2$	=C=: 207.3; C(CH$_3$)$_2$: 93.4; CH$_3$: 19.5
2,3-Pentadiene	H$_3$C(H)C=C=C(H)CH$_3$	=C=: 207.1; CH: 85.4; CH$_3$: 14.6

7.5 Substituent Increments for Aromatics

The chemical shift increments put together in Table 7.10 for the carbon atoms in the α or 1-, *ortho, meta,* or *para* position with respect to substituent X result from the differences between the ^{13}C chemical shifts of benzene (128.5 ppm) and of the carbon atoms in monosubstituted derivatives.

α-effects in benzene are also predominantly inductive. They manifest themselves as ^{13}C chemical shifts that increase with the electronegativity of substituent X, such as in the series I, Br, Cl, F (see Table 7.10).

The substituent increments for the *o*-, *m*-, and *p*-positions can be explained better with the resonance model than by resorting to inductive effects. If the substituent releases electrons into the ring, the electron density at the *o*- and *p*-positions increases as is expressed in the resonance formulas:

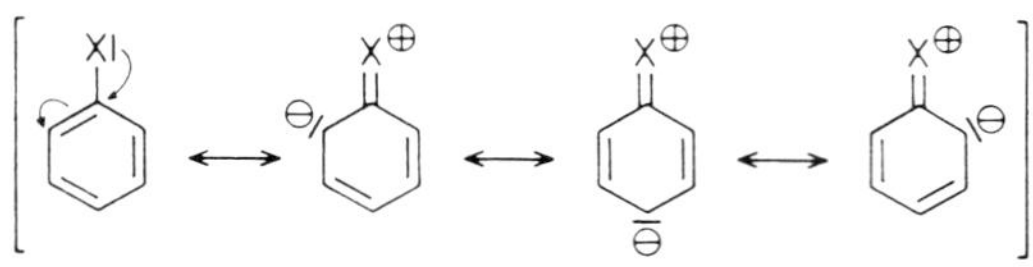

On the contrary, under the influence of an electron-attracting substituent, the *o*- and *p*-positions become electron-deficient:

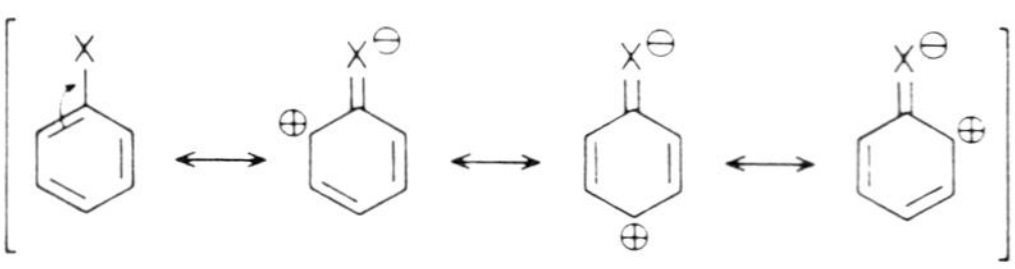

Considering these formulations, the electron density at the *m*-position changes only slightly with variations of substituent X, and the ^{13}C chemical shifts of substituted benzenes confirm this. The *meta* increments, Z_{lm}, are with few exceptions less than 1 ppm.

TABLE 7.10. EMPIRICAL SUBSTITUENT INCREMENTS IN BENZENE DERIVATES

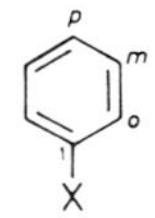

X	Z_1	Z_o	Z_m	Z_p
—CH_3	9.3	−0.1	0.7	−3.0
—CH_2CH_3	14.9	−1.3	−0.7	−3.3
—$CH_2CH_2CH_3$	14.1	−0.2	0.1	−2.7
—$CH(CH_3)_2$	20.3	−2.0	−0.1	−2.6
—$C(CH_3)_3$	21.8	−3.9	−0.9	−3.6
—Cyclopropyl (—C_3H_5)	15.1	−3.3	−0.6	−3.6
—$CH{=}CH_2$	9.1	−2.4	−0.2	−0.9
—Phenyl (—C_6H_5)	13.0	−1.1	0.5	−1.0
—$CH{\equiv}CH$	−5.8	3.9	0.1	0.4
—CF_3	−9.0	−2.2	0.3	3.2
—CH_2F	8.1	−0.9	0.1	0.2
—CH_2Cl	9.4	0.3	0.4	0.2
—CH_2Br	9.7	1.0	1.3	0.6
—CH_2OH	13.3	−0.8	0.6	−0.4
—$CH_2OCH_2C_6H_5$	10.5	−0.5	0.5	−0.5
—CH_2NH_2	15.5	−1.1	0.0	−1.9
—$CH_2NHCH_2C_6H_5$	11.9	−0.5	−0.3	−1.7
—$CH_2SCH_2C_6H_5$	10.5	0.3	0.9	−1.5
—F	35.1	−14.4	0.9	−4.4
—Cl	6.4	0.2	1.0	−2.0
—Br	−5.9	3.0	1.5	−1.5
—I	−32.3	9.9	2.6	−0.4
—OH	26.6	−12.8	1.6	−7.1
—OCH_3	31.4	−14.4	1.0	−7.8
—OC_6H_5	29.2	−9.4	1.4	−5.3
—$OCOCH_3$	23.0	−6.4	1.3	−2.3
—$OCOC_6H_5$	23.5	−6.0	1.7	−2.0
—OCN	24.6	−13.2	2.2	−1.5
—SH	2.0	0.6	0.2	−3.3
—SCH_3	10.1	−1.7	0.3	−3.5
—SC_6H_5	7.3	2.4	0.6	−1.6
—NH_2	20.2	−14.1	0.6	−9.6
—NH—NH_2	23.8	−5.7	1.4	−8.6
—$NHCH_3$	21.9	−16.4	0.6	−12.6
—$N(CH_3)_2$	22.2	−15.8	0.5	−11.8
—NHC_6H_5	14.7	−10.6	0.8	−7.6
—$N(C_6H_5)_2$	19.0	−4.6	0.9	−5.8
—$NHCOCH_3$	11.1	−9.9	0.2	−5.6
—$N{=}CHC_6H_5$	24.5	−6.9	1.0	−1.9
—$N{=}C{=}NC_6H_5$	10.8	−3.6	1.7	−2.2
—$N{=}C{=}O$	5.4	−3.7	1.2	−2.6
—$N{=}C{=}S$	17.4	−2.2	2.0	−0.4
—$P(C_6H_5)_2$	9.3	5.5	0.3	0.0

X	Z_I	Z_o	Z_m	Z_p
$—NO_2$	20.6	−4.3	1.3	6.2
$—N{=}NC_6H_5$	24.2	−5.5	1.2	3.3
$—N(CH_3)_3{}^+I^-$	8.8	−8.2	0.2	3.0
—COCl	4.8	2.9	0.6	6.9
$—COO^-Na^+$	10.3	2.8	2.2	5.1
—COOH	2.9	1.3	0.4	4.6
$—COOCH_3$	2.1	1.2	0.0	4.4
$—COOC_2H_5$	2.4	1.2	−0.1	4.3
$—COOC_6H_5$	1.9	2.3	0.8	6.0
$—CONH_2$	5.8	−1.1	−0.3	2.7
—CHO	8.2	1.2	0.5	5.8
$—COCH_3$	8.9	0.1	−0.1	4.5
$—COC_6H_5$	9.1	1.5	−0.2	3.8
—CN	−15.5	4.1	1.4	5.0
$—SO_3H$	17.2	3.3	−0.1	5.9
$—SO_2Cl$	15.8	1.4	−1.6	7.0

For electron releasing substituents such as −OR and $-NR_2$, the *p*-increments are negative without exception, in keeping with the formulated increase in electron density at the *p*-position. Conversely, the Z_{Ip} values for electron-withdrawing substituents such as $-NO_2$ and −CHO are positive (see Table 7.10), corresponding to the reduction in electron density at the *p*-position. Thus the increments show a behavior that parallels the activation of the *p*-carbon toward electrophilic substitution.

The *o*-increments behave in a less clear-cut fashion, as shown in Table 7.10, because the inductive effects and intramolecular fields owing to the substituent contribute to the shielding at the *o*-position because of the smaller distances involved. Of course, the mesomeric effect often predominates, as can be seen from the sign of the Z_o increments (e.g., $Z_{oOCH_3} = -15.0$ ppm, in contrast to $Z_{oCN} = 3.5$ ppm).

The increments discussed here can be used to predict the ^{13}C chemical shifts of polysubstituted benzenes, assuming that these effects are additive. For example, starting with the ^{13}C chemical shift of benzene (128.5 ppm) and the $COCH_3$ and OCH_3 increments, one can calculate the δ values for *m*-methoxyacetophenone in order to assign the measured chemical shifts by means of the calculated δ's, looking for the best agreement between both sets of values (see Table 7.11 A). If interactions exist between the substituents (van der Waals, dipole-dipole, hydrogen bonding), one must expect the experimental data to differ by up to 5 ppm from the values determined by increment addition, as Table 7.11 B shows for the example of vanillic acid. In such

TABLE 7.11. ESTIMATION OF THE ^{13}C CHEMICAL SHIFTS OF *M*-METHOXYACETOPHENONE (A) AND VANILLIC ACID (B) BY INCREMENT ADDITION USING THE VALUES GIVEN IN TABLE 7.10.

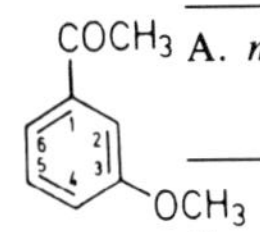

A. *m*-Methoxyacetophenone.

	C-1		C-2		C-3		C-4		C-5		C-6
	128.5		128.5		128.5		128.5		128.5		128.5
Z_{1COCH_3}	9.3	Z_{oCOCH_3}	0.2	Z_{mCOCH_3}	0.2	Z_{pCOCH_3}	4.2	Z_{mCOCH_3}	0.2	Z_{oCOCH_3}	0.2
Z_{mOCH_3}	0.9	Z_{oOCH_3}	−15.0	Z_{1OCH_3}	31.3	Z_{oOCH_3}	−15.0	Z_{mOCH_3}	0.9	Z_{pOCH_3}	−8.1
Estimated values	138.7		113.7		160.0		117.7		129.6		120.6 ppm
Measured values	137.4 (in $CDCl_3$)		111.4		158.5		118.5		128.4		118.5 ppm

($COCH_3$ 195.8 and 25.2 ppm; OCH_3 54.0 ppm)

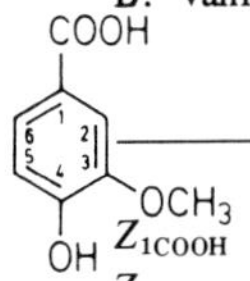

B. Vanillic acid.

	C-1		C-2		C-3		C-4		C-5		C-6
	128.5		128.5		128.5		128.5		128.5		128.5
Z_{1COOH}	2.4	Z_{oCOOH}	1.6	Z_{mCOOH}	−0.1	Z_{pCOOH}	4.8	Z_{mCOOH}	−0.1	Z_{oCOOH}	1.6
Z_{mOCH_3}	0.9	Z_{oOCH_3}	−15.0	Z_{1OCH_3}	31.3	Z_{oOCH_3}	−15.0	Z_{mOCH_3}	0.9	Z_{pOCH_3}	−8.1
Z_{pOH}	−7.6	Z_{mOH}	1.6	Z_{oOH}	−12.6	Z_{1OH}	26.9	Z_{oOH}	−12.6	Z_{mOH}	1.6
Estimated values	124.2		117.0		147.1		145.2		116.9		123.6 ppm
Measured values	123.75 (singlet)		114.85 (doublet)		151.05 (singlet)		147.1 (singlet)		112.7 (doublet)		124.15 ppm (doublet)[a]

(in $CDCl_3$); COOH 168.75 ppm; $OCH_3$55.85 ppm

[a]The multiplicities come from an off-resonance proton-decoupled ^{13}C NMR spectrum.

cases, an unequivocal assignment requires additional work (labeling, decoupling experiments, or measurement of long-range coupling constants).

7.6 Chemical Shift Increments of Heteroatoms and Substituents in Heteroaromatics

By comparing the ^{13}C chemical shifts of pyridine and benzene, the nitrogen increments can be obtained (Table 7.12 A). These heteroatom increments

TABLE 7.12. NITROGEN INCREMENTS (A) AND SUBSTITUENT EFFECTS OF SELECTED GROUPS (B) IN PYRIDINE

A. Nitrogen increments.

$Z_{2,6} = 149.7 - 128.5 = 21.2$ ppm
$Z_{3,5} = 124.2 - 128.5 = -4.3$ ppm
$Z_4 = 136.2 - 128.5 = 7.7$ ppm

$\delta_2 = 149.7 + Z_{i2}$
$\delta_3 = 124.2 + Z_{i3}$
$\delta_4 = 136.2 + Z_{i4}$
$\delta_5 = 124.2 + Z_{i5}$
$\delta_6 = 149.7 + Z_{i6}$

B. Substituent effects.

2-Substituent	Z_{22}	Z_{23}	Z_{24}	Z_{25}	Z_{26}
—CH_3	9.1	−1.0	−0.1	−3.4	−0.1
—CH_2—CH_3	14.0	−2.1	0.1	−3.1	0.2
—Br	−6.7	4.8	3.3	−0.5	1.4
—NH_2	11.3	−14.7	2.3	−10.6	−0.9
—CHO	3.5	−2.6	1.3	4.1	0.7
—$COCH_3$	4.3	−2.8	0.7	3.0	−0.2
—CN	−16.1	5.1	1.7	3.7	1.8
3-Substituent	Z_{32}	Z_{33}	Z_{34}	Z_{35}	Z_{36}
—CH_3	1.3	9.0	0.2	−0.8	−2.3
—CH_2—CH_3	0.3	15.0	−1.5	−0.3	−1.8
—Br	2.1	−2.6	2.9	1.2	−0.9
—NH_2	−11.9	21.5	−14.2	0.9	−10.8
—CHO	2.4	7.9	0.0	0.6	5.4
—$COCH_3$	0.5	−0.3	−3.7	−2.7	4.2
—CN	3.6	−14.2	4.6	0.7	4.4
4-Substituent	Z_{42}	Z_{43}	Z_{44}	Z_{45}	Z_{46}
—CH_3	0.5	0.8	10.8	0.8	0.5
—CH_2—CH_3	0.0	−0.3	15.9	−0.3	0.0
—Br	3.0	3.4	−3.0	3.4	3.0
—NH_2	0.9	−13.8	19.6	−13.8	0.9
—CHO	1.7	−0.6	5.5	−0.6	1.7
—$COCH_3$	1.6	−2.6	6.8	−2.6	1.6
—CN	1.7	2.5	15.3	2.5	1.7

show that the electron density is diminished at the α- and γ-positions of pyridine and raised at the β-positions, as was formulated in Section 6.2. For example, the chemical shift increments put together in Table 7.12 B for some substituents on the carbon atoms of the pyridine ring can be used to solve Exercise 62 (p. 00).

When benzologous heterocycles carry substituents on the benzene ring, a preliminary assignment is often possible using the substituent increments in Table 7.10. For example, the ^{13}C resonances of 1,9-dinitrophenazine can be assigned using the NO_2 increments in the table and the ^{13}C chemical shifts of phenazine (the calculated values appear in italics):

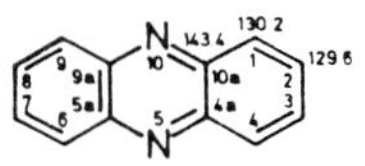

Phenazine

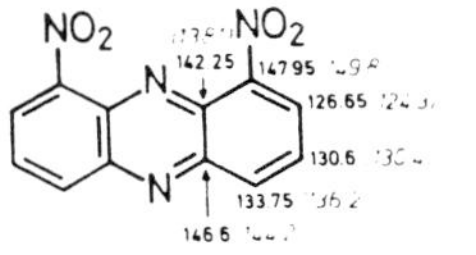

1,9-Dinitrophenazine

In the case of benzologous pyridines such as isoquinoline, naphthalene is an appropriate reference substance for the nitrogen increments (see Table 7.13 A); with its help, the ^{13}C chemical shifts of 2,6- and 2,7-naphthyridine can be estimated (Table 7.13 B).

7.7 Additivity of Substructural Shifts

In the spectral assignment and structural elucidation of more complicated molecules, the comparison of spectra often proves useful. The additivity of ^{13}C chemical shifts of whole pieces of the molecule is the basis of this method. To assign the ^{13}C NMR spectrum of a large molecule, appropriate smaller molecules are chosen in such a way that they are, or contain, parts of the larger molecule and have ^{13}C signals which have been assigned unequivocally. Comparison of the ^{13}C chemical shifts using the best fit as a criterion then leads to a preliminary assignment. Table 7.14 A illustrates this procedure for the alkaloid coniine. The ^{13}C chemical shifts of flavine-adenine dinucleotide (FAD; see Table 7.14 B) can be assembled very precisely using the data for adenosine 5′-diphosphate and riboflavine 5′-monophosphate, the building blocks of the molecule [6].

7.8 Conjugation Effects

The characteristic equalization of bond lengths upon conjugation is reflected in ^{13}C resonance spectra as a decrease in the difference between the shieldings

TABLE 7.13. ^{13}C CHEMICAL SHIFT INCREMENTS FOR NITROGEN IN ISOQUINOLINE (A), AND THEIR USE IN THE ASSIGNMENT OF THE ^{13}C RESONANCES OF 2,6- AND 2,7-NAPHTHYRIDINES (B).

A. ^{13}C chemical shift increments.

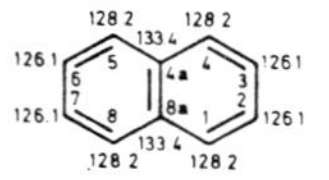

Naphthalene

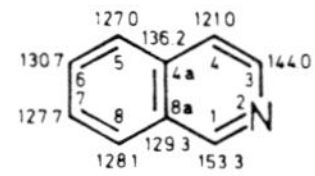

Isoquinoline

Z_1	=	153.3	−	128.2	=	25.1
Z_3	=	144.0	−	126.1	=	17.9
Z_4	=	121.0	−	128.2	=	−7.2
Z_5	=	127.0	−	128.2	=	−1.2
Z_6	=	130.7	−	126.1	=	4.6
Z_7	=	127.7	−	126.1	=	1.6
Z_8	=	128.1	−	128.2	=	−0.1
Z_{4a}	=	136.2	−	133.4	=	2.8
Z_{8a}	=	129.3	−	133.4	=	−4.1

B. Use of the increments in assigning ^{13}C resonances.

(b)

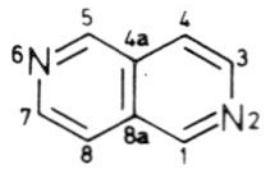

2,6-Naphthyridine

			calculated	measured
$\delta_{1,5}$	$= 128.2 + Z_1 + Z_5$	$= 128.2 + 25.1 - 1.2$	= 152.1	152.6
$\delta_{3,7}$	$= 126.1 + Z_3 + Z_7$	$= 126.1 + 17.9 + 1.6$	= 145.6	147.2
$\delta_{4,8}$	$= 128.2 + Z_4 + Z_8$	$= 128.2 - 7.2 - 0.1$	= 120.9	119.9
$\delta_{4a,8a}$	$= 133.9 + Z_{4a} + Z_{8a}$	$= 133.9 + 2.8 - 4.1$	= 132.6	131.0

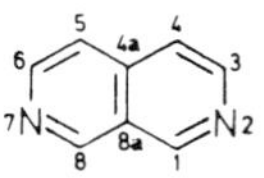

2,7-Naphthyridine

			calculated	measured
$\delta_{1,8}$	$= 128.2 + Z_1 + Z_8$	$= 128.1 + 25.1 - 0.1$	= 153.1	153.4
$\delta_{3,6}$	$= 126.1 + Z_3 + Z_6$	$= 126.1 + 17.9 + 4.6$	= 148.6	147.2
$\delta_{4,5}$	$= 128.2 + Z_4 + Z_5$	$= 128.2 - 7.2 - 1.2$	= 119.8	119.6
δ_{4a}	$= 133.9 + 2\,Z_{4a}$	$= 133.9 + 5.6$	= 139.5	138.5
δ_{8a}	$= 133.9 + 2\,Z_{8a}$	$= 133.9 - 8.1$	= 125.8	124.3

TABLE 7.14. ADDITIVITY OF THE ¹³C CHEMICAL SHIFTS OF MOLECULAR SUBSTRUCTURES AS A SIMPLE AID FOR THE ASSIGNMENT OF THE ¹³C SIGNALS OF CONIINE (A) AND FLAVINE-ADENINE DINUCLEOTIDE (B).

A = adenosine 5′-diphosphate trisodium salt, R = riboflavine 5′-monophosphate sodium salt, and A + R = flavine-adenine dinucleotide disodium salt; S = singlet, D = doublet, T = triplet, and Q = quartet, multiplicities in the off-resonance spectra.

A. Assignment of the ¹³C signals of coniine.

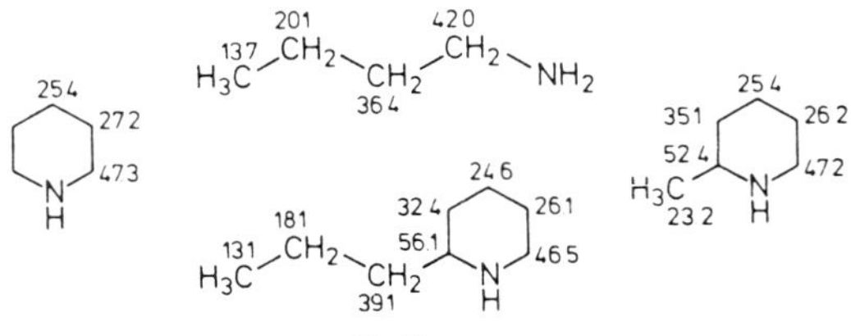

Coniine

B. Assignment of the ¹³C signals of flavine-adenine dinucleotide.

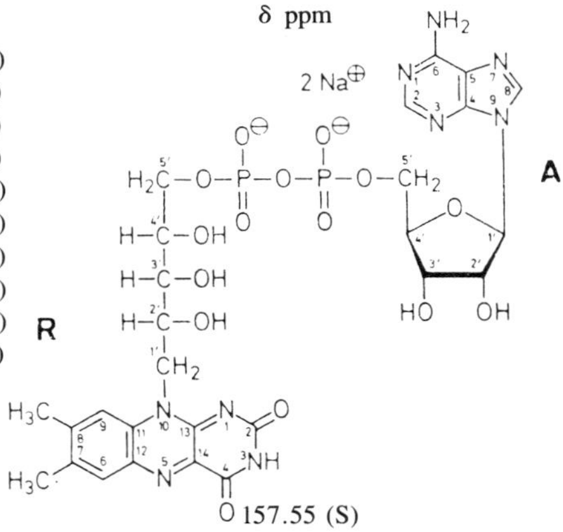

	δ ppm	δ ppm	δ ppm
A 2	152.60 (D)		152.40 (D)
A 4	148.50 (S)		148.15 (S)
A 5	118.30 (S)		117.75 (S)
A 6	155.10 (S)		154.65 (S)
A 8	139.95 (D)		139.20 (D)
A 1′	87.55 (D)		87.45 (D)
A 2′	70.15 (D)		70.25 (D)
A 3′	74.70 (D)		75.15 (D)
A 4′	83.75 (D)		83.85 (D)
A 5′	64.90 (T)		65.40 (T)
R 2		157.55 (S)	157.45 (S)
R 4		160.25 (S)	160.60 (S)
R 6		130.25 (D)	130.15 (D)
R 7		133.50 (S)	133.30 (S)
R 8		134.15 (S)	133.80 (S)
R 9		117.20 (D)	116.75 (D)
R11		131.35 (S)	131.35 (S)
R12		139.55 (S)	139.20 (S)
R13		150.85 (S)	150.65 (S)
R14		149.55 (S)	149.80 (S)
R 7 CH_3		18.70 (Q)	18.70 (Q)
R 8 CH_3		20.85 (Q)	20.85 (Q)
R 1′		47.60 (T)	47.60 (T)
R 2′		69.40 (D)	69.40 (D)
R 3′		72.65 (D)	72.65 (D)
R 4′		71.55 (D)	71.45 (D)
R 5′		65.85 (T)	67.80 (T)

of the respective olefinic carbon atoms. Examples of this are the smaller chemical shift differences of the vinyl carbon atoms in 1,3-butadiene and styrene as compared with 1-butene:

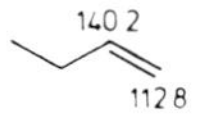

1-Butene

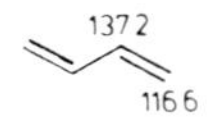

1,3-Butadiene

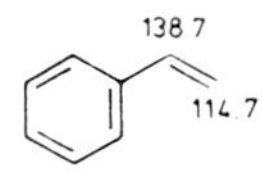

Styrene

Carbonyl carbon atoms are also shielded by about 10 ppm when the carbonyl function is conjugated with multiple bonds, as shown in the following comparison of formyl and acetyl chemical shifts:

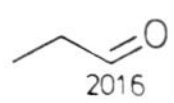

Propanal

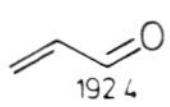

Propenal

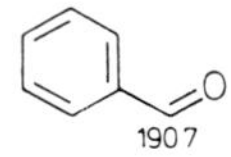

Benzaldehyde

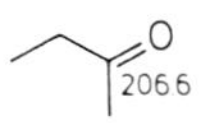

2-Butanon

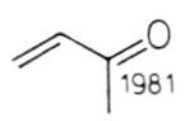

3-Butene-2-on

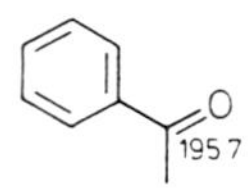

Acetophenone

For example, when conjugation is hindered by steric repulsion, the affected carbon nuclei are deshielded. Thus the ^{13}C chemical shifts of the carbonyl resonances increase in the series acetophenone, 2-methylacetophenone, 2,6-dimethylacetophenone or pivalophenone, 2-methylpivalophenone, 2,6-dimethylpivalophenone (see Fig. 7.1), and a correlation can be observed with the dihedral angle θ between the phenyl and carbonyl bond planes.

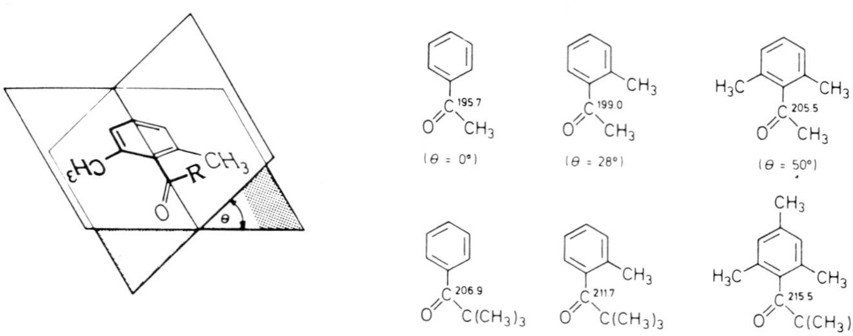

Fig. 7.1 Steric hindrance of conjugation in *o*-substituted phenylketones.

7.9 Influence of Intramolecular Interactions

The carbonyl ¹³C chemical shifts of salicylaldehyde, *o*-hydroxy, and *o*-aminoacetophenone are larger than those of the unsubstituted carbonyl compounds:

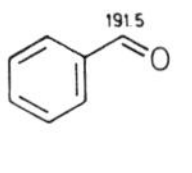

Benzaldehyde

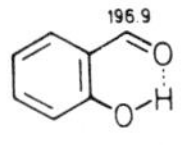

Salicylaldehyde

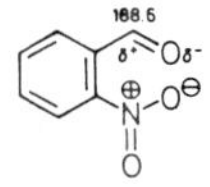

o-Nitrobenzaldehyde

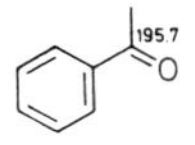

Acetophenone

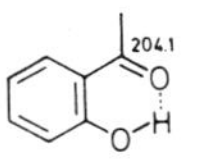

o-Hydroxyacetophenone

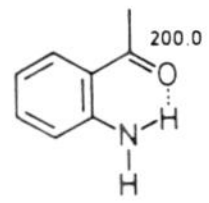

o-Aminoacetophenone

The reason for this deshielding is not so much steric hindrance of conjugation as the chelate-like intramolecular hydrogen bond, which makes the carbonyl carbon atom more strongly positive.

The influence of intramolecular dipole-dipole interactions is weaker. Thus the formyl carbon atom in *o*-nitrobenzaldehyde is shielded by 3 ppm relative to benzaldehyde. This can be interpreted as a consequence of a diminished positive charge on the carbonyl carbon atom owing to the dipole field of the *o*-nitro group.

7.10 Influence of Intermolecular Interactions (Medium Effects)

¹³C chemical shifts depend on the solvent, the concentration, and the pH of the solution being measured. Solvent and concentration shifts can be as large as 2 ppm, particularly in compounds with polar groups. For example, a hydration shift to smaller δ values is observed for the α carbon atom of pyridine upon dilution with water, whereas the β and γ carbon atoms behave in the opposite fashion (Fig. 7.2). The concentration shifts in 50% aqueous solution lie between -0.6 ppm for C-α and $+0.9$ ppm for C-γ, as compared with pure pyridine.

The effects of pH, which must be taken into account for organic bases and acids, are very much greater than those of solvent and concentration. For instance, if the ¹³C chemical shifts of pyridine are measured as functions of

Fig.7.2. Hydration (A) and pH influences (B) on the ^{13}C chemical shifts of pyridine. (Source: E. Breitmaier, G. Bauer, *Pharmazie in Unserer Zeit,* **5,** 97, 1976.)

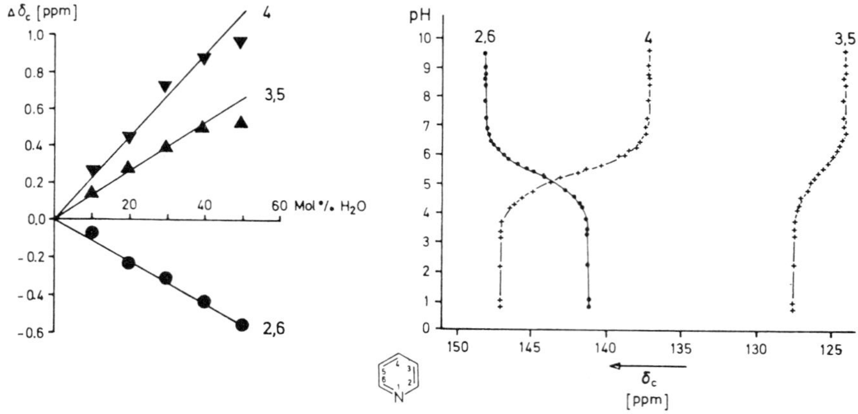

the pH value, each carbon resonance exhibits a titration curve with a point of inflection at the pH value corresponding to the pK value of pyridine (pK_a = 5.35) (see Fig. 7.2).

The protonation shifts (−7.8 ppm for C-α, 5.05 ppm for C-β, and 12.4 ppm for C-γ) have the same signs but larger values than the hydration effects (see Figs. 7.2 A and B), suggesting similar causes acting more or less effectively: Hydration and pH shifts of pyridine are related to the fact that the basic pyridine nitrogen atom forms a hydrogen bond with water and is protonated by acids. As a consequence of the coordinative N-H bond in protonated pyridine, the ring carbon atoms become poorer in electrons than is the case with a hydrogen bond, as shown by the resonance formulae of the pyridinium cation:

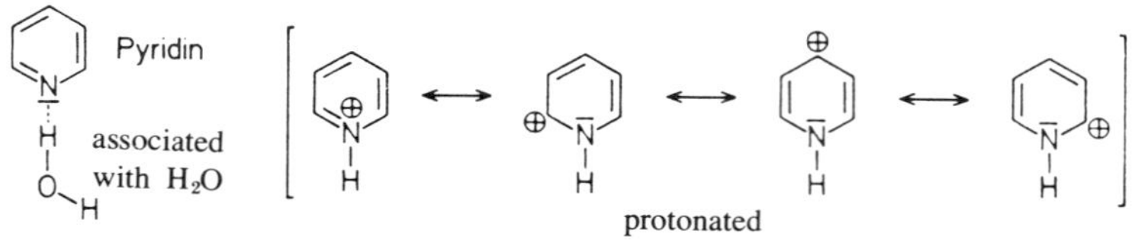

In this way, the larger protonation shifts make sense when compared with the hydrogen bonding effect. Of course, on the basis of the resonance formulae, one would expect not only C-γ but also C-α to be deshielded. The α carbon atom, however, is more strongly shielded in the pyridinium cation than in pyridine; this is attributed to the decreased double-bond character of the N-C-α bond.

References

1. E. G. Paul, D. M. Grant, *J. Am. Chem. Soc.* **85,** 1701 (1963).
2. D. M. Grant, E. G. Paul, *J. Am. Chem. Soc.* **86,** 2984 (1964).
3. J. B. Stothers, *Carbon-13 NMR Spectroscopy,* Academic Press, New York, 1972.
4. F. W. Wehrli, T. Wirthlin, *Interpretation of Carbon-13 Nuclear Magnetic Resonance Spectra,* Heyden & Son Ltd., London, 1976.
5. G. C. Levy, G. L. Nelson, *Carbon-13 Nuclear Magnetic Resonance for Organic Chemists,* Wiley-Interscience, New York, 1972.
6. E. Breitmaier, W. Voelter, *^{13}C NMR Spectroscopy: Methods and Applications,* "Monographs in Modern Chemistry," Vol. 5, H. E. Ebel, ed., Verlag Chemie GmbH, Weinheim, 2nd ed., 1978.
7. D. E. Dorman, M. Jautelat, J. D. Roberts, *J. Org. Chem.* **36,** 2757 (1971).

CHAPTER 8

Analysis of Coupling Constants

8.1 Carbon-Proton Couplings Across One Bond

Taken together with the ^{13}C chemical shifts, the coupling constants of carbon with protons, either directly bonded or several bonds away, are useful parameters. To measure them, one must record the ^{13}C NMR spectra undecoupled or use alternately pulsed double resonance (see Section 3.7). Both techniques require longer acquisition times and larger sample concentrations. Sometimes the coupling constants of directly bonded ^{13}C and ^{1}H nuclei can also be measured as the spacing of the ^{13}C satellites (0.5% intensity) of sufficiently intense signals in the PMR spectrum.

The magnitudes of the coupling constants of carbon with directly bonded protons lie between 120 and 300 Hz. Two-, three-, and four-bond C-H coupling constants ($^{2}J_{CH}$, $^{3}J_{CH}$, $^{4}J_{CH}$) are smaller than 50 Hz.

The coupling constant J_{CH} increases in the series ethane, ethylene, acetylene; or methylamine, formaldimine, hydrogen cyanide (see Fig. 8.1). On the basis of the familiar orbital hybridization model, one obtains a linear correlation between J_{CH} and the *s* character of the carbon atom's hybrid bonding orbitals, as sketched in Figure 8.1.

Conversely, the C-H coupling constant allows one to reach conclusions about the bonding state of the coupled carbon atom. Thus J_{CH} is 161 Hz in cyclopropane as compared with 128 Hz in cyclopentane and 123 Hz in cyclohexane (see Table 8.1). A similar situation is found when comparing the data for oxiran and its ring homologs (Table 8.1). The C-H bonds in three-membered rings are obviously more like those in alkenes than those in five-membered and larger rings, and one can infer that the hybrid bonding orbitals in cyclopropane and oxiran have more *s* character. The larger C-H coupling constants in three-membered rings make the assignment of the cyclopropane carbon resonances possible in polycyclic compounds, as shown below for the case of tricyclo [4.3.1.$0^{1,6}$] deca-2,4-diene [1]:

J_{CH} = 161 Hz

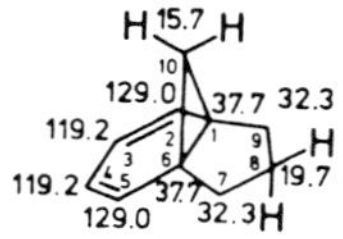

J_{CH} = 129 Hz

Decreasing electron density at the carbon atom because of bonded heteroatoms or substituents makes the C-H couplings larger. In this way, J_{CH} grows in the series ethane (125 Hz), methylamine (136 Hz), methanol (142 Hz), nitromethane (146.5 Hz), chloromethane (150 Hz). This effect becomes stronger when there are two or more substituents, such as in dichloromethane (177.6 Hz) and chloroform (210.6 Hz) (see Table 8.1 A and Fig. 1.2). Similarly, the carbon atoms in oxygen heteroalicycles show C-H couplings that are considerably larger than those in cycloalkanes of the same ring size (see Table 8.1 B). Finally, the C-H coupling constant is particularly large (up to 210 Hz) in azines and azoles when the coupled carbon atoms are flanked by two nitrogens (see Table 8.1 C), and the C-H coupling constant decreases in pyridine as the distance of the coupled carbon atom from the nitrogen atom increases.

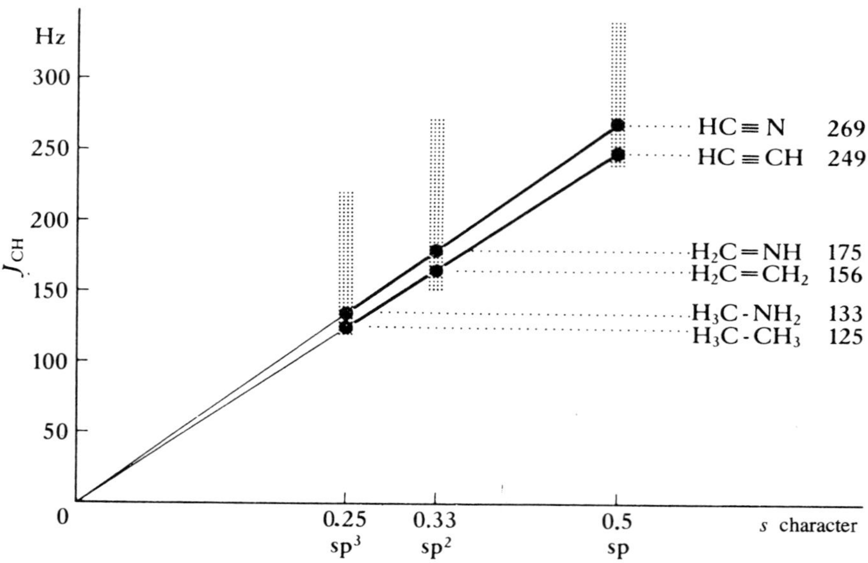

Fig. 8.1. Carbon-proton coupling constants J_{CH} and *s* character of the carbon atom's hybrid bonding orbitals. (Source: E. Breitmaier, G. Bauer, *Pharmazie in Unserer Zeit*, **5**, 97, 1976).

TABLE 8.1. ^{13}C CHEMICAL SHIFTS AND CORRESPONDING ^{13}C-^{1}H ONE-BOND COUPLINGS IN ALKANES AND SIMPLE ALKYL COMPOUNDS (A), CYCLOALKANES AND OXYGEN HETEROALICYCLICS (B), AND BENZENE, AZINES, AND AZOLES (C) (δ_C[ppm], J_{CH}[Hz])[a]

A. Alkanes and simple alkyl compounds.

CH_4: −2.3/*125* CH_3—NO_2: 61.1/*146.5*
CH_3—OH : 49.0/*142.0*
CH_3—NH_2: 28.3/*136.0*
CH_3—Cl : 25.1/*150.0;* CH_2Cl_2 : 53.85/*177.6;*
$CHCl_3$: 78.05/*210.6*

58.0/*140.5*

CH_3——CH_3 CH_3——CH_2——OH

5.9/*125* 18.0/*125.2*

B. Cycloalkanes and oxygen heteroalicyclics.

(b)

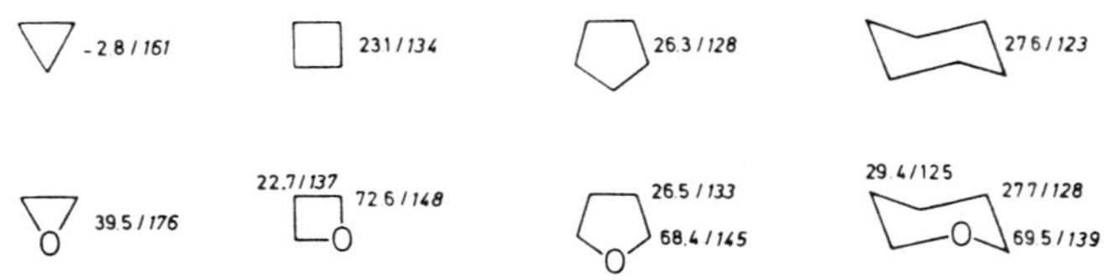

C. Benzene, azines, and azoles.

(c)

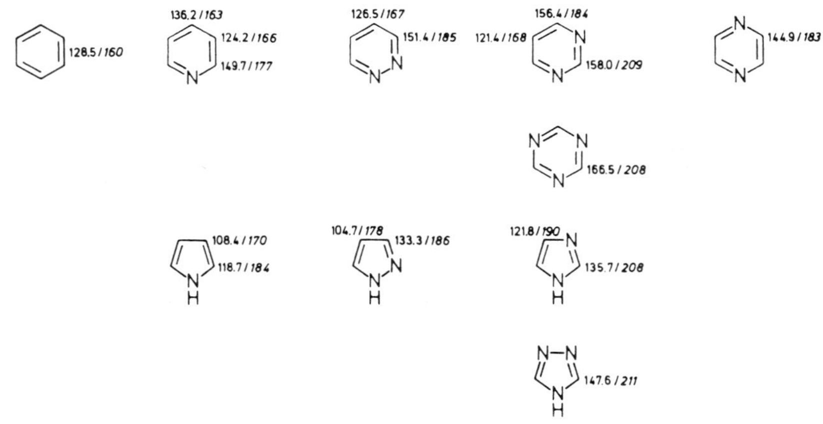

[a]E. Breitmaier, W. Voelter, *^{13}C NMR Spectroscopy: Methods and Applications,* "Monographs in Modern Chemistry," Vol. 5, H. E. Ebel, ed., Verlag Chemie GmbH, Weinheim, 2nd ed., 1978; J. T. Clerc, E. Pretsch, S. Sternhell, *^{13}C-Kernresonanzspektroskopie,* Akademische Verlagsgesellschaft, Frankfurt am Main, 1973; G. C. Levy, G. L. Nelson, *Carbon-13 Nuclear Magnetic Resonance for Organic Chemists,* Wiley-Interscience, New York, 1972; J. B. Stothers, *Carbon-13 NMR Spectroscopy,* Academic Press, New York, 1972; and F. W. Wehrli, T. Wirthlin, *Interpretation of Carbon-13 Nuclear Magnetic Resonance Spectra,* Heyden & Son Ltd., London, 1976.

8.2 Carbon-Proton Long-Range Couplings

Geminal carbon-proton couplings ($^{2}J_{CH}$) in open-chained aliphatic compounds are smallest for tetracoordinated carbon (alkanes) and largest for dicoordinated carbon (alkynes) (see Table 8.2 A), as is the case for direct couplings. *Vicinal* C-H coupling constants ($^{3}J_{CH}$) are larger for nuclei in a *trans* configuration, when the conformation is fixed, than the corresponding *gauche* or *cis* couplings (Table 8.2 A). In benzene derivatives and pyridine (see Tables 8.2 B and C), three-bond C-H couplings ($^{3}J_{CH}$) are the largest. The C-H long-range couplings for some heteroaromatics (Tables 8.2 C and D) reflect in part the influence of the electron-attracting heteroatoms. Thus the coupling of C-3 to H-2 in furan (14 Hz) is clearly larger than in pyrrole (7.8 Hz) or thiophene (4.7 Hz; see Table 8.2 D), and the largest long-range coupling in pyridine is that of C-2(6) to H-6(2) across the nitrogen atom (12 Hz; see Table 8.2 C).

Figure 8.2 illustrates the analysis of all the C-H coupling constants in pyrimidine. The C-2 resonance at δ = 158.4 ppm is split into a doublet of long-range triplets as a consequence of direct coupling to H-2 (203 Hz) and long-range coupling to H-4 and H-6 (10.4 Hz). Carbon atoms C-4 and C-6 appear, because of their chemical equivalence, as a double signal (156.4 ppm), which is split in the first place into a doublet by direct coupling with H-4 or H-6 (181.5 Hz). The fine structure of both doublet signals shows an eightfold splitting owing to the couplings $^{3}J_{C-4(6),\ H-2}$ = 9.0 Hz; $^{3}J_{C-4,\ H-6}$ or $^{3}J_{C-6,\ H-4}$ = 5.3 Hz; and $^{2}J_{C-4,\ H-5}$ = 3.0 Hz. The remaining signal at 121.8 ppm corresponds to C-5. Because of the direct coupling with H-5 (168.0 Hz) and the long-range couplings $^{2}J_{C-5,\ H-4(6)}$ = 7.8 Hz and $^{4}J_{C-5,\ H-2}$ = 1.6 Hz, it appears as a doublet of long-range double triplets.

The long-range multiplicities of carbon resonances can be decisive assignment criteria if enough adequately soluble substance is present to record the coupled ^{13}C NMR spectrum. In this way, one can distinguish between the long-range splittings of carbons C-2(6) and C-3(5) in monosubstituted aromatics such as nitrobenzene (see Fig. 8.3): Whereas the C-3(5) signal essentially shows only a *m*-coupling of 8 Hz with H-5(3), the C-2(6) resonances are split into long-range double doublets owing to two *m*-couplings with H-6(2) and H-4 (8 and 6 Hz), showing an additional fine structure because of *o*-coupling with H-3(5). The conclusion is that the signal at 123.3 ppm corresponds to carbon atoms C-2 and C-6 (see Fig. 8.3).

The long-range C-H couplings are of the same order of magnitude as *geminal* and *vicinal* H-H coupling constants, as Table 8.2 shows [2-4]. The relationship $J_{CH(trans)} > J_{CH(cis)}$ in alkenes and the dependence of the C-H vicinal coupling constant on the dihedral angle formed with the coupling partner

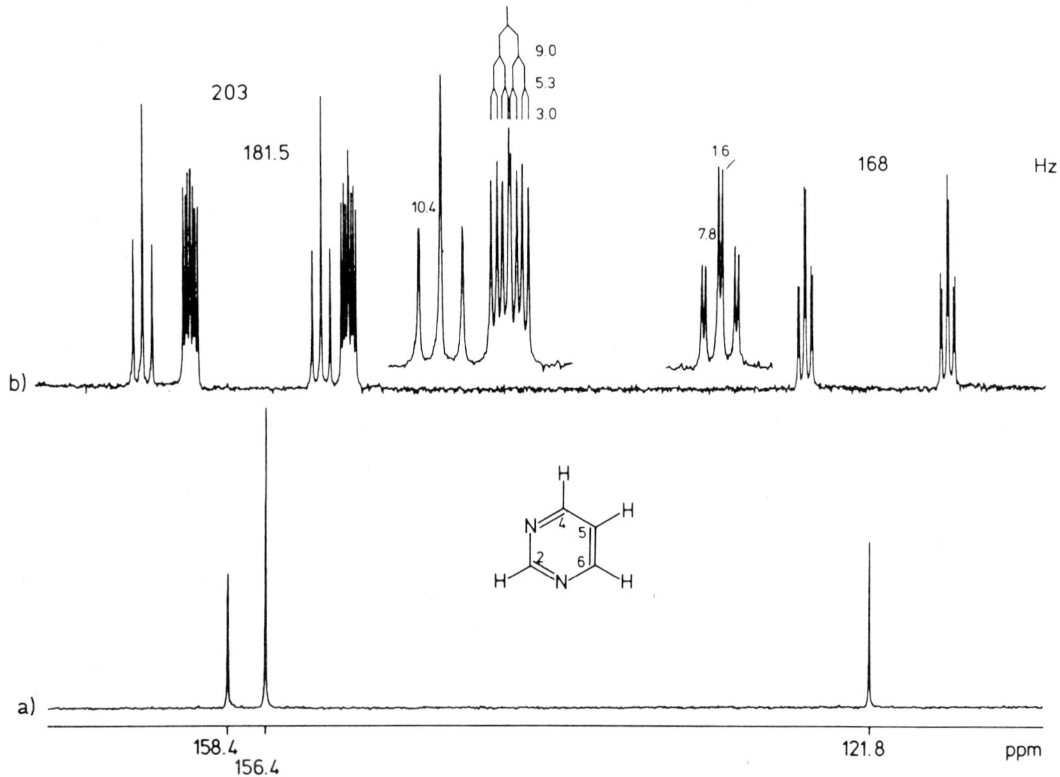

Fig. 8.2. 20 MHz ^{13}C NMR spectra of pyrimidine, 75% v/v in $CDCl_3$ at 25° C. A: Broad-band proton-decoupled, single scan. B: Undecoupled, 750 accumulated scans.

TABLE 8.2. LONG-RANGE C-H COUPLINGS $^{2,3,4}J_{CH}$[Hz] IN (A) OPEN-CHAIN COMPOUNDS, (B) AROMATICS, (C) PYRIDINE, AND (D) FIVE-MEMBERED HETEROAROMATICS.[a]

A. Geminal and vicinal C-H couplings.

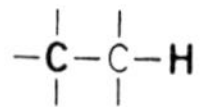	$^2J_{CH} = 1 - 6$
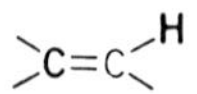	$^2J_{CH} = 1 - 16$
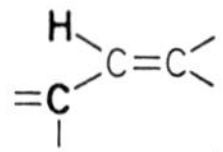	$^2J_{CH} = 8$
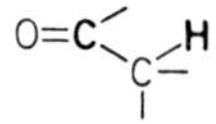	$^2J_{CH} = 5 - 8$
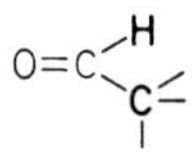	$^2J_{CH} = 20 - 25$
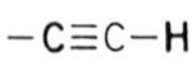	$^2J_{CH} \approx 40$

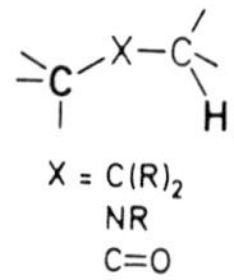	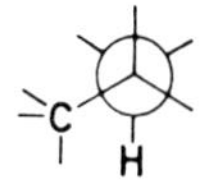	$^3J_{CH} \approx 0$ (*gauche-*)
	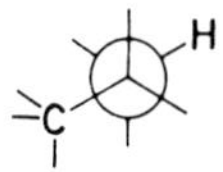	$^3J_{CH} = 5 - 7$ (*trans-*)
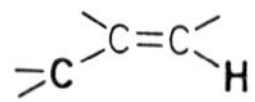		$^3J_{CH} \leqq 12$ (*cis-*)
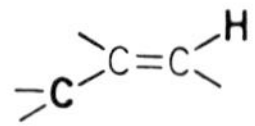		$^3J_{CH} \leqq 18$ (*trans-*)

B. Aromatics.

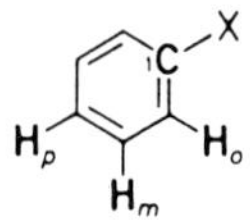

$^2J_{CH_o} = 1 - 4$
$^3J_{CH_m} = 7 - 10$
$^4J_{CH_p} = 0.5 - 2$

C. Pyridine.

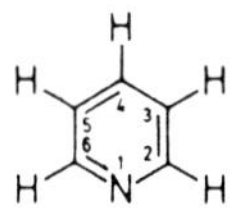

$^2J_{C\text{-}2,H\text{-}3} = 2$
$^3J_{C\text{-}2,H\text{-}4} = 7$
$^3J_{C\text{-}2,H\text{-}6} = 12$
$^2J_{C\text{-}3,H\text{-}2} = 7$
$^2J_{C\text{-}3,H\text{-}4} = 2$
$^3J_{C\text{-}3,H\text{-}5} = 6$
$^3J_{C\text{-}4,H\text{-}2} = 4$

D. Five-membered heteroaromatics.

	C	$^{2}J_{CH-2}$	$^{2}J_{CH-3}$	$^{2,3}J_{CH-4}$	$^{3,4}J_{CH-5}$
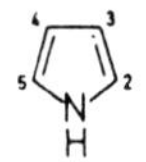	2	—	7.6	7.6	7.6
	3	7.8	—	4.6	7.8
	2	—	7.0	10.8	7.0
	3	14.0	—	4.0	5.8
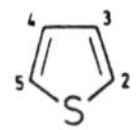	2	—	7.3	10.0	5.1
	3	4.7	—	5.9	9.5
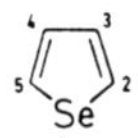	2	—	7.0	10.0	3.5
	3	4.5	—	6.0	10.4

[a]E. Breitmaier, W. Voelter, *^{13}C NMR Spectroscopy: Methods and Applications,* "Monographs in Modern Chemistry," Vol. 5, H. E. Ebel, ed., Verlag Chemie GmbH, Weinheim, 2nd ed., 1978; J. T. Clerc, E. Pretsch, S. Sternhell, *[13]C-Kernresonanzspektroskopie,* Akademische Verlagsgesellschaft, Frankfurt am Main, 1973; G. C. Levy, G. L. Nelson, *Carbon-13 Nuclear Magnetic Resonance for Organic Chemists,* Wiley-Interscience, New York, 1972; J. B. Stothers, *Carbon-13 NMR Spectroscopy,* Academic Press, New York, 1972; and F. W. Wehrli, T. Wirthlin, *Interpretation of Carbon-13 Nuclear Magnetic Resonance Spectra,* Heyden & Son Ltd., London, 1976.

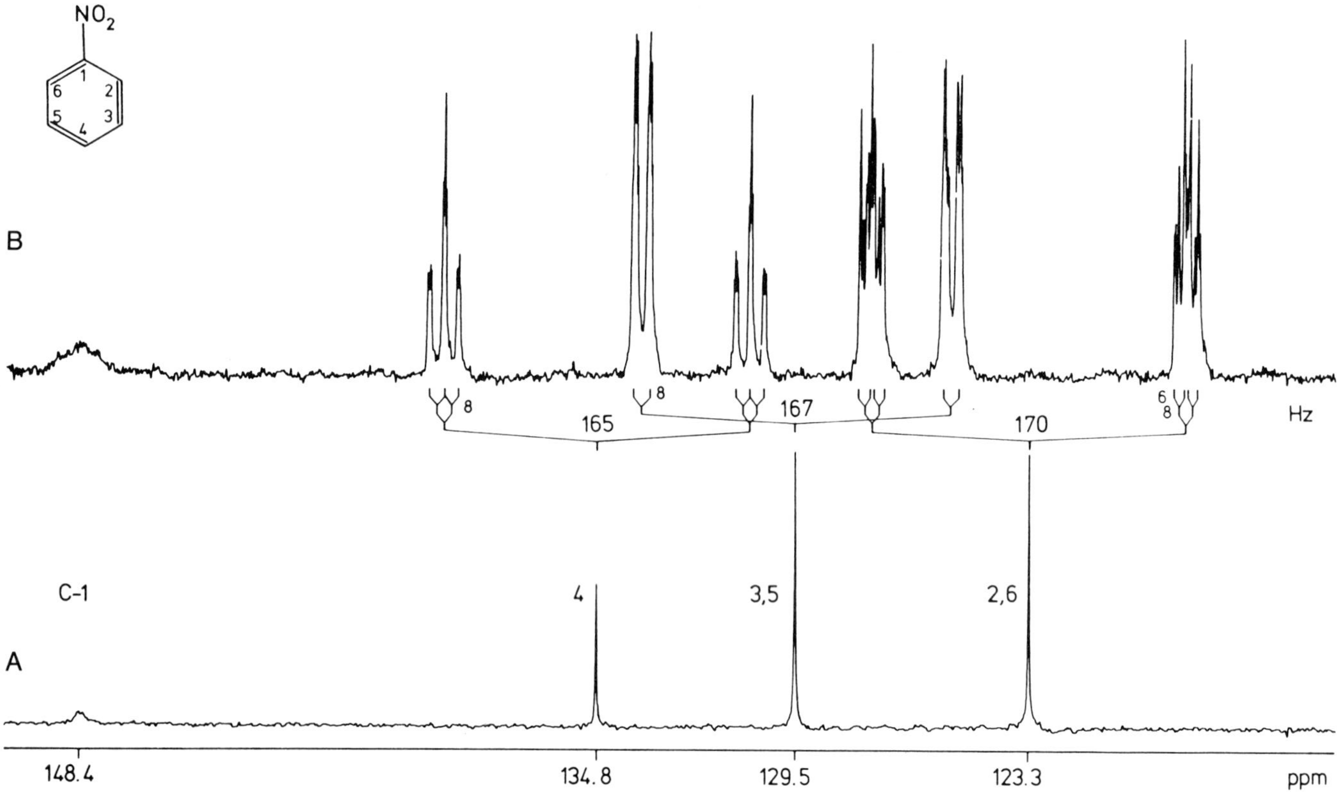

Fig. 8.3. 20 MHz ^{13}C NMR spectra of nitrobenzene, 90% v/v in acetone-d_6 at 25° C. A: Proton noise-decoupled, single scan. B: Undecoupled, 400 accumulated scans. The *o*-couplings (2 Hz) are not shown in the coupling diagrams.

($J_{trans} > J_{gauche}$; see the Karplus relationships [2]) are a striking parallel with analogous situations in proton magnetic resonance.

The systematic analysis of C-H coupling is often *not* possible using the (n + 1) rule. On the contrary, **higher order splittings** may appear in the coupled ^{13}C NMR spectrum as a consequence of the mixing of energetically similar spin states in such a way that, for example, the high- and low-field portions of a multiplet may show different fine structures. Higher order splittings may also complicate the interpretation of carbon-carbon couplings discussed in the following paragraph.

8.3 Carbon-Carbon Couplings

Carbon-carbon couplings usually remain lost in the baseline noise when ^{13}C NMR spectra are recorded at natural ^{13}C abundance. The prerequisite for homonuclear C-C coupling is a direct bond between two ^{13}C nuclei. At the natural ^{13}C abundance of 10^{-2} (1.1%), the probability that this situation will arise is only 10^{-4}. Therefore, C-C couplings are observed only as weak satellites of intense signals when analyzing concentrated samples of simple compounds. These signals are stronger when ^{13}C-enriched compounds are used. On the other hand, the extremely low intensity of C-C multiplets simplifies the interpretation of ^{13}C NMR spectra of samples with natural ^{13}C abundance.

Largely similar relationships are valid for C-C and C-H coupling constants. Thus the J_{CC} values increase in the sequence ethane (34.6 Hz, sp^3-sp^3 bond), ethylene (67.6 Hz, sp^2-sp^2 bond), and acetylene (171.5 Hz, sp-sp bond). If substituents or heteroatoms are joined to the coupled carbon atom, the C-C coupling constants are also affected by the charge density and bond polarity. For instance, the C-C coupling constant increases in the series thiophene, pyrrole, furan (see Table 8.3), a situation that can be explained by the growing polarization of the C-X bonds with increasing electronegativity of the heteroatom.

TABLE 8.3. C-C COUPLING CONSTANTS IN SELECTED HETEROCYCLES.

$J_{2,3}$ = 64.2 Hz	$J_{2,3}$ = 65.6 Hz	$J_{2,3}$ = 69.1 Hz
		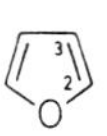

$J_{2,3}$ = 53.8 Hz	$J_{3,4}$ = 53.5 Hz	$J_{2,5}$ = 14.0 Hz

8.4 Carbon-Heteroatom Couplings

Heteroatoms endowed with nuclear spin are not affected by proton decoupling, as has already been shown for deuteriated compounds (see Section 4.1). Consequently, ^{13}C-heteronuclear couplings are observable in broad-band proton-decoupled ^{13}C NMR spectra: e.g., with ^{19}F or ^{31}P in trifluoroethanol or phosphoenolpyruvic acid (see Figs. 8.4 A and B). Some C-F and C-P coupling constants are listed in Table 8.4. ^{13}C-^{14}N couplings are usually not resolved owing to line-broadening caused by the quadrupole moment of ^{14}N. ^{13}C-^{15}N couplings are only observed in ^{15}N-enriched compounds—as shown in Figure 8.4 C for formamide—because of the low natural abundance (0.4%) of ^{15}N.

TABLE 8.4. MAGNITUDES OF SOME C-X COUPLING CONSTANTS IN Hz

A. ^{13}C—^{14,15}N

Compound	(J_{CN})
$R—^{14}\overset{+}{N}(CH_3)_3$	4 – 10
$R—^{15}\overset{+}{N}(CH_3)_3$	4 – 6
$H_2{}^{15}N—CH_2—COOH$	6.2
$H_3{}^{15}\overset{+}{N}—CH_2—COOH$	4.9
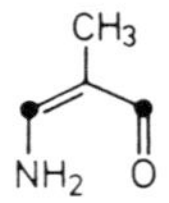	14.0 3.5 ($^3J_{CN}$;$^2J_{CN}$ < 2)

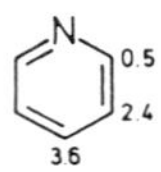

B. ^{13}C—^{19}F

Compound	Carbon	J_{CF}
1-Fluorohexane	CH_2F	166.6
	CH_2	19.9
	CH_2	5.3
	CH_2	<2.0
	CH_2	0
	CH_3	0
$FCH_2—COOH$	CH_2	181
$F—CH-F—COOH$	CH-F	254
$CF_3—COOH$	CF_3	283
	COOH	44
$CF_3—CH_2OH$	CF_3	278
	CH_2OH	35

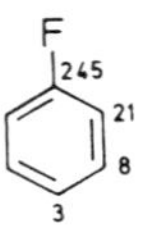

C. ^{13}C—^{31}P

Compound	J_{CP}
$P(—CH_2—CH_2—CH_2—CH_3)_3$	14, 15, 11, 0
$OP(—O—CH_2—CH_2—CH_2—CH_3)_3$	6, 7, 0, 0

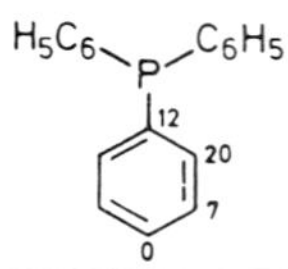

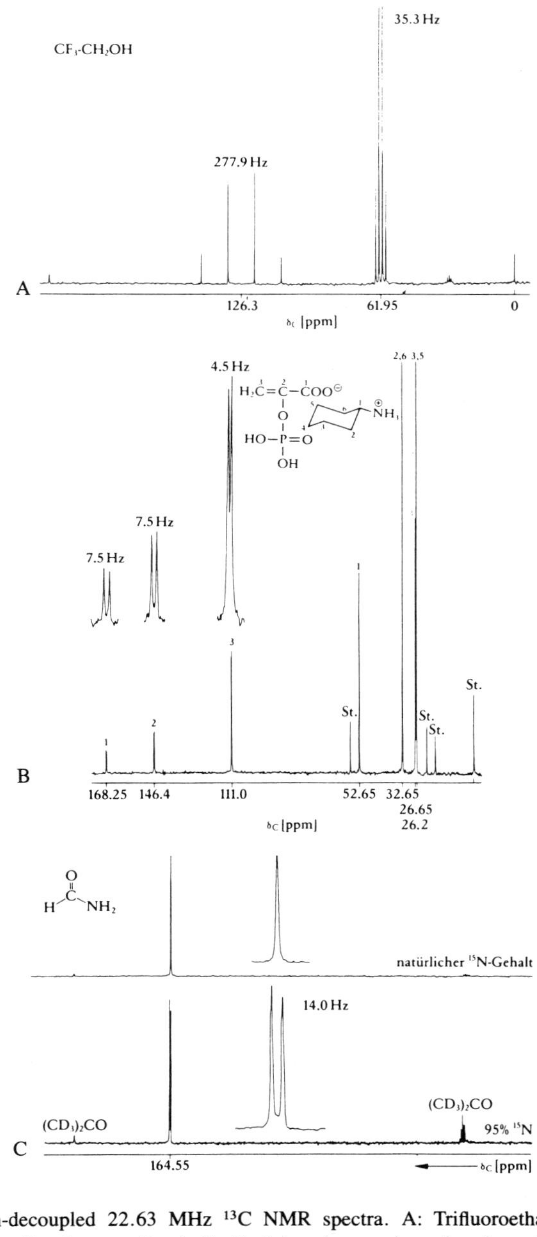

Fig. 8.4. Proton-decoupled 22.63 MHz ^{13}C NMR spectra. A: Trifluoroethanol, 95% v/v in acetone-d_6 (carbon-fluorine couplings). B: Cyclohexylammonium phosphoenolpyruvate, 10% in D_2O, sodium trimethylsilypropanesulfonate as internal standard labeled "St." (carbon-phosphorus couplings). C: Formamide, 95% v/v in acetone-d_6; *top:* sample with natural ^{15}N abundance (0.365%), *bottom:* sample enriched in ^{15}N (95% ^{15}N, ^{13}C − ^{15}N coupling). (Source: E. Breitmaier, G. Bauer, *Pharmazie in Unserer Zeit,* **5,** 97, 1976.)

References

1. J. B. Stothers, *Carbon-13 NMR Spectroscopy,* Academic Press, New York, 1972.
2. H. Friebolin et al., *NMR-Spektroskopie, Eine Einführung mit Übungen,* Verlag Chemie GmbH, Weinheim, 1974, Taschentext Bd. 15.
3. H. Günther, *NMR-Spektroskopie, Eine Einführung,* Georg Thieme Verlag, Stuttgart, 1973.
4. H. Suhr, *Anwendungen der Kernmagnetischen Resonanz in der Organischen Chemie,* Springer Verlag, New York, 1965.

CHAPTER 9

^{13}C Spin-Lattice Relaxation Times: Measurement and Range of Application

9.1 Methods

To measure the spin-lattice relaxation times T_1 described in Section 1.3, the time course of the relaxation process must be recorded. This can be done using 180°, τ, 90°, or 90°, τ, 90° pulse sequences, varying the interval τ between pulses. The first pulses (180° or 90°) disturb the equilibrium of the nuclei in the magnetic field and initiate the relaxation process. The second pulses (90°) allow one to record the partially relaxed magnetization, which increases with the interval τ.

9.1.1 180°, τ, 90° or Inversion Recovery Method

In the inversion recovery experiment, the Boltzmann distribution of the nuclear spins (see Fig. 1.3C), and thus the resulting magnetization vector $\vec{M}_0$, is inverted by a 180° pulse (see Figs. 9.1A and B). The spin-lattice relaxation process follows, and the magnetization recovers from $-\vec{M}_0$ through zero to $+\vec{M}_0$ (see Figs. 9.1B-E). After a variable time τ, a 90° pulse rotates the partially relaxed magnetization $\vec{M}_\tau$ in the x,y plane, and this induces the FID signal in the receiver coil of the spectrometer (see Section 2.2).

If the partially relaxed magnetization $\vec{M}_\tau$ is still negative ($M < 0$ after a short time τ, a 90° pulse with its field vector $\vec{H}_1$ along the x axis rotates it into the negative y axis (see Fig. 9.1F). The negative transverse magnetization gives rise to a nuclear induction signal with negative initial amplitude in the receiver, and the corresponding Fourier transform shows an inverted NMR signal (see Fig. 9.1H).

After a certain time τ_0, the spin-lattice relaxation has proceeded to an equal distribution of the nuclei between both states of nuclear spin. The resulting magnetization is then zero, and no FID signal will be recorded in the receiver (**zero transition;** see Fig. 9.1D).

After a time $\tau > \tau_0$ has elapsed, the partially relaxed magnetization is positive again (see Fig. 9.1E). The 90° pulse then generates a positive transverse magnetization (see Fig. 9.1G), giving rise to an FID signal with positive

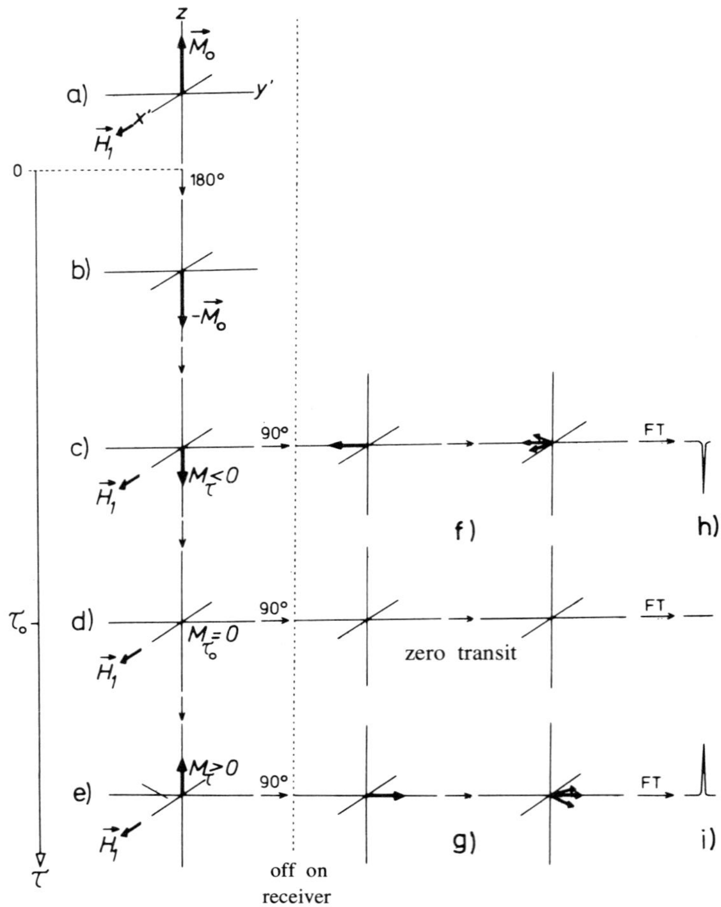

Fig. 9.1. Pulse-driven motion of the magnetization vector $\vec{M}_0$ and sign of the NMR signals in the inversion recovery or 180°, τ, 90° method of T_1 measurement ("FT" means Fourier transformation). (Source: E. Breitmaier, K.-H. Spohn, S. Berger, *Angew. Chem.* **87,** 152, 1975.)

initial amplitude. Fourier transformation leads to a "normal" NMR signal with an intensity that is attenuated relative to the equilibrium value corresponding to complete relaxation (see Fig. 9.1I).

To interpret the computer-controlled series of measurements shown in Figure 9.2A for adamantane, Eq. (1.6) (see Section 1.3) must be solved by integration between the limits t = 0 and τ and $M_z = -M_o$ and M_τ:

$$\ln (M_o - M_\tau) - \ln 2M_o = -\frac{\tau}{T_1} \tag{9.1}$$

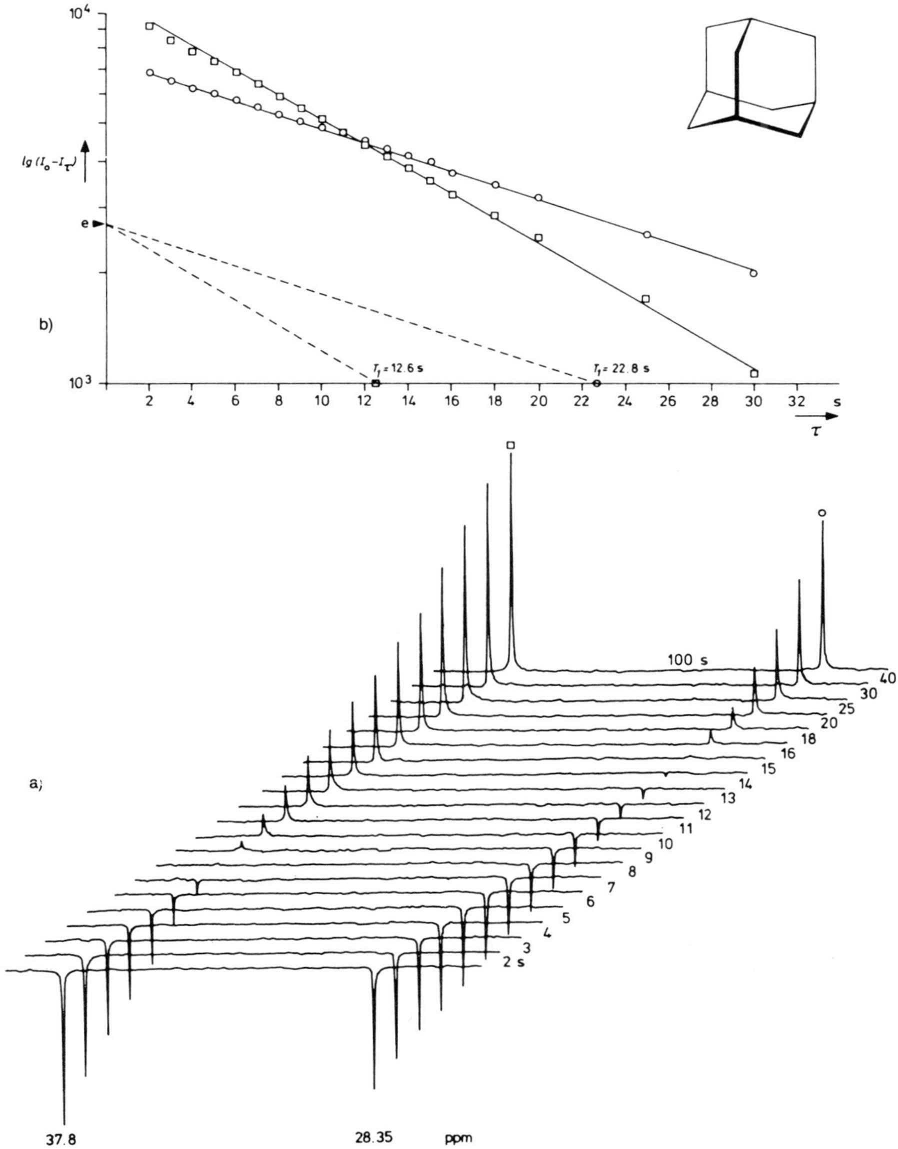

Fig. 9.2. Measurement of ^{13}C spin-lattice relaxation times by the inversion-recovery method; sample: adamantane, saturated solution in $CDCl_3$ at 25° C and 25 MHz; 50 accumulated scans for each experiment; A: Computer-controlled series of 21 experiments; the interval τ = 100 s was chosen to measure the equilibrium amplitude I_0; B: Graphic determination of the T_1 values from series A (CH_2: 12.6 s, CH: 22.8 s).

The intensity differences $I_0 - I_\tau$ of the signals in Figure 9.2A, responding to the magnetization differences $M_0 - M_\tau$, are then plotted semilogarithmically versus the pertinent pulse intervals τ, as shown in Figure 9.2B. Two straight lines are thus obtained for the two kinds of carbon nuclei in adamantane. Their reciprocal slopes, evaluated graphically or by computation, are the spin-lattice relaxation times of both kinds of carbon nuclei.

The zero transition times τ_0 for $M_\tau = 0$ also permit a rapid determination of T_1: The solution of Eq. (9.1) for these parameters produces Eq. (9.2):

$$T_1 = \frac{\tau_0}{\ln 2} = \frac{\tau_0}{0.69} \tag{9.2}$$

Of course, the T_1 values obtained in this way are based on a single measurement, namely the zero transition, so that experimental errors (up to 10%) escape evaluation.

9.1.2 90°, τ, 90° Methods

One can also initiate spin-lattice relaxation of a nuclear-spin species using a 90° pulse (Fig. 9.3B) and rotate the magnetization $M_\tau < M_0$—partially relaxed along the z axis after time τ has elapsed (Fig. 9.3D)—onto the y axis by means of a second 90° pulse (Fig. 9.3E). Naturally, when doing this, the transverse magnetization induced by the first pulse must be eliminated before firing the second pulse so that the signal intensity is not altered. This is achieved by using what is called a **homospoil pulse** in the lock channel of the spectrometer. The pulse disrupts the field homogeneity briefly and thus triggers a rapid decay of the transverse magnetization.

The experiment, controlled by the spectrometer's computer, produces the spectrum series shown in Figure 9.4A. Note that the signal intensity approaches an equilibrium value I_0 asymptotically, corresponding to the equilibrium magnetization M_0, as the pulse interval τ increases and the spin-lattice relaxation proceeds. T_1 is obtained from Eq. (1.6), again by plotting the intensity differences $I_0 - I_\tau$ semilogarithmically versus the pulse interval τ as the reciprocal slope of the straight regression lines (see Fig. 9.4B).

When the variation of the 90°, τ, 90° method known as **progressive saturation** is used, the signal sequence following a series of 90° pulses is accumulated. One thus proceeds in the same way as when accumulating interferograms for Fourier NMR spectroscopy but waits a certain interval τ between the 90° pulses, which is varied in a series of 10 to 20 measurements to follow the spin-lattice relaxation. When this is done, the signals must not be acquired by the receiver and summated in the computer before the dynamic equilibrium between excitation and relaxation is reached. Therefore, one begins accumulating signals for each individual measurement corresponding to a

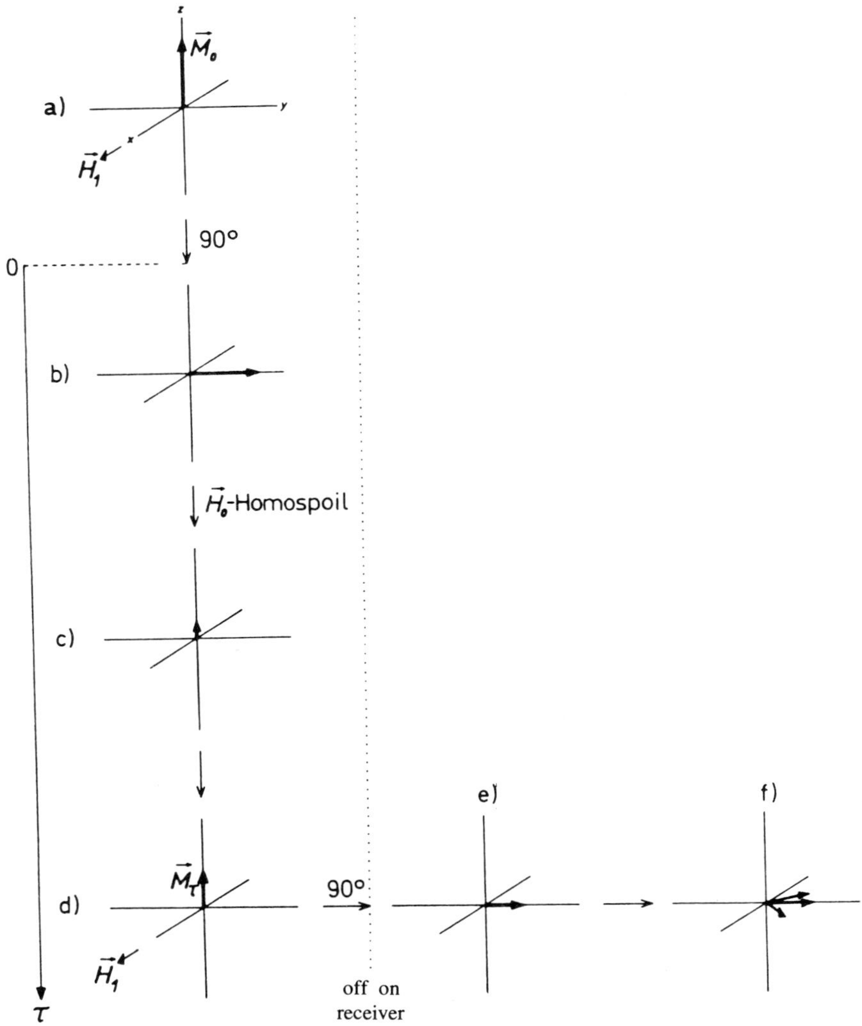

Fig. 9.3. Pulse-driven motion of the magnetization vector and spin-lattice relaxation during a 90°, τ, 90° sequence for the measurement of T_1: the first 90° pulse rotates M_0 (a) onto the y axis (b) and initiates the spin-lattice relaxation (b-d). Immediately afterwards the homospoil pulse in the lock channel briefly disrupts the H_0 field homogeneity, so that the transverse magnetization decays (b-c). After a chosen time τ, the partially relaxed magnetization M_τ (d) is rotated onto the y axis by means of a second 90° pulse (e), and it decays as a result of spin-spin relaxaton (f). The corresponding FID signal is recorded by the receiver and is Fourier-transformed by the computer into the NMR signal. The intensity of the NMR signal is a measure of the magnitude of the partially relaxed magnetization M_τ.

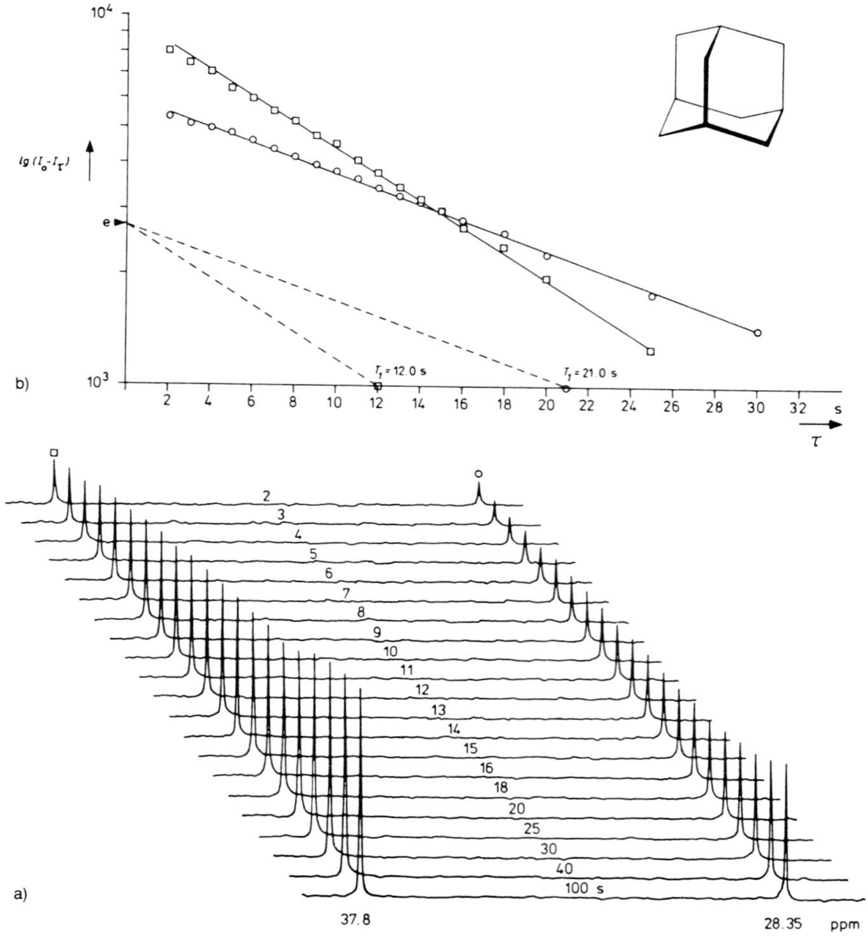

Fig. 9.4. Measurement of ^{13}C spin-lattice relaxation times using the 90°-*(homospoil)*, τ, 90° method; sample and conditions are the same as outlined in Fig. 9.2. A: Computer-controlled series of 21 experiments. B: Graphic determination of the T_1 values from series A (CH_2: 12.0 s, CH: 21.0 s).

certain interval τ *only* after the fourth or fifth 90° pulse. Again, the measurement of T_1 by progressive saturation and its evaluation are illustrated using the example of adamantane (see Figs. 9.5A and B).

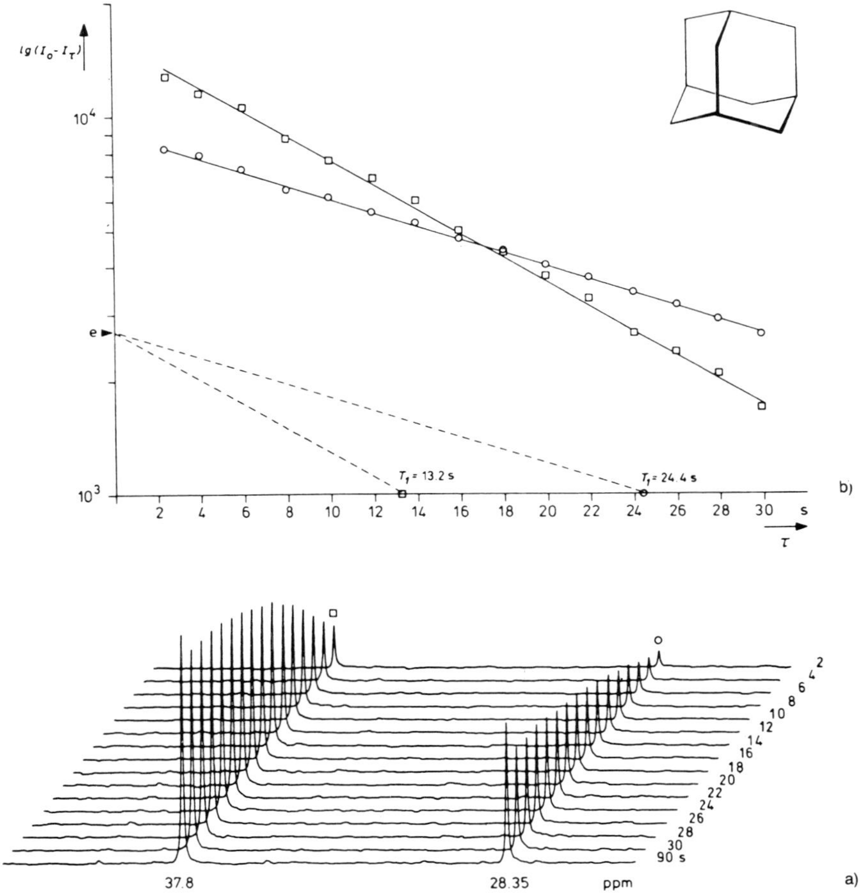

Fig. 9.5. Measurement of ^{13}C spin-lattice relaxation times by progressive saturation; sample and conditions are the same as in Fig. 9.2 but at 20 MHz. A: Computer-controlled series of 16 experiments; the interval τ = 90 s was chosen to measure the equilibrium amplitude I_0. B: Graphic determination of the T_1 values from series A (CH_2: 13.2 s, CH: 24.4 s).

9.1.3 Comparison of the Methods

The partially relaxed signals that follow the pulse sequences usually must be accumulated to improve the signal:noise ratio. In such cases, one must wait for all the nuclei in a molecule to be relaxed after each pulse sequence. This waiting time should amount to at least $3T_{1max}$, where T_{1max} is the spin-lattice relaxation time of the slowest nuclei in the molecule. Under these conditions, T_1 measurements are time consuming, especially when dilute solutions or larger molecules are involved. In these cases, the progressive saturation variant is used, particularly for determinating T_1 values for slow nuclei. On the other hand, in the progressive saturation method, the smallest possible interval is equal to the acquisition time of the interferogram (1 to 4 seconds); thus it is impossible to measure small T_1 values by this method.

9.2 The Dipole-Dipole Relaxation Mechanism

Spin-lattice relaxation is the exchange of energy between a nuclear spin system and its environment (see Section 1.3). Similarly to the NMR experiment, this energy exchange involves interactions of the nuclear spins with oscillating local electromagnetic fields arising from molecular translations, vibrations, and rotations. Five relaxation mechanisms can be distinguished according to the molecular property producing these fluctuating local fields. Unpaired electrons **(paramagnetic relaxation),** spatially oriented intramolecular shielding fields **(chemical shift anisotropy),** rapidly relaxing bonded nuclei **(scalar coupling),** or a molecule's magnetic moment resulting from electron motion **(spin rotation)** can set up magnetic fields at the point occupied by the observed nucleus that fluctuate with molecular motion.

The mechanism that usually predominates for the C-H carbon atoms in diamagnetic organic molecules is **dipole-dipole relaxation** (DD mechanism), based on the internuclear dipole-dipole interaction of ^{13}C and ^{1}H nuclei. The fluctuating local field at a ^{13}C nucleus arises because a bonded proton—by virtue of its magnetic moment, which is the largest in comparison with all other nuclei—creates a magnetic field that varies with the molecule's motion.

9.3 Nuclear Overhauser Effect and Dipole-Dipole Relaxation

During proton decoupling in a ^{13}C experiment (see Chapter 3), the protons discharge their excitation energy to the lattice by means of the dipole-dipole relaxation mechanism, and the lower energy ^{13}C spin states (see Fig. 1.3B) become more densely populated through coupled ^{1}H-^{13}C relaxation. In this way, the decoupled ^{13}C signals are strengthened more than one would expect from the multiplet intensities in the coupled spectrum. This change in intensity that accompanies decoupling is known as the nuclear Overhauser effect [1].

If the ^{13}C relaxation proceeds exclusively by the DD mechanism, the nuclear Overhauser factor, η_c, which measures the intensity loss of the ^{13}C signal, is given by Eq. (9.3) as the half-quotient of the gyromagnetic ratios of ^1H and ^{13}C [4]:

$$\eta_c = \frac{\gamma H}{2\gamma_c} = 1.988 \tag{9.3}$$

The nuclear Overhauser factors η_c are measured by comparing the signal intensities of proton-coupled and decoupled ^{13}C spectra with the same baseline noise amplitudes. Of course, the multiplets in coupled spectra of complex molecules are often too weak, which makes their integration too inexact and their interpretation more difficult owing to multiplet overlap. In these cases, a variation of alternately pulsed double resonance is recommended (see Section 3.7.2.), which in practice can easily be combined with T_1 measurements. To do this, the signal intensity is measured (1) with the decoupler turned on for a long time and then (2) with a short decoupling pulse only while the interferogram is being recorded (see Fig. 9.6A). Both measurements yield proton-decoupled spectra, the first with, and the second practically without the NOE (the reason is explained in Section 3.7). The quotient of both intensities, I_1/I_2, is η_c (see Fig. 9.6B).

Nuclear Overhauser factors of $\eta_c = 1.988 \approx 2.0$ are therefore indicative of the DD mechanism. If $\eta_c < 2$, other spin-lattice relaxation mechanisms are operating for the ^{13}C nuclei bonded to hydrogen. The percentage contribution of the DD mechanism can be calculated from η_c using Eq. (9.4):

$$\%\ DD = \frac{\eta_c}{1.988}\ . \tag{9.4}$$

The time constant T_{1DD} for the dipole-dipole mechanism can be determined accordingly using Eq. (9.5):

$$T_{1DD} = T_1 \frac{1.988}{\eta_c} \tag{9.5}$$

If several relaxation mechanisms are acting in conjunction, the measured relaxation rate $1/T_1$ is the sum of the rate constants $1/T_{1i}$ for all participating mechanisms (i):

$$\frac{1}{T_1} = \Sigma \frac{1}{T_{1\,i}} \tag{9.6}$$

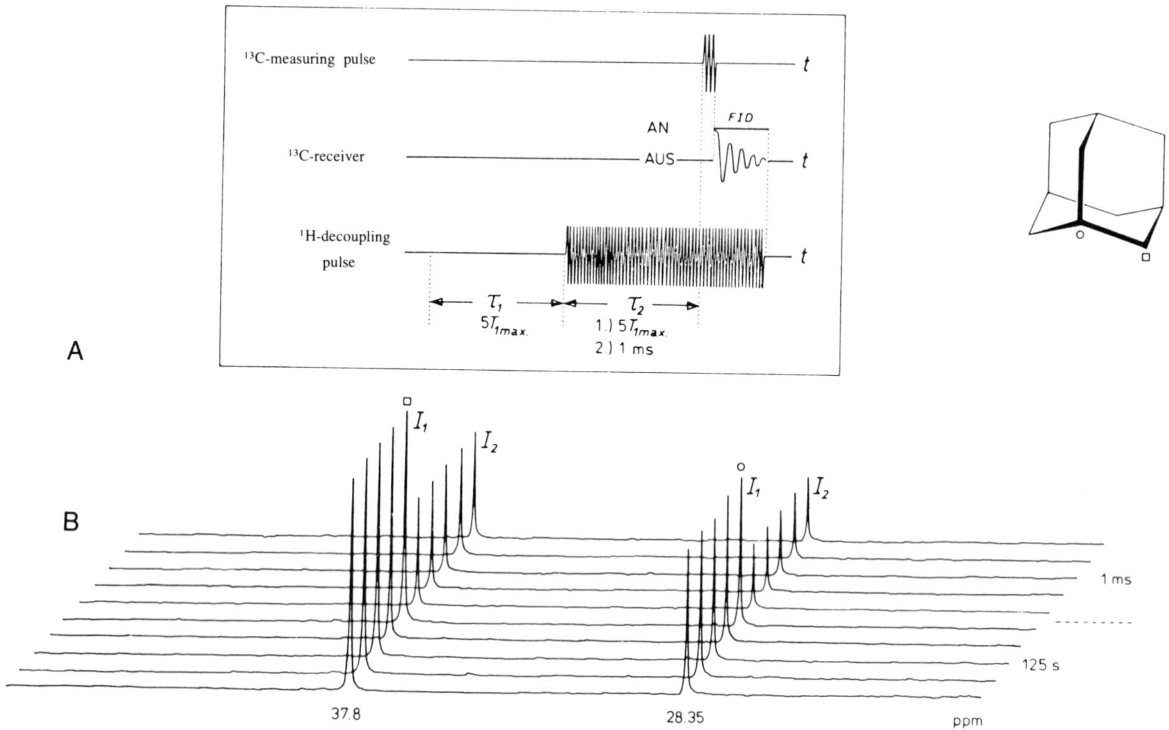

Fig. 9.6. Measurement of the nuclear Overhauser effect using pulsed $^{13}C\ \{^1H\}$ double resonance; sample: adamantane, saturated solution in $CDCl_3$ at 25°C and 20 MHz; 16 accumulated scans for each experiment. A: Timing of excitation pulses, receiver and decoupler for each individual measurement (Experiment 2, with $\tau_2 = 1$ ms, corresponds to the pulse sequence shown in Fig. 14a); B: Computer-controlled series of 5 experiments with $\tau_2 = 1255$ ($I_{1\ max} = 255$, see Figure 9.5), and 5 measurements with $\tau_2 = 1$ ms. The nuclear Overhauser factor is seen to be $I_1/I_2 \approx 2$, corresponding to dipolar spin-lattice relaxation.

In addition to the DD mechanism, spin rotation (SR) contributes to spin-lattice relaxation of C-H carbon atoms, especially in extremely small symmetrical organic molecules. For example, cyclobutane has the experimental values $\eta_c = 1.4$ and $T_1 = 35.7$ s [2,3]. The contribution of dipole-dipole relaxation can be calculated from these data using Eq. (9.4); the dipole-dipole relaxation time T_{1DD}, using Eq. (9.5); and the time constant for spin rotation, T_{1SR}, using Eq. (9.6), as follows:

$$\% \, DD = \frac{1.4}{1.988} \cdot 100 = 70.5, \text{ and} \tag{9.4}$$

$$T_{1DD} = \frac{1.988}{1.4} \cdot 35.7 = 50.7 \text{ s.} \tag{9.5}$$

$$\text{From } \frac{1}{T_1} = \frac{1}{T_{1DD}} + \frac{1}{T_{1SR}}, \tag{9.6}$$

$$\text{it follows that } T_{1SR} = \frac{T_1 \, T_{1DD}}{T_{1DD} - T_1} = \frac{35.7 \cdot 50.7}{50.7 - 35.7} = 121 \text{ s.}$$

9.4 Relationship Between Molecular Motion and Dipole-Dipole Relaxation.

As happens in NMR experiments, energy exchange between a nuclear spin and a fluctuating local field is possible only when the fluctuation frequency contains components corresponding to the Larmor frequency of the nuclear spin. The fluctuation frequency, however, depends on the frequencies of the molecular motions; thus a relationship exists between the relaxation time T_1 and the molecular mobility.

It is difficult to classify a molecule's motions in solution into translations, vibrations, and rotations. In conclusion, the mean time elapsed between two molecular reorientations is taken as a measure of the molecule's mobility. As in the physical processing of NMR operations, the angular frequency $\omega = 2\pi\nu$ is commonly used instead of the frequency ν, and the reorientation time is referred to the unit circle so that it can be correlated with the Larmor frequency $\omega_0 = 2\pi\nu_0$. Therefore, the reorientation time is divided by 2π and then designated as the **effective molecular correlation time** τ_c.

As mentioned earlier, only molecular motions with "frequencies" that lie in the range of ^{13}C precession frequencies participate in the dipole-dipole relaxation of C-H nuclei. For example, at a magnetic field strength of $H_0 \approx 2.1$ T, ^{13}C nuclei precess with a frequency of $\nu_0 \approx 22.6$ MHz $= 2.26 \cdot 10^7$ Hz or $\omega_0 \approx 2\pi \, .2.26 \cdot 10^7 \approx 1.42 \cdot 10^8$ rad/s. The most effective, and therefore the fastest dipole-dipole relaxation is made possible by molecular motions with correlation times of

$$\tau_c \approx \frac{1}{1.42 \cdot 10^8} \approx 7 \cdot 10^{-9} \text{ s/rad}.$$

For these motions, the correlation function $T_1 = f(\tau_c)$ goes through a minimum at $H_0 \approx 2.1$ T (see Fig. 9.7). At stronger magnetic fields, the minimum is reached with shorter correlation times, as shown in the figure. Rapid dipole-dipole relaxation then requires even faster molecular motions.

Relatively slow molecular motions with $\tau_c > 10^{-9}$ s/rad (at $H_0 \approx 2.1$ T) contribute less effectively to dipole-dipole relaxation and cause an increase in T_1 while the spin-spin relaxation time T_2 (see Section 2.2) decreases (see Fig. 9.7), so the signal broadens as a consequence of the relationship $\Delta\nu_{1/2} \sim 1/T_2$. For this reason, the "slower" macromolecules often yield poorly resolved ^{13}C NMR-band spectra.

On the other hand, small- to medium-sized molecules move too fast in low viscosity solutions for effective dipole-dipole relaxation to be achieved ($\tau_c < 10^{-9}$ s at $H_0 \approx 2.1$ T). In these cases, T_1 and T_2 increase ($T_1 \sim T_2$) so that the signals become sharper. One then measures in the range of the molecular motion frequency, which is the most attractive frequency for high resolution NMR spectroscopy ("motional narrowing"). Under these conditions, the dipole-dipole relaxation time of a ^{13}C nucleus is inversely proportional to the number N of directly bonded protons and to its effective correlation time, according to Eq. (9.7) (h is Planck's constant and $h/2\pi \equiv \hbar$; γ_C *and* γ_H are

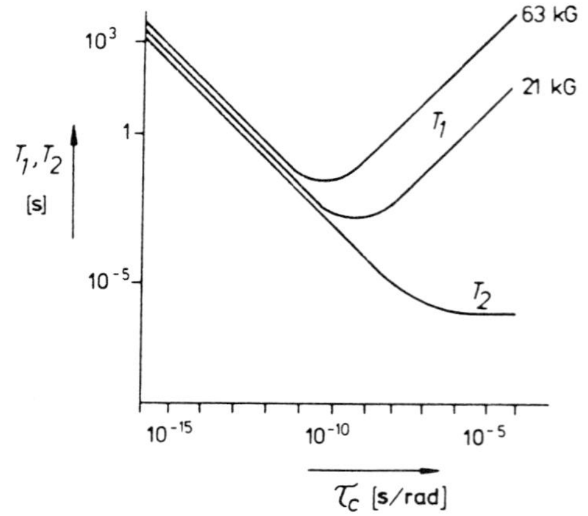

Fig. 9.7. Relationship between relaxation times T_1 and T_2 and correlation times τ_c. (Source: E. Breitmaier, K.-H. Spohn, S. Berger, *Angew. Chem.* **87,** 152, 1975.)

the gyromagnetic ratios of ^{13}C and ^{1}H; r_{CH} is the ^{13}C-^{1}H internuclear distance, usually 0.109 nm):

$$\frac{1}{T_{1DD}} = \hbar^2 \gamma_c^2 \gamma_H^2 r_{CH}^{-6} N \tau_c \qquad (9.7)$$

Many applications of ^{13}C spin-lattice relaxation times are based on this relationship.

9.5 Selected Chemical Applications

9.5.1 Differentiation Between CH_2, CH, and Quaternary Carbon Atoms

In rigid molecules without any motionally preferred orientation, all carbon atoms are strongly affected by molecular motions in the medium. Therefore, the correlation time τ_c is the same for all carbons. The C-H bond lengths are also approximately equal (0.109 nm) in organic compounds, with the exception of those in ethynyl groups (0.106 nm). Under these conditions, Eq. (9.7) shows that a carbon atom's dipole-dipole relaxation time depends only on the number of directly bonded hydrogen atoms. Because $1/T_1 \sim r_{CH}^{-6}$, more distant protons have virtually no influence:

$$T_{1DD} = const./N \qquad (9.7a)$$

Accordingly, the T_1 values of CH and CH_2 carbon atoms in the same molecule should be in the ratio 2:1, and quaternary carbon atoms with $N = 0$ should relax considerably more slowly:

$$(T_{1CR_4} \gg)\ T_{1CH} : T_{1CH_2} = 2:1 \qquad (9.8)$$

For example, this relationship can be recognized in the case of adamantane, with $T_{1CH} \approx 24.4$ s and $T_{1CH_2} \approx 13.2$ s (see Fig. 9.5). It is a valuable aid in assigning the ^{13}C signals of the rigid carbon skeletons of larger molecules, particularly when signal crowding in the coupled or off-resonance proton-decoupled ^{13}C NMR spectra preclude any unequivocal distinction between triplets, doublets, and singlets.

Methyl carbon atoms have shorter correlation times than do CH and CH_2 nuclei of the rigid molecular skeleton because of their methyl group rotations—which are usually faster than overall molecular motions—and thus their T_1 values are larger.

9.5.2 Anisotropy of Molecular Motion

The relationship in Eq. (9.7) is valid only when the motion of the molecule under study is **isotropic,** i.e., when it has no preferred direction. This is true for spherically symmetrical molecules such as adamantane. Ellipsoidal or more disc-shaped molecules such as substituted benzenes rotate and move in such a way that they displace as few other molecules as possible. Therefore, the motions of these molecules have preferred directions, and they are **anisotropic.** The T_1 values in the direction of the preferred molecular rotations can be estimated for simple molecules without using mathematical relationships between nuclear relaxation and anisotropic molecular motions [4]. Thus the C-H nuclei at the *p* position of monosubstituted benzenes relax faster than those in *ortho* or *meta* positions relative to the substituent (see Table 9.1).

The reason for *p* C-H nuclei relaxing more rapidly is a preferred rotation around the molecular axis passing through the substituent X and the *p* carbon atom. When the molecule rotates in this manner, the *p* C-H bond does not alter its orientation with regard to the external magnetic field vector $\vec{H}_0$. Fluctuating local fields at the *p* carbon nucleus arise, then, only by molecular rotations perpendicular to the preferred axis, but these less favored rotations still have large enough frequencies and therefore small enough correlation times to relax the *p* carbon atom rapidly. However, the preferred, and thus the fastest rotation alters the orientation of the *o* and *m* C-H bonds with regard to $\vec{H}_0$ *too rapidly* for the *o* and *m* carbon atoms to relax by the dipole-dipole mechanism.

9.5.3 Steric Effects

It was pointed out in Section 9.5.1 that the rotation of methyl groups is usually faster than the overall motion of the molecule, so that $\tau_{c(CH_3)} \ll \tau_{c(skeleton)}$. In conclusion, the T_1 values are usually larger for methyl carbon atoms than one would expect from Eq. (9.7) for the DD mechanism:

$$T_{1(CH)}:T_{1(CH_2)}:T_{1(CH_3)} = 6:3:2.$$

The methyl group of toluene is an example (see Table 9.1), relaxing more slowly (16.3 s) than the *p* carbon atom (15 s).

Steric interactions, however, can hinder the methyl rotation and thus accelerate relaxation of the methyl carbon nucleus. Therefore, the T_1 values of methyl groups can serve as probes for steric interactions. In this way, the T_1 values of 2-butanone oxime permit the conclusion that the CH_3 group *syn* to

TABLE 9.1 ^{13}C SPIN-LATTICE RELAXATION T_1 IN BENZENE AND MONOSUBSTITUTED DERIVATIVES.[a]

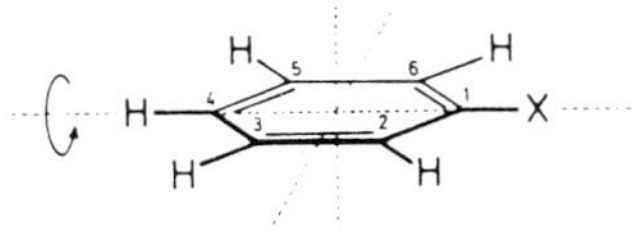

X	C Atom	T_1[s]
—H	1 – 6	29.3[b]
—CH_2—CH_3	1	36.0
	2,6	18.0
	3,5	18.0
	4	13.0
	—CH_2	13.0
	—CH_3	9.0
—CH=CH_2	1	75.0
	2,6	14.8
	3,5	13.5
	4	11.9
	—CH=	17.0
	=CH_2	7.8
—C≡CH	1	107.0[b]
	2,6	14.0
	3,5	14.0
	4	8.2
	—C≡	132.0
	≡CH	9.3
—CH_3	1	58.0
	2,6	20.0
	3,5	21.0
	4	15.0
	CH_3	16.3
—OH	1	21.5
	2,6	4.4
	3,5	3.9
	4	2.4

[a]E. Breitmaier, K.-H. Spohn, S. Berger, *Angew. Chem.* **87,** 152 (1975); and G. C. Levy, Accounts Chem. Res. **6,** 161 (1973).
[b]Degassed, i.e., O_2-free sample.

the OH group in the stable *anti* isomer rotates more rapidly (T_1 = 6.1 s) than the one *anti* to the OH group in the more labile *syn* isomer (T_1 = 2.8 s):

anti- *syn-*

2-Butanone oximes

In the *anti* isomer, van der Waals interactions between the methyl and the hydroxyimino and methylene groups are clearly of the same order of magnitude, and the methyl rotation is still too fast to enable rapid relaxation of the methyl carbon nucleus. On the other hand, the methyl group in the *syn* isomer interacts unilaterally with the methylene group and prefers an energetically favorable staggered conformation, which restricts its rotation.

In linalool, the 1,2-dimethylvinyl methyl carbon *trans* to the large alkyl group relaxes considerably more slowly (8.0 s) than that in *cis* position (3.5 s). The *cis* methyl rotation is hindered as a result of the interaction between the *cis* methyl and methylene groups, leading again to a more rapid relaxation of the methyl carbon nucleus.

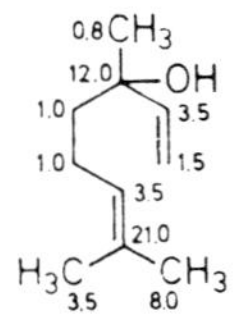

(cis) *(trans)*

Linalool

Substituents can also hinder intramolecular rotation in biphenyl derivatives and related compounds. Because the rotation of the phenyl group is anisotropic (see Section 9.5.2), so that $T_{1(o,m)} > T_{1(p)}$, the ratio $T_{1(o,m)}/T_{1(p)}$ decreases with increasing hindrance of rotation, as can be seen by comparing the data reported for 3-bromo and 2,2′,6,6′-tetramethylbiphenyl with those for the parent compound:

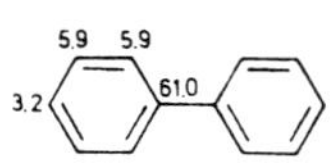

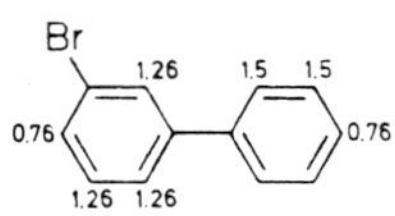

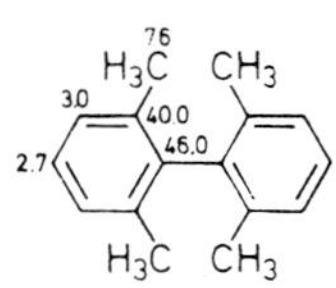

$_{1(o,m)}:T_{1(p)}$ = 1.84 1.66 1.97 1.1

Biphenyl 3-Bromobiphenyl 2,2′,6,6′-Tetramethylbiphenyl

9.5.4 Molecular Flexibility

The different mobility of distinct parts of a molecule **(segment mobility)** can result in considerable deviations from the T_1 values given by Eq. (9.7a). In such cases, the correlation time τ_c can be different for each carbon atom. The product NT_1 is then no longer constant— as it is for the carbon atoms in rigid molecules—but inversely proportional to the correlation time according to Eq. (9.7b):

$$NT_1 \sim \frac{1}{\tau_c} \tag{9.7b}$$

This product can be taken as a measure of intramolecular mobility.

In long-chain alkanes and their substituted derivatives, the carbon nuclei already have such different T_1 values that one can conclude that these molecules are relatively flexible. In this way, the spin-lattice relaxation times and their corresponding products NT_1 go through a minimum in decane, which is displaced somewhat toward the heavy end of the molecule in 1-bromodecane. Therefore, the ends of the molecule are more mobile than the center and light molecular segments are more mobile than heavy ones:

Decane

	CH_3 —	CH_2 —	CH_2 —	CH_2 —	CH_2 —	CH_2 —	CH_2 —	CH_2 —	CH_2 --	CH_3	
T_1 =	8.74	6.64	5.71	4.95	4.36	4.36	4.95	5.71	6.64	8.74	s
NT_1 =	26.2	13.3	11.4	9.9	8.7	8.7	9.9	11.4	13.3	26.2	s

1-Bromodecane

Br—	CH_2 —	CH_2 —	CH_2 —	CH_2 —	CH_2 —	CH_2 —	CH_2 —	CH_2 —	CH_2 —	CH_3	
T_1 =	2.8	2.7	1.9	2.0	2.1	2.1	2.2	3.1	3.9	5.3	s
NT_1 =	5.6	5.4	3.8	4.0	4.2	4.2	4.4	6.2	7.8	15.9	s

In cholesteryl chloride, the T_1 values of the ring CH and CH_2 carbon atoms are in the ratio 2:1. Therefore, the androstene skeleton is relatively rigid. In the side chain, on the other hand, mobility increases with distance from the steroidal residue. The C-26 and C-27 methyl carbon nuclei relax more slowly than the C-21 nucleus, and the C-24 methylene carbon relaxes more slowly than the C-20 one:

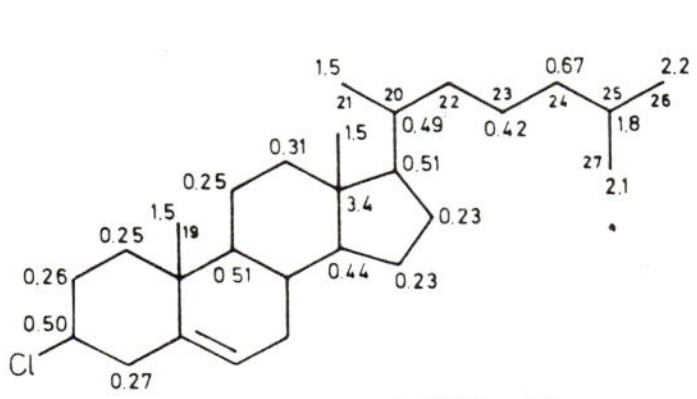

Cholesteryl Chloride

9.5.5 Association Effects on Molecular Mobility

Hydrogen bonds and ion-dipole interactions hinder the mobility of molecules and affected portions of molecules [2,3]. Thus they raise the correlation times τ_c of the ^{13}C-H nuclei and favor their dipole-dipole relaxation. For example, the formyl carbon relaxes more rapidly in formic acid than in methyl formate [2,3]. According to this, the methyl ester is more mobile than the dimeric carboxylic acid. Furthermore, one can see from the nuclear Overhauser factors of the carbon atoms of both molecules that in formic acid, the ^{13}C relaxation proceeds by a purely dipole-dipole mechanism (η = 2.0), whereas spin rotation also participates in methyl formate (η_c = 1.55):

Formic acid

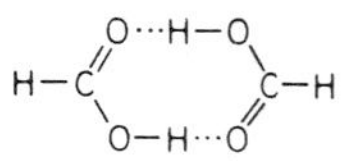

T_1 = 10.1 ± 1 s
η_C = 2.0 ± 0.15 s
(25 MHz)

Methyle formate

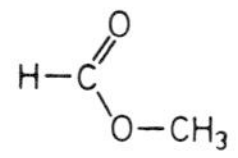

T_1 = 14.5 ± 1.5 s; 18.0 ± 2.5
η_C = 1.55 ± 0.15s; 0.8 ± 0.1
(25 MHz)

The ^{13}C spin-lattice relaxation times increase on going from the associated groups in 1-decanoic acid and 1-decanol ($-COOH$ and $-CH_2OH$) toward the hydrophobic ends of both molecules. As the hydrocarbon chain becomes more mobile toward the hydrophobic extremity, the molecule is anchored by hydrogen bonds at its polar end:

1-Decanoic acid

	CH_3 —	CH_2 —	CH_2 —	CH_2—	$(CH_2)_3$-	CH_2 —	CH_2 —	C
T_1	= 3.0	1.9	1.4	1.2	0.8	0.6	0.4 s	
NT_1	= 9.0	3.8	2.8	2.4	1.6	1.2	0.8 s	

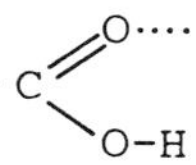

1-Decanol

	CH_3 —	CH_2 —	CH_2 —	CH_2—	$(CH_2)_3$-	CH_2 —	CH_2 —	CH_2— O
T_1	= 3.1	2.2	1.6	1.1	0.84	0.77	0.77	0.65 s
NT_1	= 9.3	4.4	3.2	2.2	1.68	1.54	1.54	1.3 s

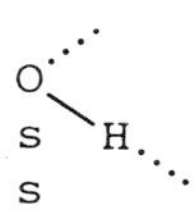

^{13}C spin-lattice relaxation times are also useful in some cases to establish the presence of hydrogen bonds. The increasing mobility in a large molecule from the associated to the hydrophobic end, and the concomitant rise of T_1 or NT_1, can also be useful for assigning the ^{13}C signals. In phytol, for example, the diterpene alcohol component of chlorophyll, assignment of the signals for the CH_2 carbon pairs C-4 and C-14, C-6 and C-12, and C-9 and C-13 is uncertain because of their similar chemical shifts and identical multiplicities. The assignment can be made, however, after measuring the spin-lattice relaxation times: carbon atoms 12, 13 and 14, which are more removed from the hydroxyl group, move more rapidly and therefore have larger NT_1 values.

Phytol

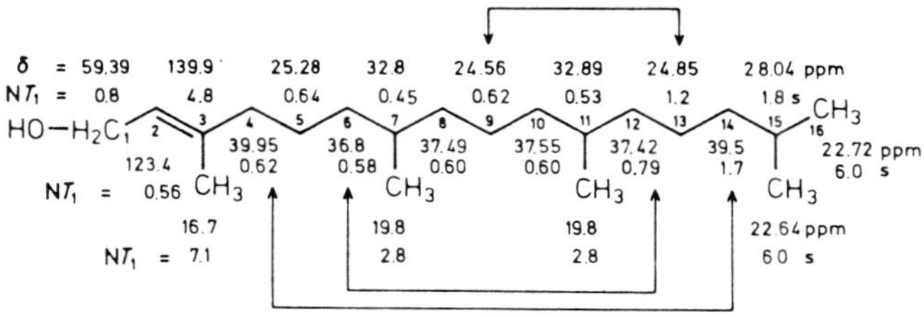

9.6 Paramagnetic Relaxation

As a consequence of their considerably larger magnetic moments (μ_e:μ_H = 1836:1), unpaired electrons in radicals and other paramagnetic compounds produce much stronger and more far-reaching fluctuating fields than protons do. As a result, relaxation processes in the presence of unpaired electrons can occur by the electron spin-nucleus dipole-dipole interaction, which is analogous to the dipole-dipole mechanism but more effective. For this reason, smaller T_1 values are observed in paramagnetic compounds or in their presence, which also happens in solutions containing dissolved oxygen. To measure reproducible relaxation times, paramagnetic impurities—including oxygen, which is almost always dissolved in liquid samples—must be removed. **(degassing).** The influence of dissolved oxygen is particularly large for slowly relaxing carbon atoms ($T_1 > 20$ s). For example, the quaternary carbon atoms of phenylacetylene relax much more rapidly without degassing, as shown below (the figures appear in parentheses):

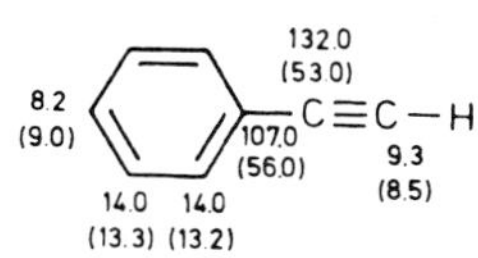

Phenylacetylene
T_1 [s]

It is possible to make use of this particularly large speeding up of the relaxation of slow ^{13}C nuclei in PFT ^{13}C NMR spectroscopy. When a rapid succession of excitation pulses is used, which is usually the case when accumulating signals, the spin-lattice relaxation of slow ^{13}C nuclei can no longer follow the excitation. The corresponding ^{13}C signals then become attenuated. Adding small quantities of paramagnetic relaxation accelerators such as radicals (*t*-butyl nitroxide) or transition metal salts (cobalt(II)acetate) to the sample will enhance these signals. The reagents also suppress the nuclear Overhauser effect in proton decoupling because paramagnetic relaxation is faster than dipole-dipole relaxation. This balancing out of signal intensities by adding paramagnetic substances permits quantitative interpretation of ^{13}C NMR spectra (see, also, Section 3.7.2).

References

1. J. H. Noggle, R. E. Schirmer, *The Nuclear Overhauser Effect: Chemical Applications,* Academic Press, New York, 1971.
2. E. Breitmaier, K.-H. Spohn, S. Berger, *Angew. Chem.* **87,** 152 (1975).
3. G. C. Levy, *Accounts Chem. Res.* **6,** 161 (1973).
4. D. E. Woessner, *J. Chem. Phys.* **37,** 647 (1962).

PART II

Section A

Exercises in the Interpretation of ^{13}C NMR Spectra and Their Application to Structural Problems in Organic Chemistry

1

What compound with the molecular formula C_8H_{18} fits ^{13}C NMR Spectrum 1?

Recording Conditions

Solvent:	Deuteriochloroform (triplet at 77.1 ppm, not shown).
Concentration:	75% v/v.
Temperature:	30° C.
Operating frequency:	22.63 MHz.
Decoupling:	*Top:* Proton off-resonance decoupled (400 scans). *Bottom:* Proton noise-decoupled (200 scans).

C_8H_{18}

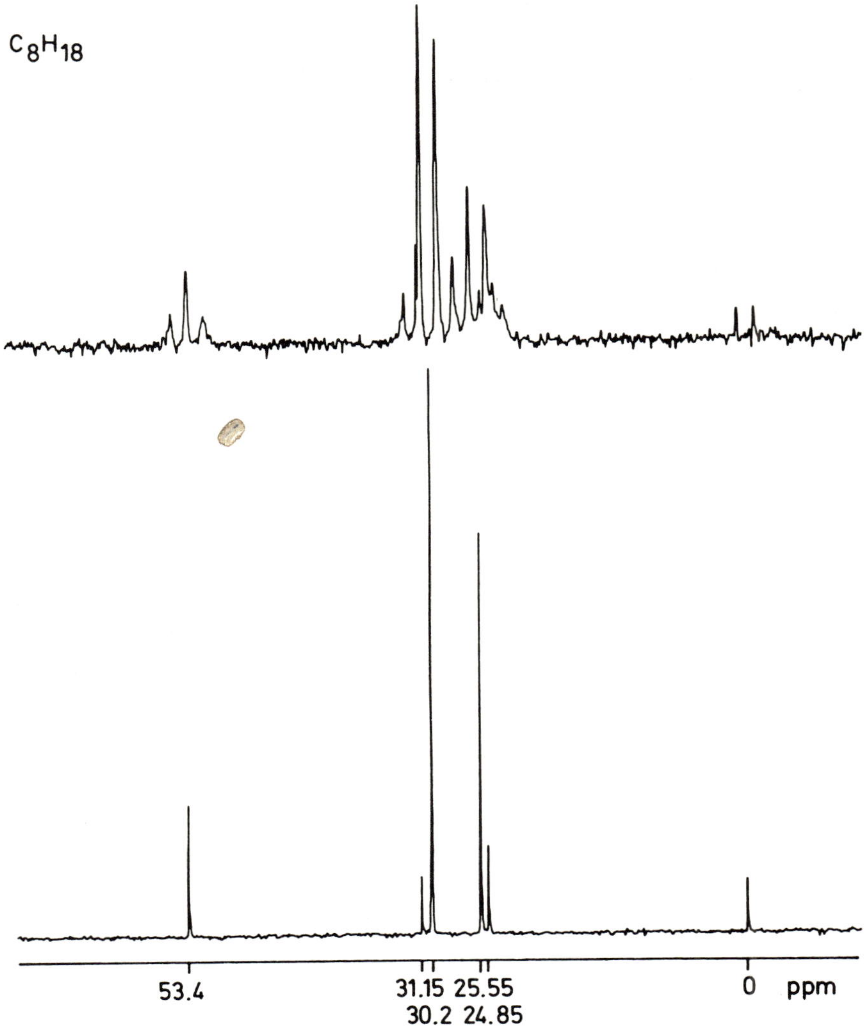

(1)

2

What compound with the molecular formula $C_4H_{10}S$ fits ^{13}C NMR Spectrum 2? Assign each signal.

Recording Conditions

Solvent:	Deuteriochloroform (triplet at 77.25 ppm, not shown).
Concentration:	50% v/v.
Temperature:	28° C.
Operating frequency:	22.63 MHz.
Decoupling:	*Top:* Proton off-resonance decoupled (200 scans). *Bottom:* Proton noise-decoupled (100 scans).

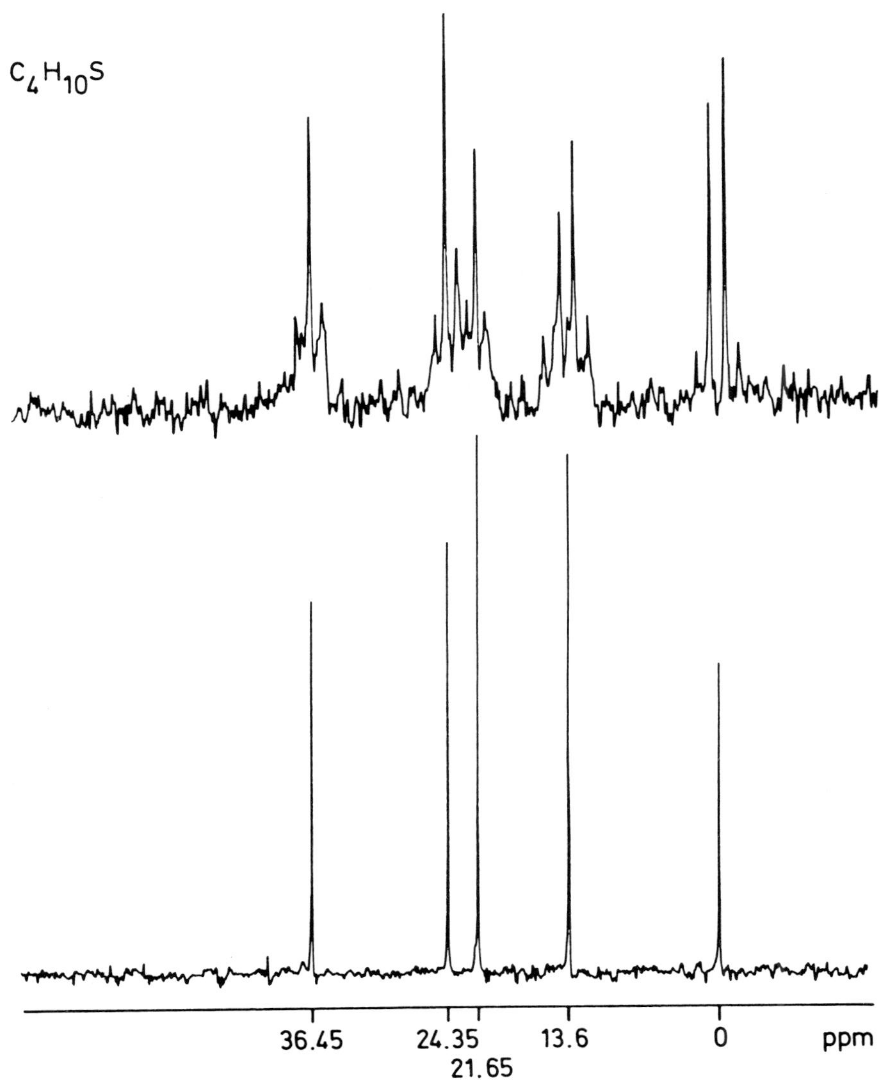

2

3

What primary amine with the molecular formula $C_5H_{13}N$ corresponds to ^{13}C NMR Spectra 3A-C? Analyze all the spectra and use the literature in Footnote 1 to interpret Spectrum B:

Recording Conditions

Solvent:	Tetradeuteriomethanol (septet at 49.3 ± 0.2 ppm with J_{CD} = 21.5 Hz).
Concentration:	A. 10% v/v. B. 10% v/v + 0.1 ml concentrated HCl. C. 10% v/v + 30 mg tris (1,1,1,2,2,3,3-hepta-fluoro-7,7-dimethyl-4,6-octanedionato) europium (III).
Temperature:	30° C.
Operating frequency:	20 MHz.
Decoupling:	Proton noise-decoupled (1,500 scans for each spectrum).

[1] J. G. Batchelor, J. Feeney, G. C. K. Roberts, *J. Magn. Reson.* **20,** 19 (1975); and J. G. Batchelor, *J. Am. Chem. Soc.* **97,** 3410 (1975).

VR = Shift reagent

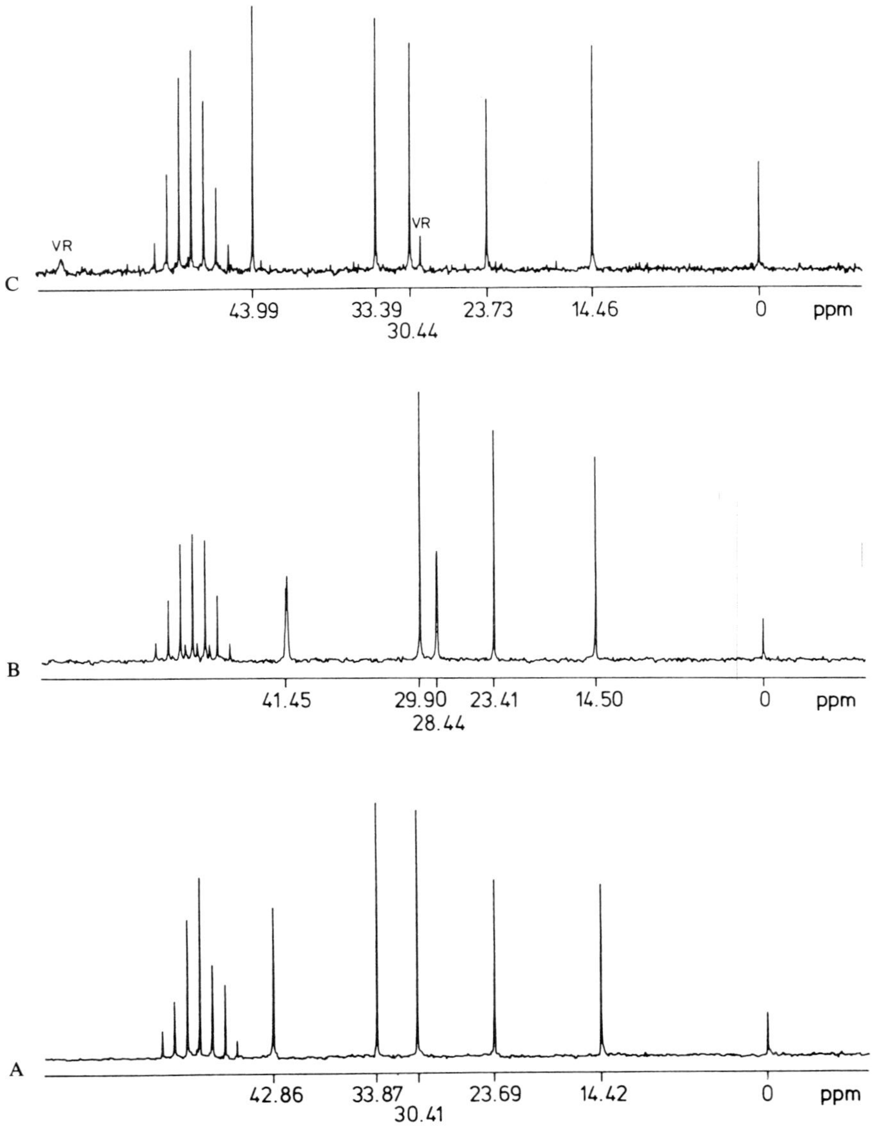

3

4

The series of spectra in Figure 4 is the result of a T_1 measurement on 2-octanol using the inversion recovery method. First, assign the ^{13}C chemical shifts of 2-octanol using the data for octane and the OH increments (Table 7.2, p. 52). **Second, determine the T_1 values for all the ^{13}C nuclei from their zero transition times. Interpret the relaxation behavior and check your assignment based on increments with the following results[1]:**

^{13}C Chemical Shifts of Octane

CH_3 —	CH_2 —	CH_2 —	CH_2 —	CH_2 —	CH_2 —	CH_2 —	CH_3	
14.0	23.0	32.4	29.7	29.7	32.4	23.0	14.0	ppm

Recording Conditions

Solvent:	Hexadeuterioacetone (extremely weak septet at 29.85 ppm; singlet at 206 ppm not shown).
Concentration:	75% v/v.
Temperature:	30° C.
Operating frequency:	20 MHz.
Decoupling:	Proton noise-decoupled (30 scans).
T_1 experiment using 180°, τ, 90° pulse sequence:	Spectrum series for τ = 0.4, 0.6, 0.8, 1.0, 1.2, 1.4, 1.6, 1.8, 2.0, 2.2, 2.4, 2.6, 2.8, 3.0, 3.2, 3.4, 3.6, 3.8, 4.0, and 32.0 s (interval for fairly complete relaxation).

[1]See J. B. Stothers, *Carbon-13 NMR Spectroscopy,* Academic Press, New York, 1972, p. 56.

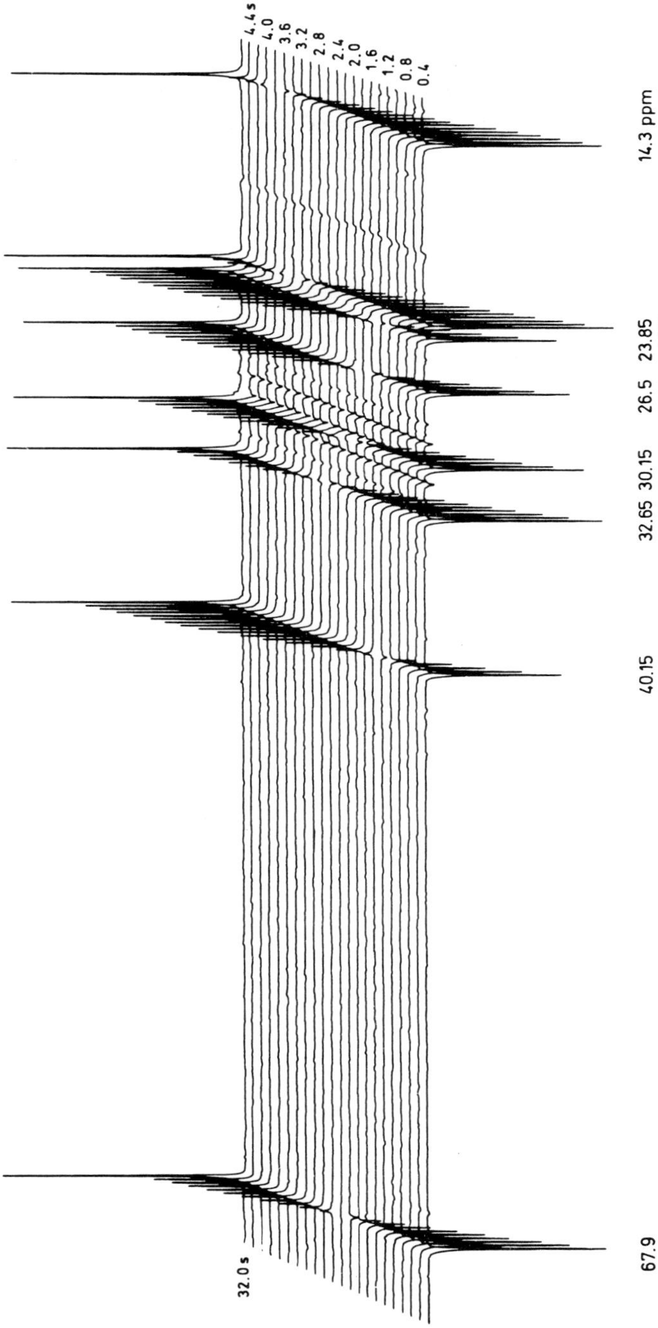
4.4 s
4.0
3.6
3.2
2.8
2.4
2.0
1.6
1.2
0.8
0.4
32.0 s
14.3 ppm
23.85
23.2
26.5
30.15
32.65
40.15
67.9

4

5

A commercial sample of isoamyl nitrite was found to be analytically pure but gave ^{13}C NMR Spectrum 5. What compounds make up the sample? For the assignment, begin with the ^{13}C chemical shifts of an appropriate hydrocarbon and the necessary shift increments.[1]

Recording Conditions

Solvent:	Hexadeuterioacetone (signals not visible owing to high sample concentration).
Concentration:	90% v/v.
Temperature:	30° C.
Operating frequency:	20 MHz.
Decoupling:	*Top:* Proton off-resonance decoupled (1,000 scans). *Bottom:* Proton noise-decoupled (50 scans).

---

[1]See Table 7.2, p. 52; J. B. Stothers, *Carbon-13 NMR Spectroscopy,* Academic Press, New York-London, 1972; and E. Breitmaier, W. Voelter, *^{13}C NMR: Methods and Applications,* "Monographs in Modern Chemistry," Vol. 5, H. E. Ebel, ed., Verlag Chemie GmbH, Weinheim, 2nd ed., 1978.

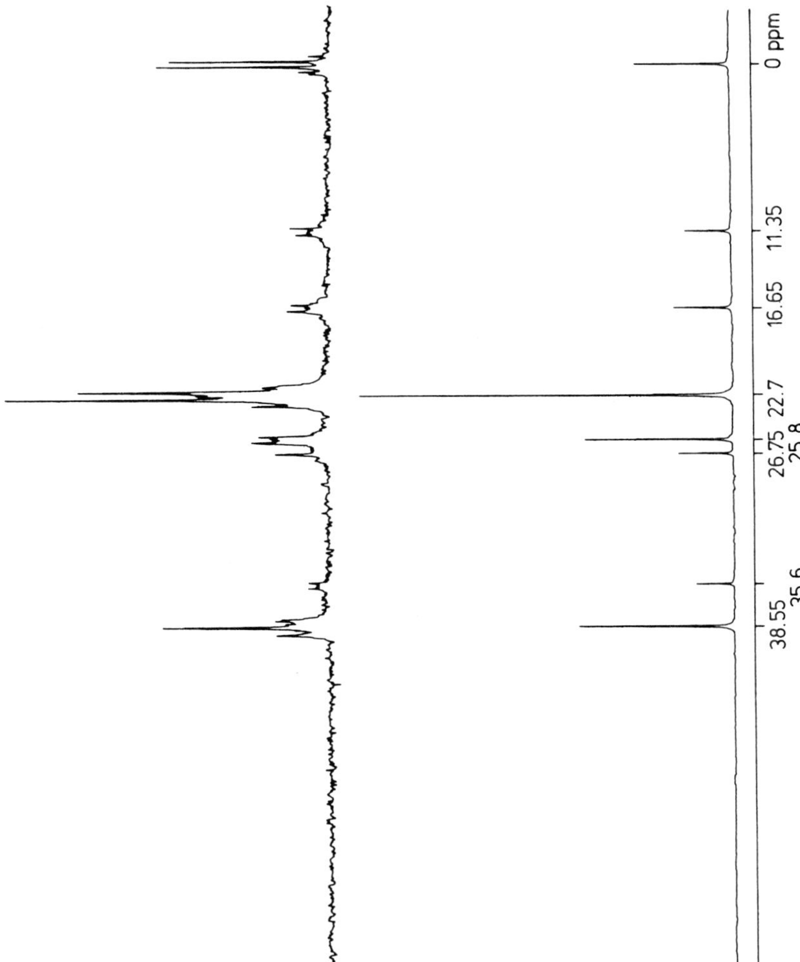

5

E

6

What compound with the molecular formula $C_5H_7NO_2$ fits ^{13}C NMR Spectrum 6?

Recording Conditions

Solvent:	Deuteriochloroform (not visible because the solution is concentrated).
Concentration:	75% v/v.
Temperature:	28° C.
Operating frequency:	22.63 MHz.
Decoupling:	*Top:* Proton off-resonance decoupled (1,000) scans). *Bottom:* Proton noise-decoupled (200 scans).

$C_5H_7NO_2$

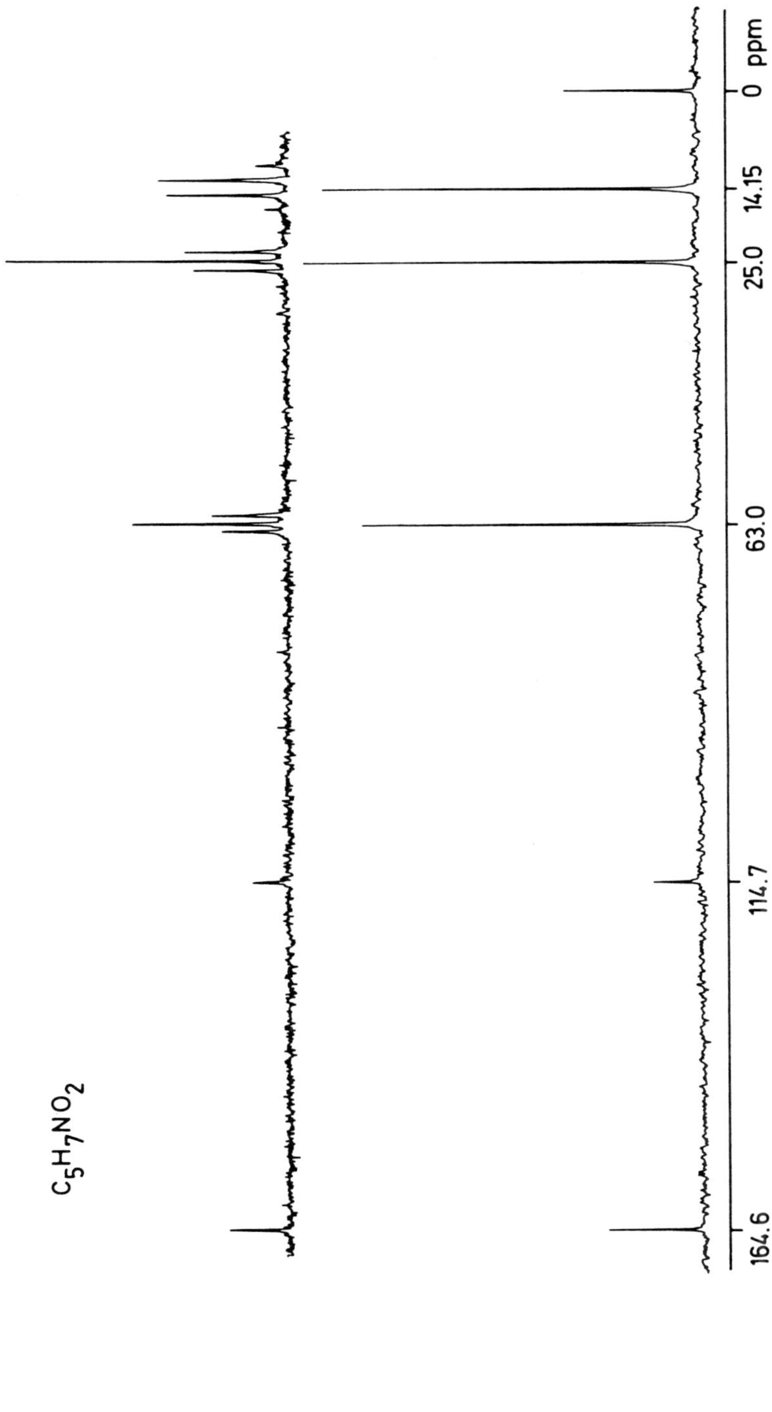

6

7

What compound with the molecular formula $C_3H_3F_3O_2$ fits ^{13}C NMR Spectrum 7?

Recording Conditions

Solvent:	Hexadeuterioacetone (septet at 29.85 ppm, J_{CD} = 19.1 Hz, singlet at 206.3 ppm).
Concentration:	50% v/v.
Temperature:	28° C.
Operating frequency:	22.63 Hz.

Coupled spectrum.

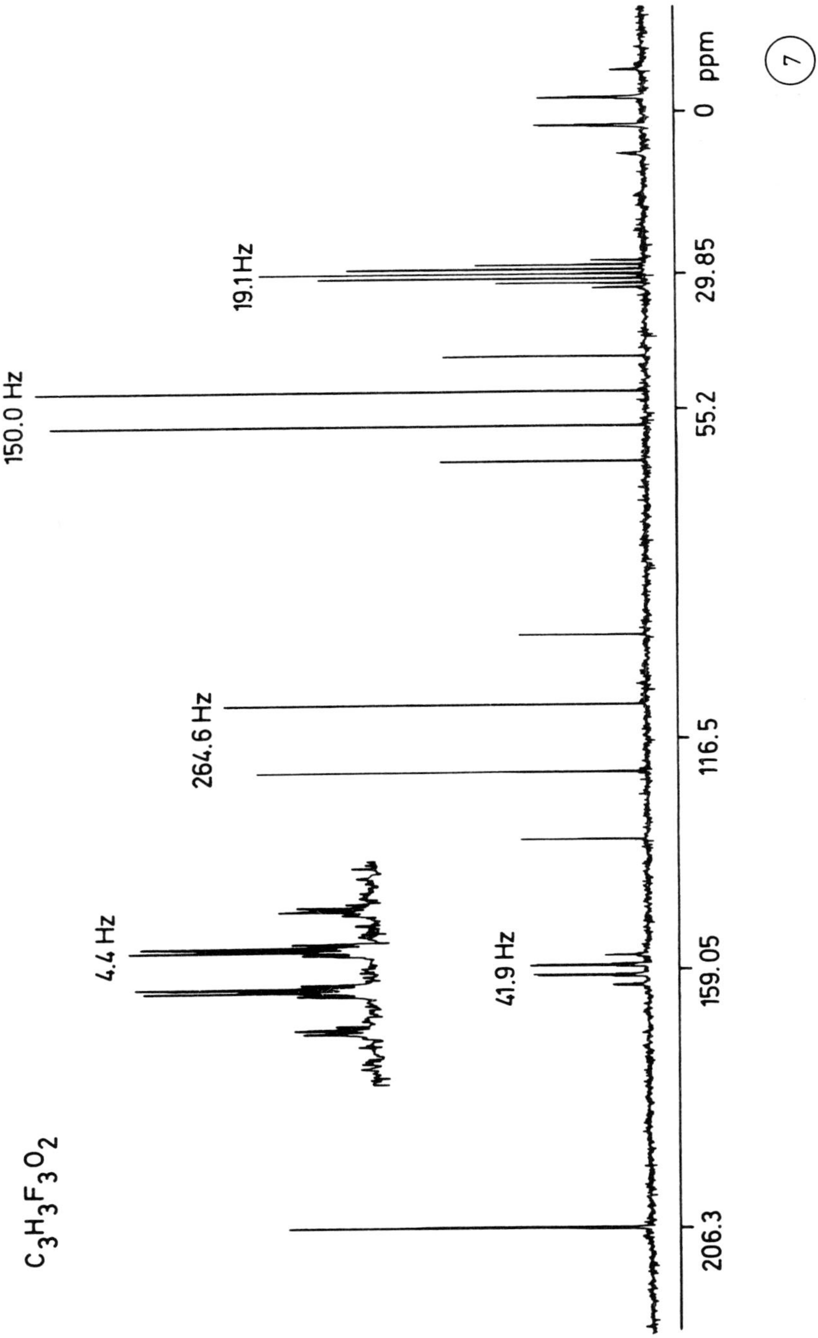

$C_3H_3F_3O_2$
150.0 Hz
19.1 Hz
264.6 Hz
4.4 Hz
41.9 Hz
0 ppm
29.85
55.2
116.5
159.05
206.3
7

8

δ-Hydroxylevulinic acid could be prepared by acid hydrolysis and decarboxylation of bromoacetylsuccinic acid diethyl ester:

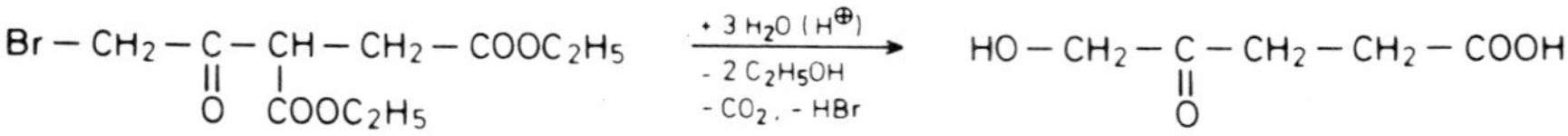

When the reaction was carried out, a compound with the molecular formula $C_6H_6O_5$ was obtained, as established by its elemental analysis and molecular weight determination. What is the compound's structure according to its ^{13}C NMR Spectrum 8?

Recording Conditions

Solvent:	Hexadeuteriodimethylsulfoxide (septet at 39.7 ppm) with a little hexadeuterioacetone (septet at 29.8 ppm, partly visible; singlet at 206.8 ppm not shown).
Concentration:	200 mg in 1 ml.
Temperature:	30° C.
Operating frequency:	15.08 MHz.
Decoupling:	*Top:* Coupled spectrum (25,000 scans). *Bottom:* Proton noise-decoupled (2,000 scans).

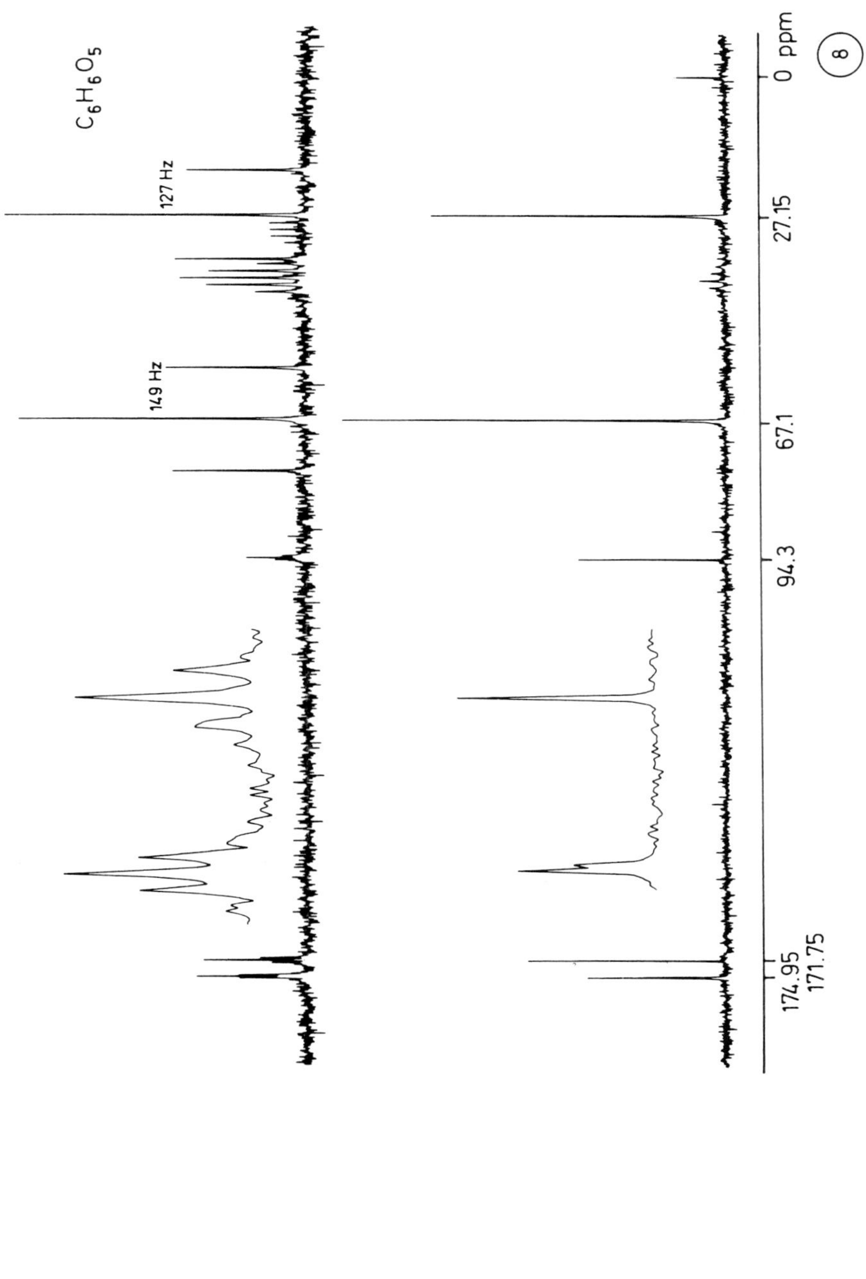

$C_6H_6O_5$
127 Hz
149 Hz
0 ppm
27.15
67.1
94.3
174.95
171.75
8

9

Identify the compound with the molecular formula C_4H_9NO on the basis of its ^{13}C NMR Spectrum 9. How do you justify the appearance of four signals?

Recording Conditions

Solvent:	Deuteriochloroform (not visible, as the solution is concentrated).
Concentration:	75% v/v.
Temperature:	30° C.
Operating frequency:	15.08 MHz.
Decoupling:	*Top:* Proton off-resonance decoupled (2,000 scans). *Bottom:* Proton noise-decoupled (500 scans).

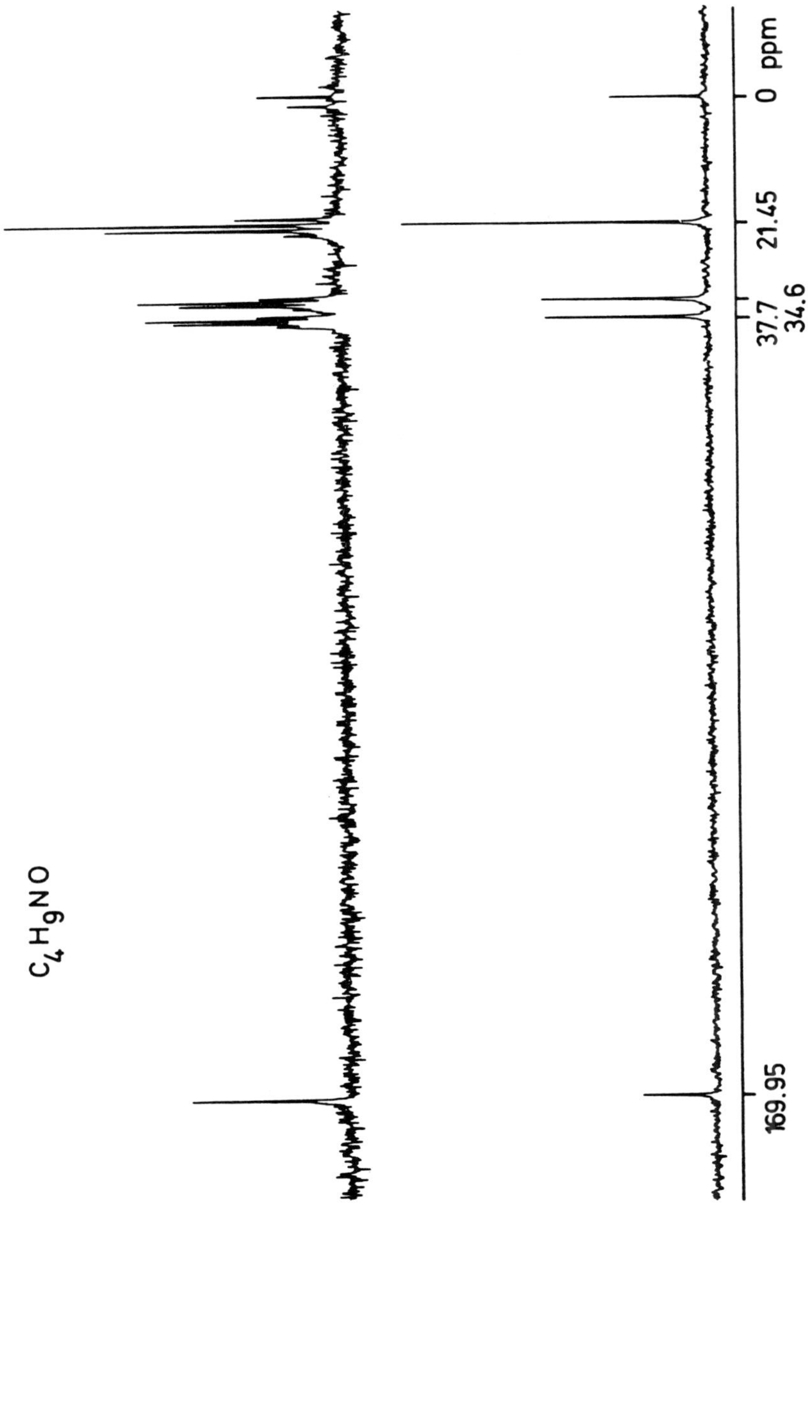

9

10

What compound with the molecular formula $C_5H_{13}NO_2$ has ^{13}C NMR Spectrum 10?

Recording Conditions

Solvent:	Deuteriochloroform (triplet at 77.1 ppm with J_{CD} = 32.5 Hz).
Concentration:	50% v/v.
Temperature:	28° C.
Operating frequency:	22.63 MHz.
Decoupling:	*Top:* Coupled spectrum (2,500 scans). *Bottom:* Proton noise-decoupled (100 scans).

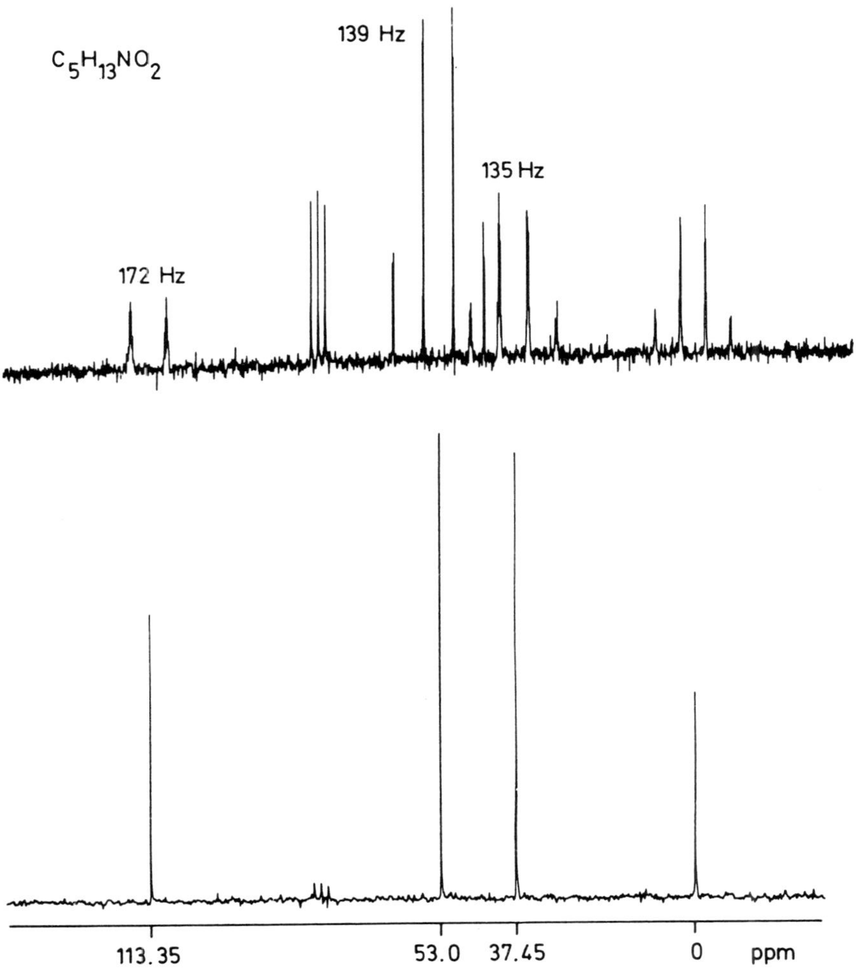

10

11

The reaction of 3-methyl-1-butenyl ether with ethyl orthoformate in the presence of boron trifluoride etherate gives a compound with the elementary composition $C_{14}H_{30}O_4$ for which ^{13}C NMR Spectrum 11 was recorded. What compound is this? How do you explain the appearance of two signals around 63 and 15.7 ppm?

Recording Conditions

Solvent:	Hexadeuterioacetone (signals not visible owing to high sample concentration).
Concentration:	90% v/v.
Temperature:	25° C.
Operating frequency:	20 MHz.
Decoupling:	*Top:* Proton off-resonance decoupled (200 scans). *Bottom:* Proton noise-decoupled (100 scans).

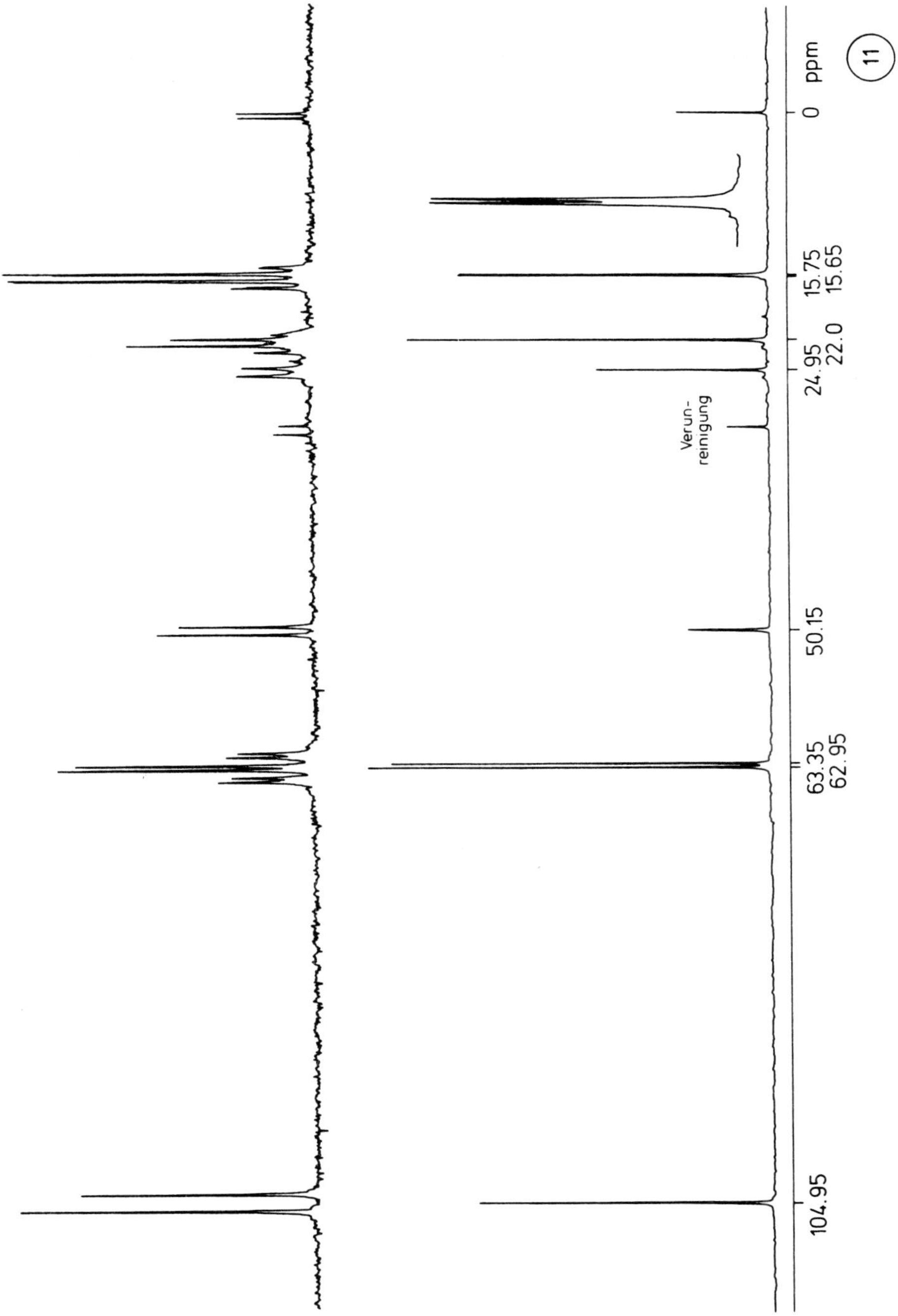
11
ppm
0
15.75
15.65
22.0
24.95
Verun-
reinigung
50.15
63.35
62.95
104.95

12

What compound with the molecular formula $C_4H_8O_2$ fits ^{13}C NMR Spectrum 12?

Recording Conditions

Solvent:	Deuteriochloroform (triplet at 77.8 ppm).
Concentration:	50% v/v.
Temperature:	28° C.
Operating frequency:	22.63 MHz.
Decoupling:	*Top:* Proton off-resonance decoupled (500 scans). *Bottom:* Proton noise-decoupled (100 scans).

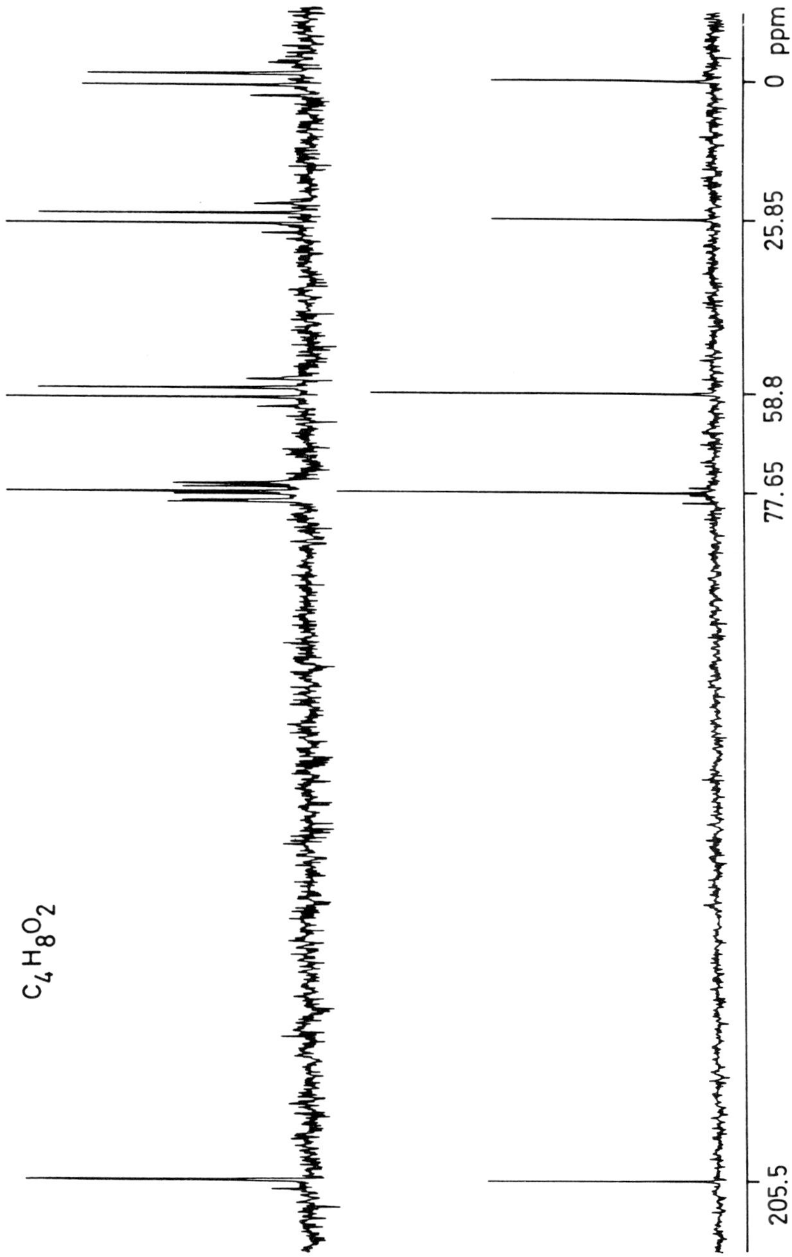

12

13

The treatment of methyl vinyl ketone with nitromethane affords reaction product A, which is derivatised with 1,2-ethanethiol to A′ having ^{13}C NMR Spectrum 13. Assign the signals and use the spectrum to establish the structure of derivative A′.

Recording Conditions

Solvent:	Hexadeuterioacetone (septet at 29.8 ppm partly visibly; singlet at 206.2 not shown).
Concentration:	50% v/v.
Temperature:	30° C.
Operating frequency:	15.08 MHz.
Decoupling:	*Top:* No decoupling (5,000 scans). *Bottom:* Proton noise-decoupled (500 scans).

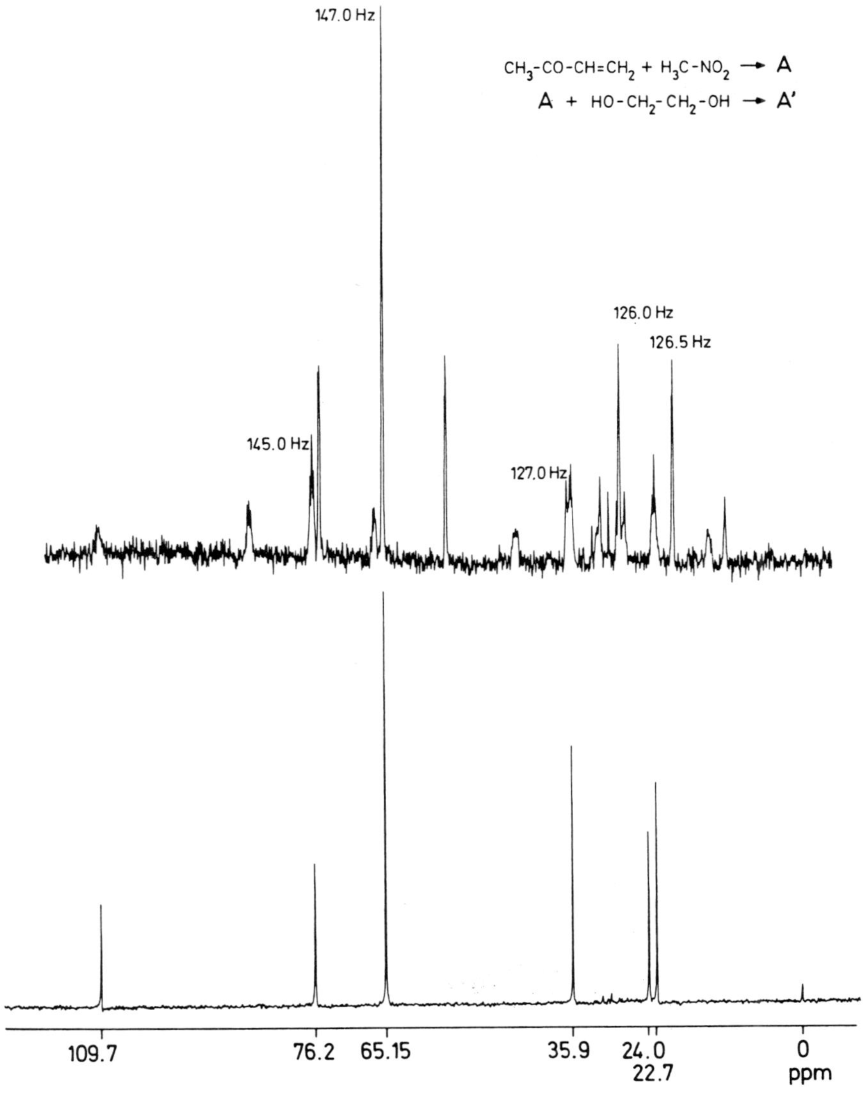
147.0 Hz
CH_3-CO-CH=CH_2 + H_3C-NO_2 → A
A + HO-CH_2-CH_2-OH → A'
126.0 Hz
126.5 Hz
145.0 Hz
127.0 Hz
109.7
76.2
65.15
35.9
24.0
22.7
0
ppm

14

Quantitatively analyse ^{13}C NMR Spectrum 14 of ethyl acetoacetate, using the signal intensities tabulated below. How do you interpret your results?

Recording Conditions

Solvent:	Hexadeuterioacetone (septet at 30.1 ppm, singlet at 207.5 ppm).
Concentration:	95% v/v.
Temperature:	25° C.
Operating frequency:	20 MHz.
Decoupling:	*Top:* Proton off-resonance decoupled (500 scans). *Bottom:* Pulsed proton decoupling as shown in Fig. 3.7A, p. 27 (300 scans).
Remarks:	In the pulsed decoupled spectrum, the waiting time between pulses was 120 s.

Measurement protocol:

Address	Frequency	ppm	Integral	Intensity
528	4,150.085	207.504	21,178.368	4,056.576
642	4,045.715	202.285	99,429.376	56,021.504
1182	3,551.330	177.566	7,125.504	4,723.200
1253	3,486.328	174.316	5,728.256	4,126.208
1373	3,376.464	168.823	94,940.160	39,467.520
3083	1,810.913	90.545	2,161.664	2,987.520
3714	1,233.215	61.660	100,802.560	44,512.768
3734	1,214.904	60.745	5,819.904	2,671.104
3958	1,009.826	50.491	102,142.464	56,429.056
4385	618.896	30.944	7,815.680	4,511.232
4403	602.417	30.120	112,625.664	57,795.072
4427	580.444	29.022	7,125.504	3,429.888
4600	422.058	21.102	6,320.640	3,190.272
4744	290.222	14.511	12,967.424	5,723.648
4747	287.475	14.373	98,109.440	53,208.576

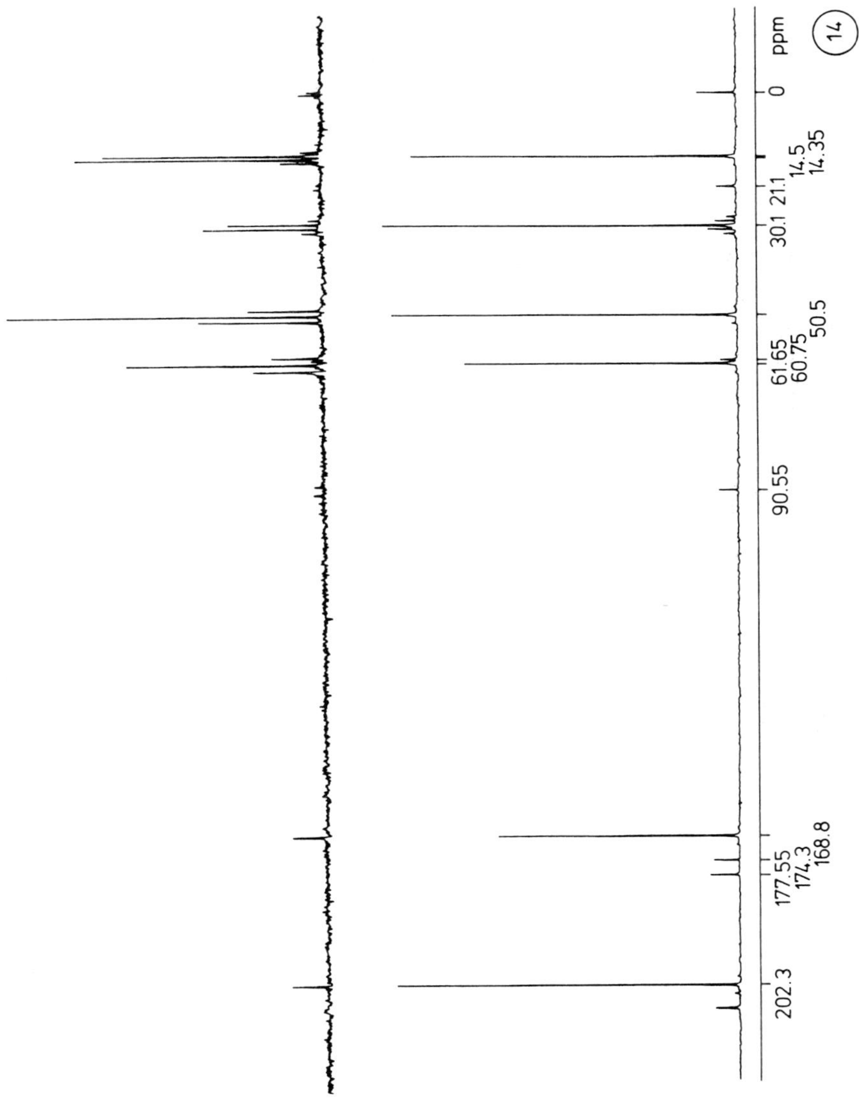
ppm
0
14.5
14.35
21.1
30.1
50.5
61.65
60.75
90.55
177.55
174.3
168.8
202.3
14

15

Ethyl trimethylsilylcarbodiimide gave ^{13}C NMR Spectrum 15 in deuteriochloroform. Try to interpret the spectrum.[1]

Recording Conditions

Solvent:	Deuteriochloroform (triplet at 77.5 ppm).
Concentration:	50% v/v.
Temperature:	25° C.
Operating frequency:	20 MHz.
Decoupling:	A. Pulsed proton decoupling, as in Fig. 3.7A; waiting time between pulses: 5 s (500 scans). B. Part of spectrum A expanded. C-D. Pulsed proton decoupling as in Fig. 3.6A; waiting time between pulses: 5 s (1,000 scans); expanded parts of the spectrum.
Remarks:	Half the pulse width required for a 90° pulse was chosen. The numbers on spectra C and D are the C-H coupling constants (long-range couplings in small type).

[1] See I. Ruppert, *Angew. Chem.* **89**, 336 (1977).

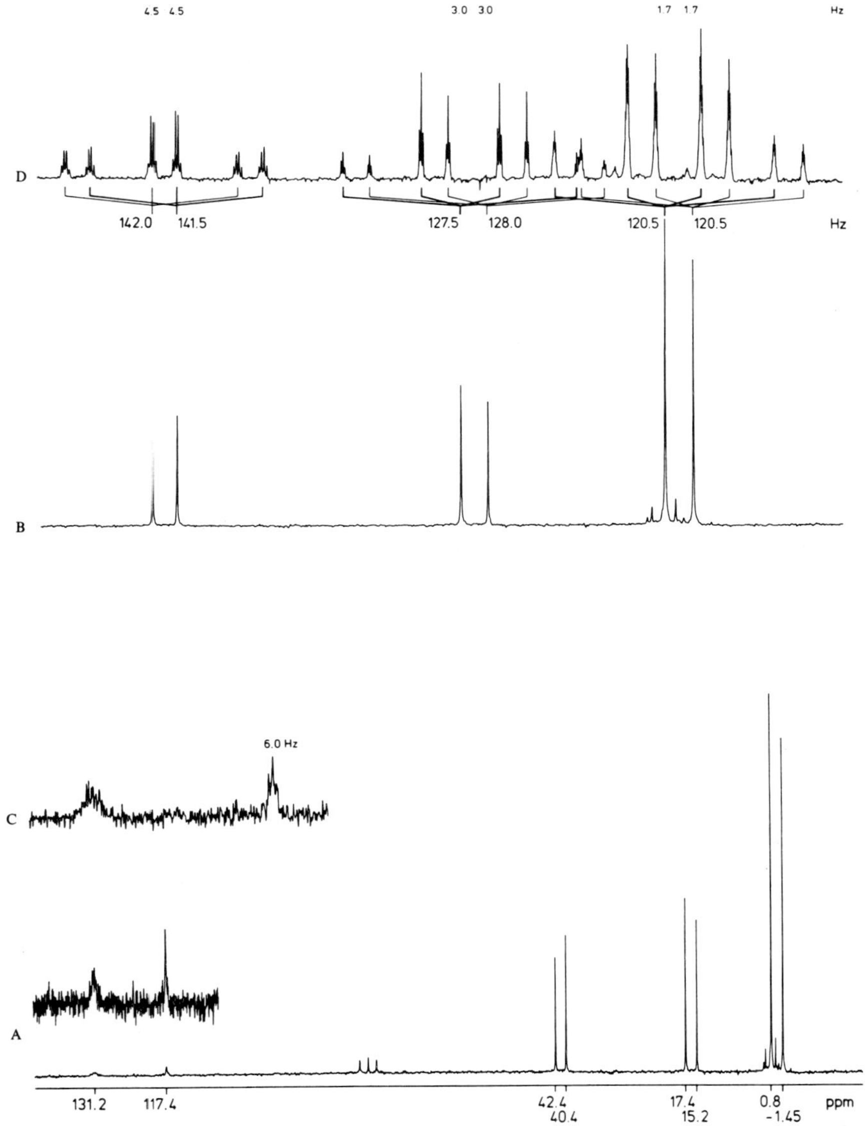
4.5 4.5
3.0 3.0
1.7 1.7
Hz
D
142.0
141.5
127.5
128.0
120.5
120.5
Hz
B
6.0 Hz
C
A
131.2
117.4
42.4
40.4
17.4
15.2
0.8
-1.45
ppm

15

16

^{13}NMR Spectra 16A and B were recorded for natural rubber and gutta-percha. Assign the signals produced by both polymers.

Recording Conditions for Both Spectra

Solvent:	Deuteriochloroform (triplet at 77.2 ppm).
Concentration:	A. 150 mg in 1 ml. B. 100 mg in 1 ml.
Temperature:	28° C.
Operating frequency:	22.63 MHz.
Decoupling:	Proton noise-decoupled. A: 5,000 scans, B: 8,000 scans.
Remarks:	The central signal of the chloroform triplet was used as an internal standard. The chemical shifts were recalculated on the TMS scale using the value of 77.2 ppm for $CDCl_3$. (When this is done, however, errors of up to one ppm are possible.)

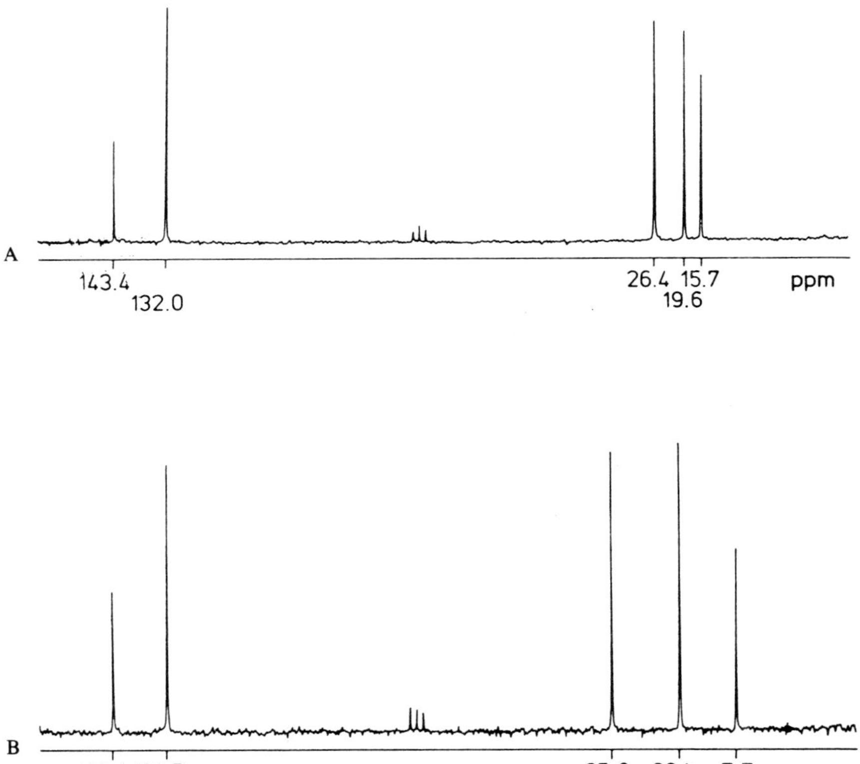

16

17

The following ^{13}C chemical shifts were measured for cyclohexylammonium phosphoenolpyruvate:

168.25 ppm (weak)	32.65 ppm (strong)
146.4 ppm (weak)	26.65 ppm (medium)
111.0 ppm (medium)	26.2 ppm (strong)
52.65 ppm (medium)	

Assign these chemical shifts to the salt's carbon atoms. The coupled ^{13}C partial spectrum (17A) and the decoupled ^{31}P NMR spectrum (17B) were recorded for a sample in which one of the salt's carbon atoms was labeled with ^{13}C (95% enrichment). Which is the labeled carbon atom? Assign the couplings.

Recording Conditions for Both Spectra

Solvent:	Deuterium oxide.
Concentration:	100 mg in 1 ml.
Standards:	^{13}C: 20 mg sodium trimethylsilylpropanesulfonate in 1 ml, internal. ^{31}P: 1 N phosphoric acid in deuterium oxide, external.
Temperature:	28° C.
Operating frequency:	22.63 MHz.
Decoupling:	Spectrum A: Coupled (2,000 scans). Spectrum B: Proton noise-decoupled (1,000 scans).

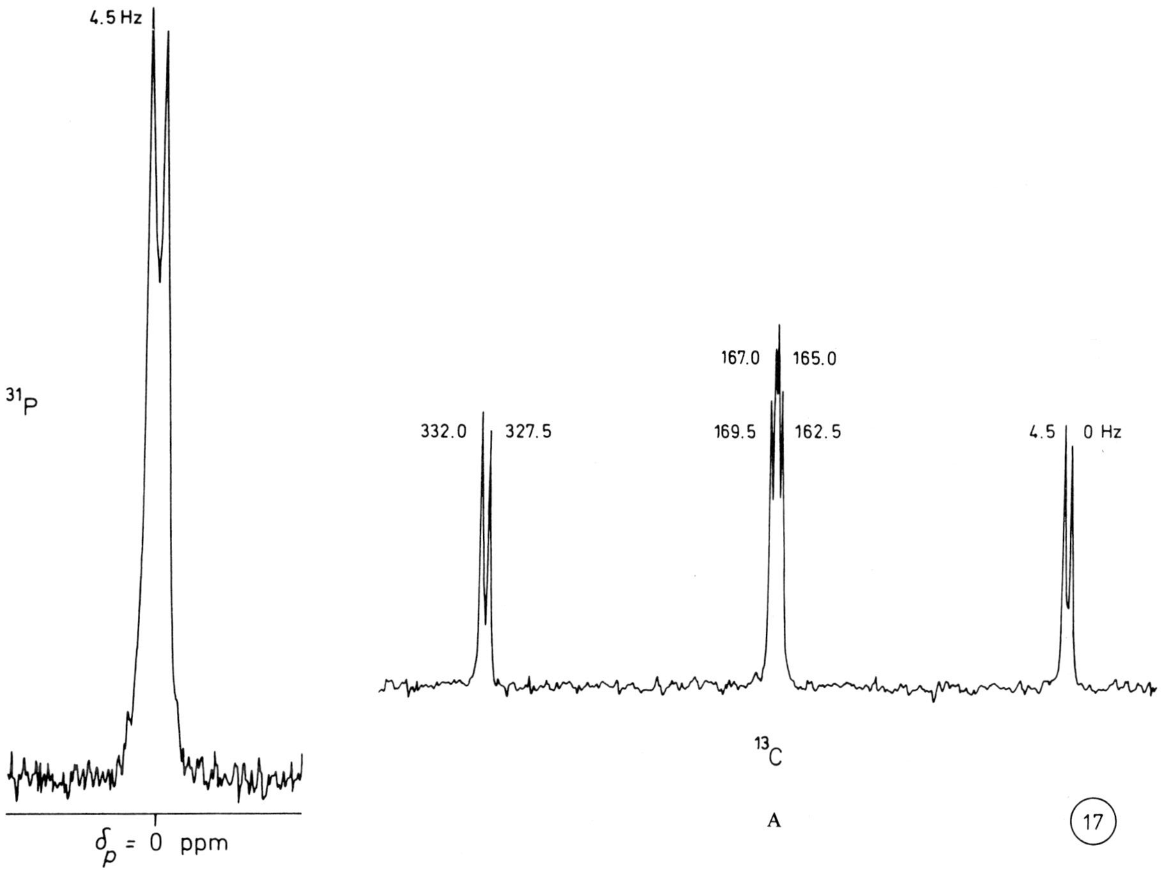

B

A

18

^{13}C NMR Spectrum 18 was recorded for an analytically pure aliphatic compound with the elementary composition $C_5H_{10}O$. What is the compound's structure, and what is its state?

Recording Conditions

Solvent:	Hexadeuterioacetone (septet at 29.8 ppm, singlet at 206.8 ppm folded back to about 143 ppm).
Concentration:	90% v/v.
Temperature:	30° C.
Operating frequency:	20 MHz.
Decoupling:	*Top:* Proton off-resonance decoupled (500 scans). *Bottom:* Pulsed noise decoupling as in Fig. 3.7A, p. 27, with a waiting time of 60 s between successive 90° pulses (100 scans).
Remarks:	The signal intensities can be estimated under these conditions. In addition to the direct C-H coupling of 127 Hz, the two upfield signals in the coupled spectrum show long-range doublet-of-doublet splittings of 12 and 7 Hz (12.75 ppm) and of 18

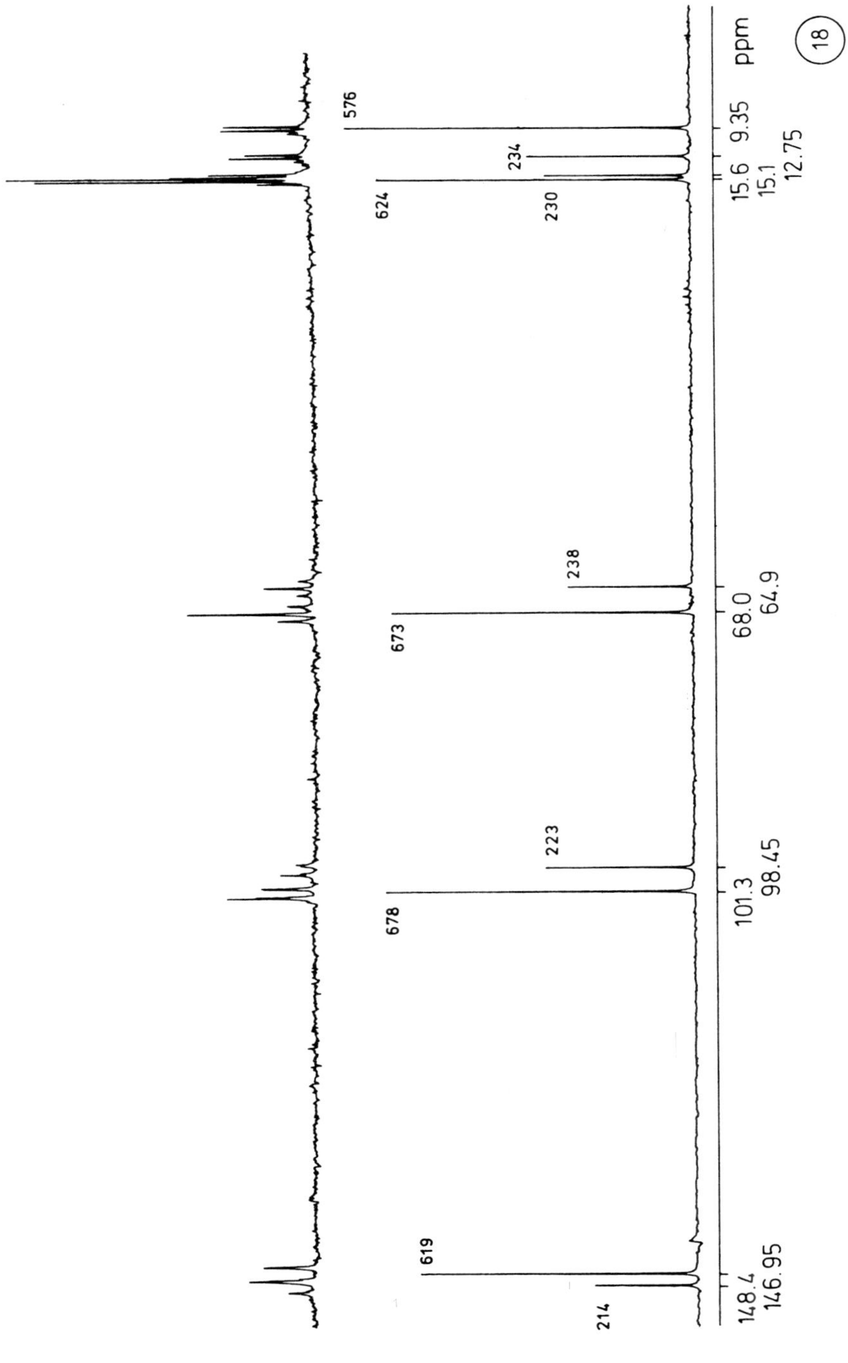

576
624
234
230
ppm
9.35
15.6
15.1
12.75
18
238
673
68.0
64.9
223
678
101.3
98.45
619
214
148.4
146.95

19

What is the structure of the compound with the molecular formula $C_6H_9O_2N$ on the basis of its ^{13}C NMR spectrum 19?

Recording Conditions

Solvent:	Hexadeuteriodimethylsulfoxide (septet at 39.7 ppm).
Concentration:	200 mg in 1 ml.
Temperature:	30° C.
Operating frequency:	15.08 MHz.
Decoupling:	*Top:* ^{1}H off-resonance decoupled (2,000 scans). *Bottom:* ^{1}H noise-decoupled (500 scans).

$C_6H_9O_2N$

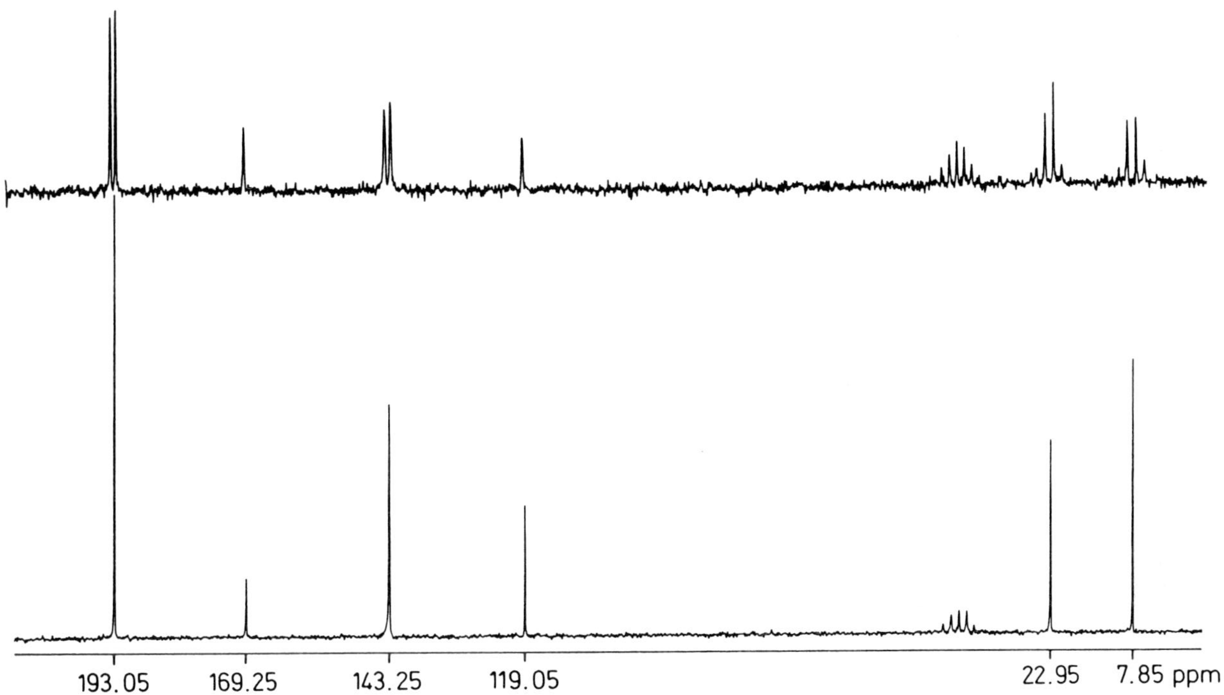

19

and 7 Hz (9.35 ppm), respectively.**20**

^{13}C NMR Spectrum 20 was recorded for technical ketene. What is the state of the compound? Analyze the spectrum.

Recording Conditions

Solvent:	Hexadeuterioacetone (septet at 29.8 ppm, singlet at 206.8 ppm not shown).
Concentration:	75% v/v.
Temperature:	30° C.
Operating frequency:	20 MHz.
Decoupling:	Alternately pulsed ^{1}H decoupling as in Fig. 3.6A, p. 26 (500 scans).
Remarks:	Sections B and C of Spectrum 20 are expanded by a factor of 2. The numbers above the multiplets in A are the direct C-H coupling constants; in B and C, they are the long-range C-H coupling constants.

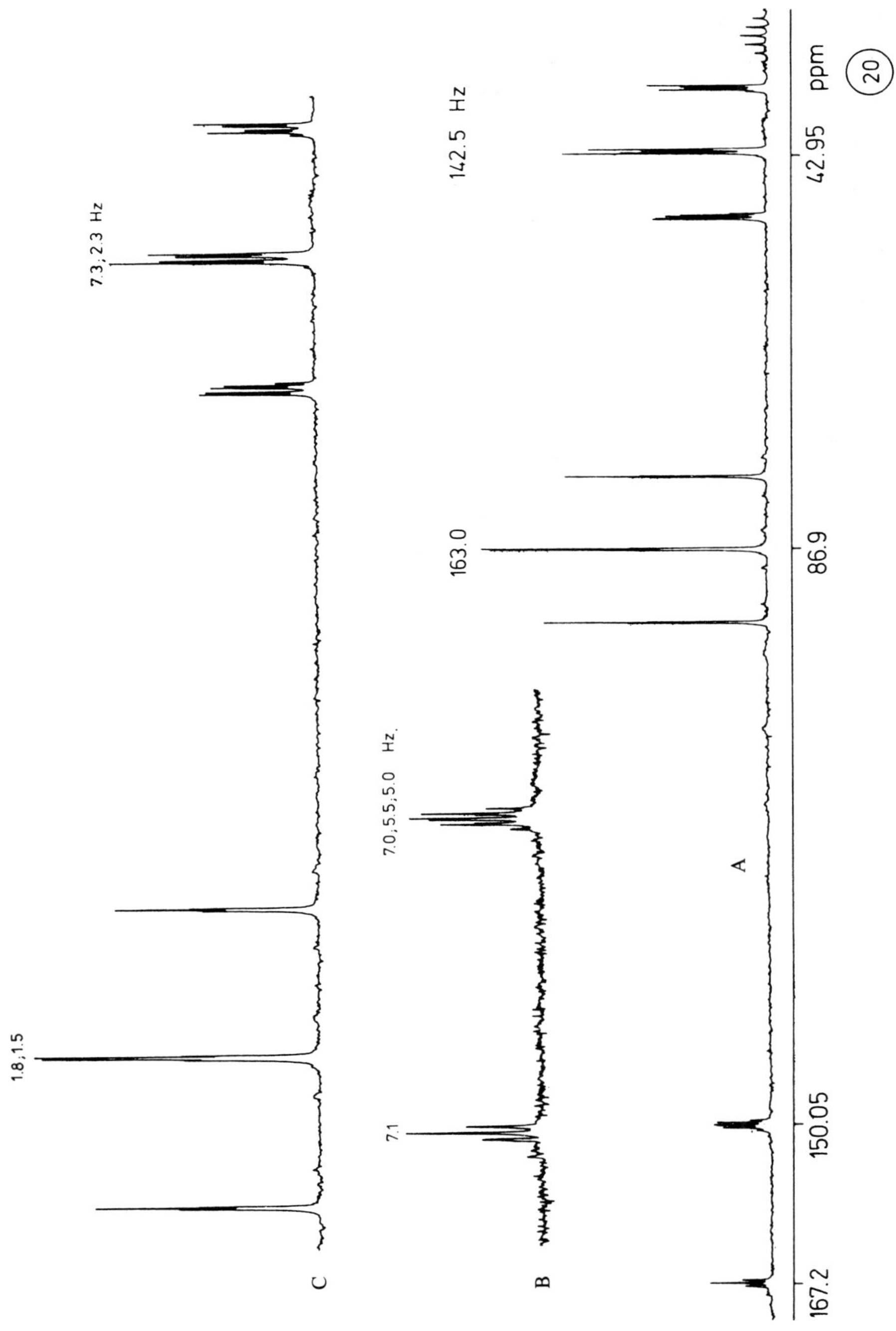

7.3; 2.3 Hz
142.5 Hz
163.0
7.0; 5.5; 5.0 Hz
7.1
1.8; 1.5
A
B
C
167.2
150.05
86.9
42.95
ppm
20

21

A sample of cyclopentadiene gave ^{13}C NMR Spectrum 21. What other compound is present?

Recording Conditions

Solvent:	Hexadeuterioacetone (septet at 29.8 ppm, singlet at 206.5 ppm not shown).
Concentration:	95% v/v.
Temperature:	25° C.
Operating frequency:	20 MHz.
Decoupling:	A. ^{1}H noise-decoupled (50 scans). B. ^{1}H off-resonance decoupled (200 scans). C. Pulsed proton decoupling as in Fig. 3.6A, p. 26 (300 scans, waiting time between pulses, 10 s).
Remarks:	The following sources can be used to assign the signals (as far as possible) and the configuration: J. B. Stothers, *Carbon-13 NMR Spectroscopy,* Academic Press, New York, 1972, p. 334; G. C. Levy, G. L. Nelson, *Carbon-13 NMR for Organic Chemists,* Wiley Interscience, New York, 1972, p. 61.

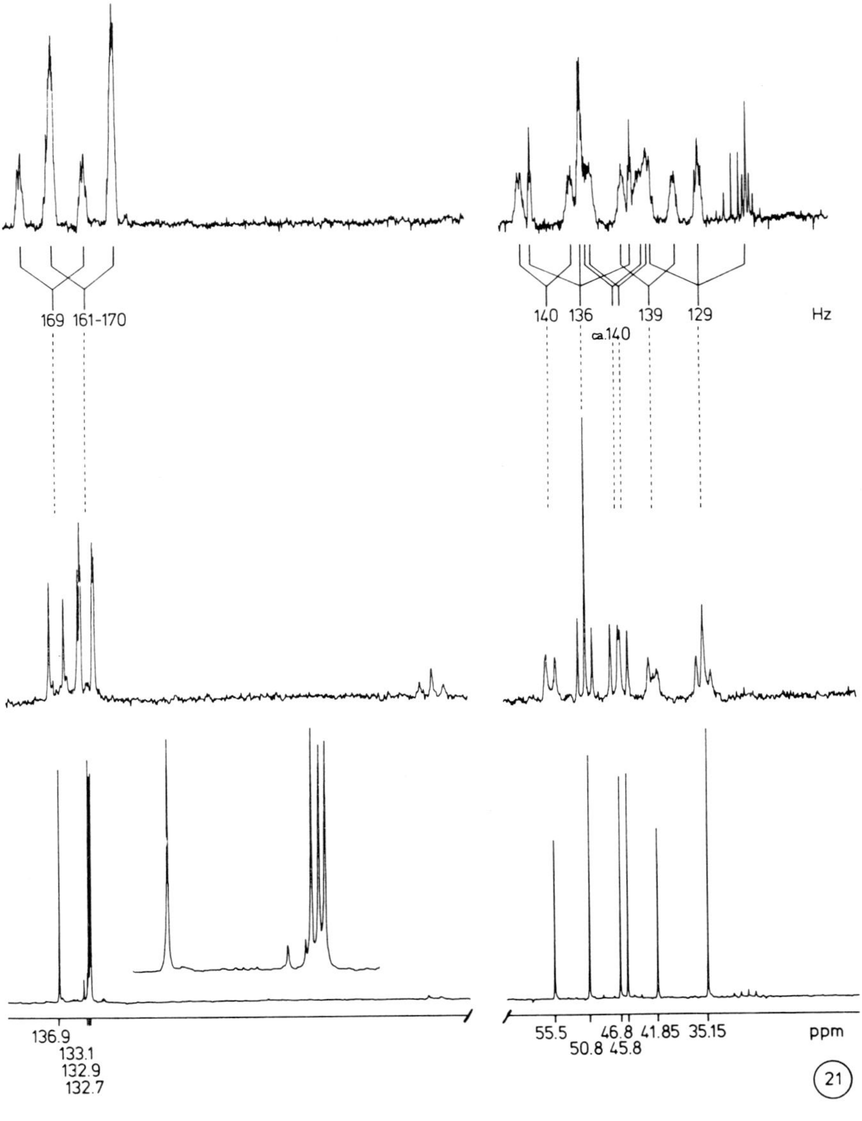
169 161-170
140 136
ca.140
139 129
Hz
136.9
133.1
132.9
132.7
55.5
50.8 45.8
46.8 41.85 35.15
ppm

21

22

What adamantane derivative with the molecular formula $C_{12}H_{19}NO$ fits ^{13}C NMR Spectrum 22? Determine the substituent increments $Z_{\alpha-\delta}$ for *Cl, Br, I, OH,* and *CH_3* from the chemical shifts in the following table:

^{13}C Chemical Shifts of Adamantanes in Deuteriochloroform
(δ values in ppm with reference to TMS)

X	C—1	C—2,8,9	C—3,5,7	C—4,6,10	CH_3
—H	28.35	37.8	28.35	37.8	
—CH_3	29.85	44.65	31.45	37.0	28.9
—OH	68.25	45.2	30.8	36.2	
—Cl	68.6	47.8	31.7	35.6	
—Br	66.0	49.2	32.35	35.4	
—I	50.5	52.4	33.0	35.6	

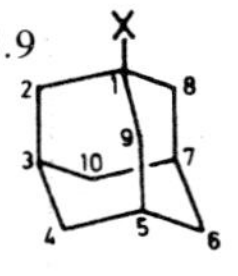

Recording Conditions

Solvent: Deuteriochloroform (triplet at 77.25 ppm).

Concentration: 200 mg/ml.

Temperature: 30° C.

Operating frequency: 20 MHz.

Decoupling:
A. ^{1}H noise-decoupled (100 scans).
B. Without decoupling (1000 scans).
C. Expanded part of Spectrum B.

Remarks: The numbers above the signals in Spectrum C indicate the frequency shift with reference to TMS.

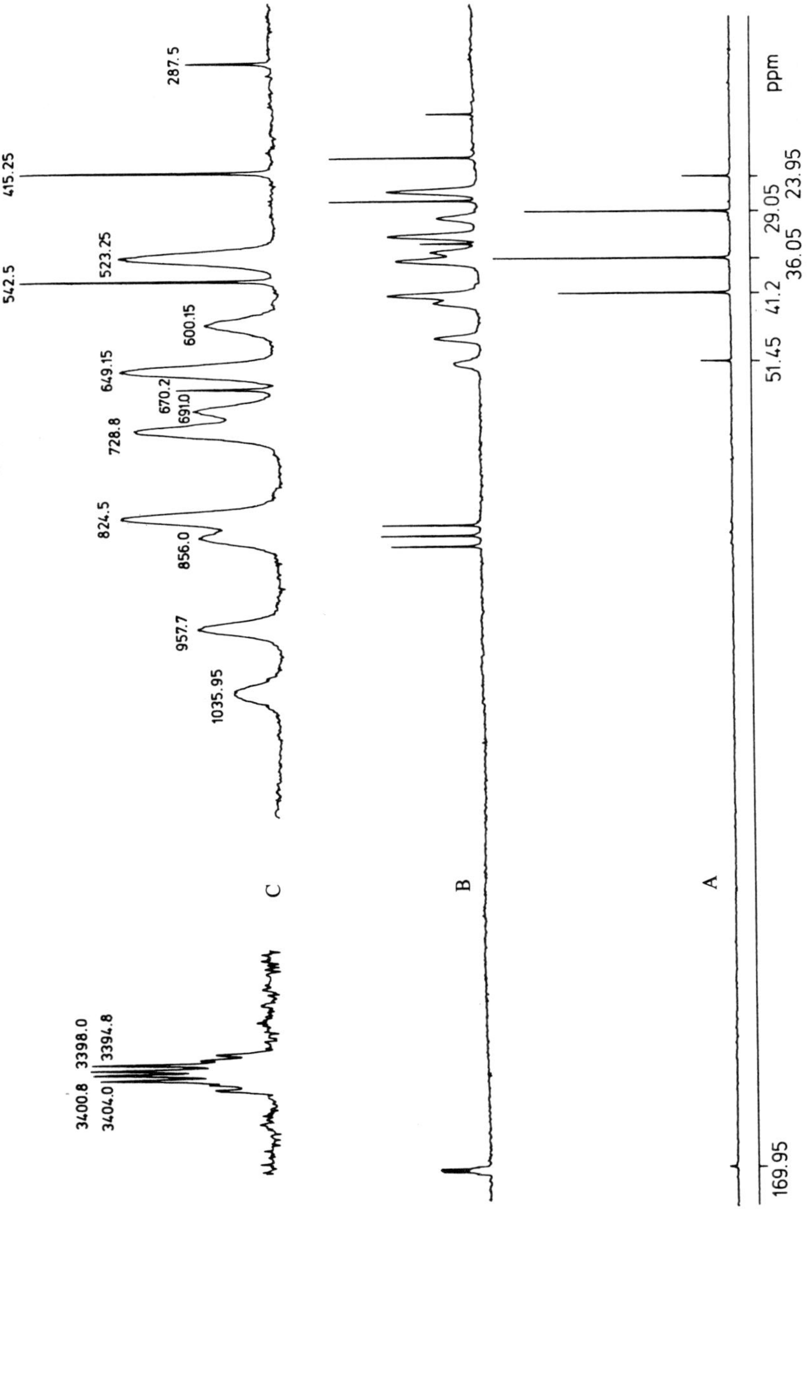

A
B
C
ppm
169.95
51.45
41.2
36.05
29.05
23.95
3400.8
3404.0
3398.0
3394.8
1035.95
957.7
856.0
824.5
728.8
691.0
670.2
649.15
600.15
542.5
523.25
415.25
287.5

22

23

Series 23 of ^{13}C NMR spectra of cis-1,2-dimethylcyclohexane was recorded at temperatures between −100 and 40° C. How do you explain the observed temperature dependence of the spectra? Using the increments in Table 7.5, assign the signals at −100 and +40° C. Determine the free energy of activation $\Delta G^{\ddagger}$ of the ring inversion from the coalescence temperature of the basic splitting $\Delta\nu$ of appropriate pairs of signals.[1]

Recording Conditions

Solvent:	Trichlorofluoromethane/hexadeuterioacetone/tetramethylsilane (75:20:5). The hexadeuterioacetone septet centered around 30 ppm is visible.
Concentration:	50% v/v.
Temperature:	−100, −60, −50, −40, −30, and +40° C.
Operating frequency:	20 MHz.
Decoupling:	^{1}H noise-decoupled (−100 and +40° C: 50 scans each; −60 to −30°C: 100 scans each).
Remarks:	All recorded chemical shifts are referred to tetramethylsilane as an internal reference. The waiting time between interferograms acquired using a pulse-width equaling one-third of the 90° pulse, was 5 s. Under these conditions, relaxation effects on the intensities are suppressed. In this way, considering the assignment, equal intensities are observed in the −100° C spectrum for the pairs of signals at 35.5 and 33.45 and at 20.3 and 11.75 ppm.

[1]H. Kessler, *Angew. Chem.* **82,** 237 (1970); and H. J. Schneider, R. Price, T. Keller, *Angew. Chem.* **83,** 759 (1971).

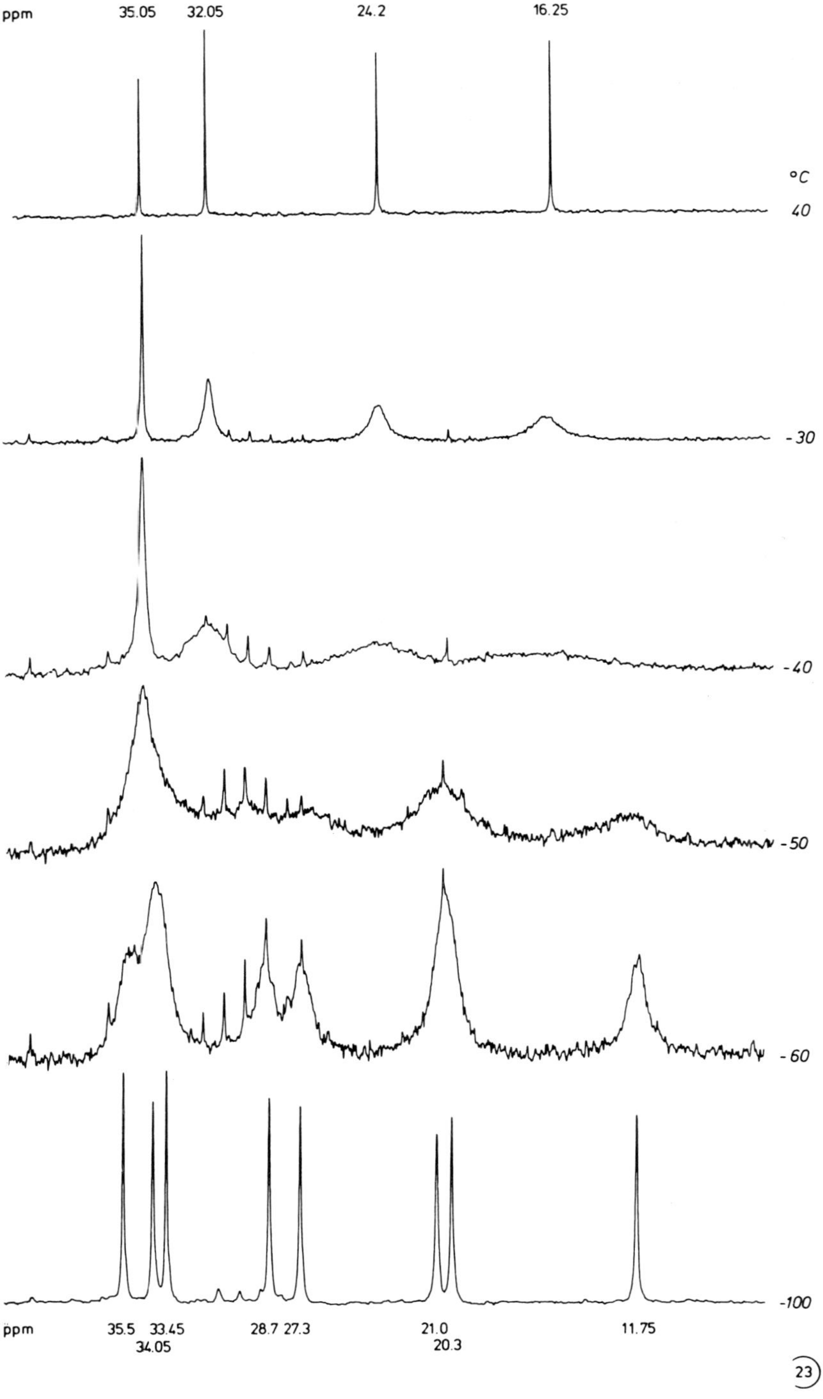
ppm
35.05
32.05
24.2
16.25
°C
40
-30
-40
-50
-60
-100
ppm
35.5
33.45
34.05
28.7
27.3
21.0
20.3
11.75

23

24

4-t-Butylcyclohexane and $LiAlH_4$ in ether give a product with ^{13}C NMR Spectrum 24. What compound is formed, and what is its state? Assign the signals and analyze the spectrum quantitatively.

Recording Conditions

Solvent:	Hexadeuterioacetone (septet at 29.8 ppm, singlet at 206.8 ppm not shown).
Concentration:	Saturated solution.
Temperature:	28° C.
Operating frequency:	22.63 MHz.
Decoupling:	^{1}H noise-decoupled (1,000 scans).
Remarks:	Owing to the presence of quaternary carbon atoms, the interferograms were acquired using a strongly reduced pulse angle (3 μs instead of 18μs for 90° pulses). In this fashion, the relaxation effect on the signal intensities is suppressed, although the intensities are not strictly comparable (nuclear Overhauser effect).

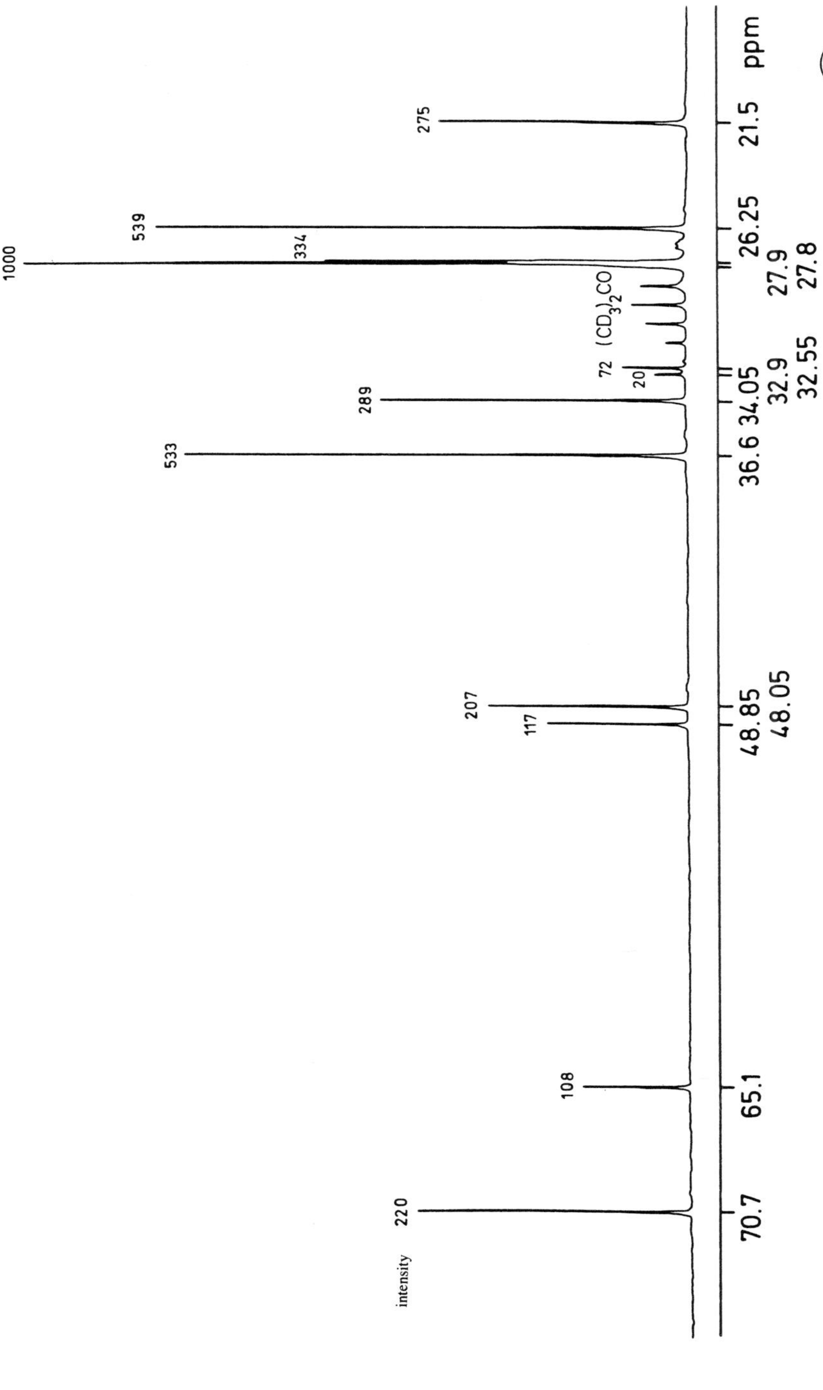

ppm
21.5
26.25
27.9
27.8
32.9
32.55
34.05
36.6
48.85
48.05
65.1
70.7
275
539
334
1000
$(CD_3)_2CO$
72
20
289
533
207
117
108
220
intensity

24

25

What compound with the molecular formula C_6H_8O fits ^{13}C NMR Spectrum 25?

Recording Conditions

Solvent:	Deuteriochloroform (triplet at 77.2 ppm in the upper spectrum).
Concentration:	50% v/v.
Temperature:	28° C.
Operating frequency:	22.63 MHz.
Decoupling:	*Top:* Off-resonance proton decoupled (500 scans). *Bottom:* ^{1}H noise-decoupled (100 scans).

(25)

C_6H_8O

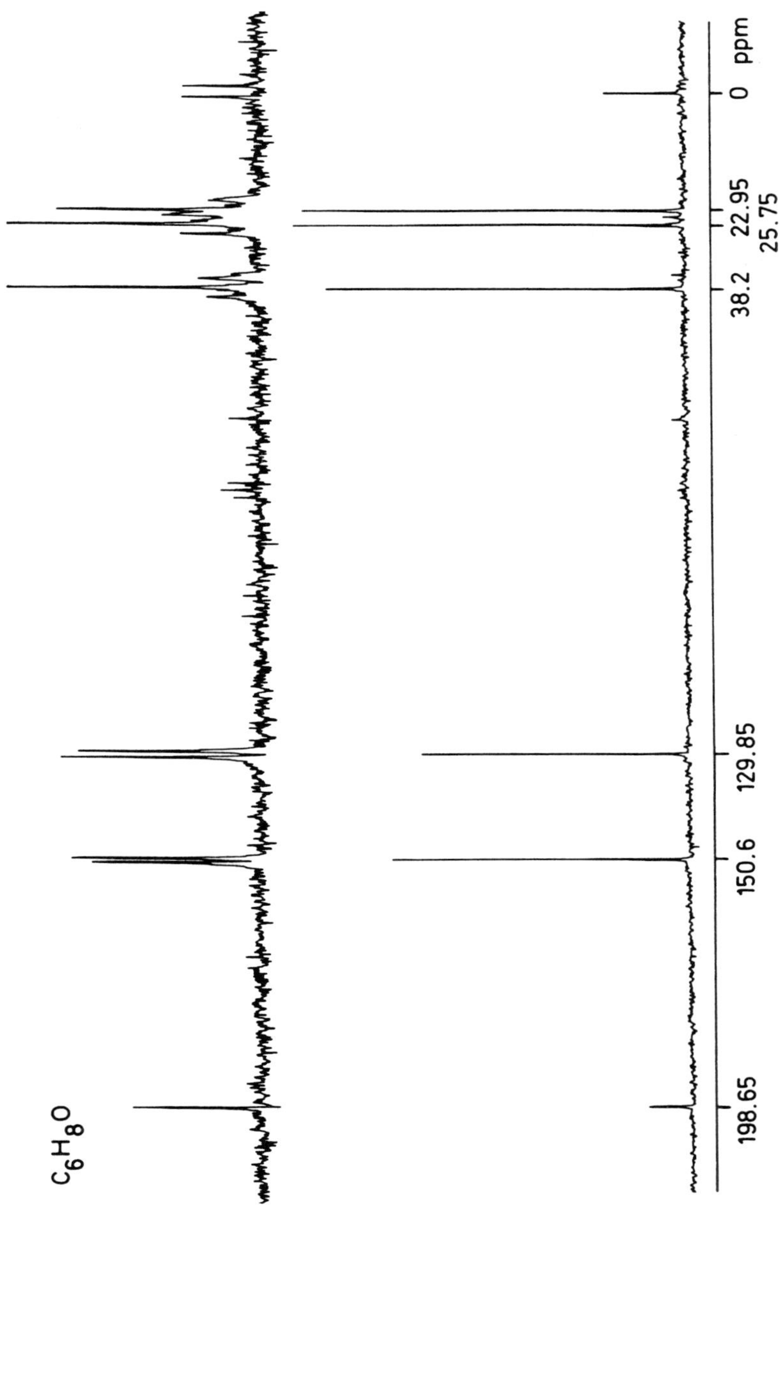

26

^{13}C NMR Spectrum 26A was recorded for cyclooctatetraene. To what compound do the additional signals at δ = 133.85, 129.25, 128.55, and 43.15 ppm correspond? ^{1}H NMR Spectrum 26B and ^{13}C NMR Spectra 26C-F were recorded for the adduct of tetracyanoethylene and cyclooctatetraene. What is the adduct's structure? Assign all ^{13}C chemical shifts.[1]

Recording conditions

Solvents:	A.	Hexadeuterioacetone (weak septet at 29.9 ppm, singlet at 206.8 ppm not shown).
	B.	Hexadeuteriodimethylsulfoxide.
	C.	Hexadeuteriodimethylsulfoxide (septet at 39.7 ppm)
	D-F.	^{12}C-Hexadeuteriodimethylsulfoxide.
Concentration:	A.	50% v/v.
	B-F.	Saturated solutions.
Temperature:		30° C.
Operating frequency:		20 MHz.
Decoupling:	A.	Pulsed ^{1}H decoupling according to Fig. 3.7A, p. 27; waiting time between the approximately 90° pulses: 20 s (100 scans)
	B.	No decoupling (single experiment)
	C.	^{1}H Noise-decoupled (5 Watts, 400 scans)
	D-F.	Selective ^{1}H decoupling without noise modulation (0.2 Watts, 200 scans each) at 3.21 ppm (D), 5.95 ppm (E), and 6.38 ppm (F).

[1] K. Grohman, J. B. Grutzner, J. D. Roberts, *Tetrahedron Lett.* **1969,** 917; R. Huisgen, F. Mietzsch, *Angew. Chem.* **76,** 36 (1964) (^{1}H chemical shifts); D. E. Dorman, M. Jautelat, J. D. Roberts, *J. Org. Chem.* **36,** 2757 (1971) (^{13}C chemical shifts of cycloalkenes).

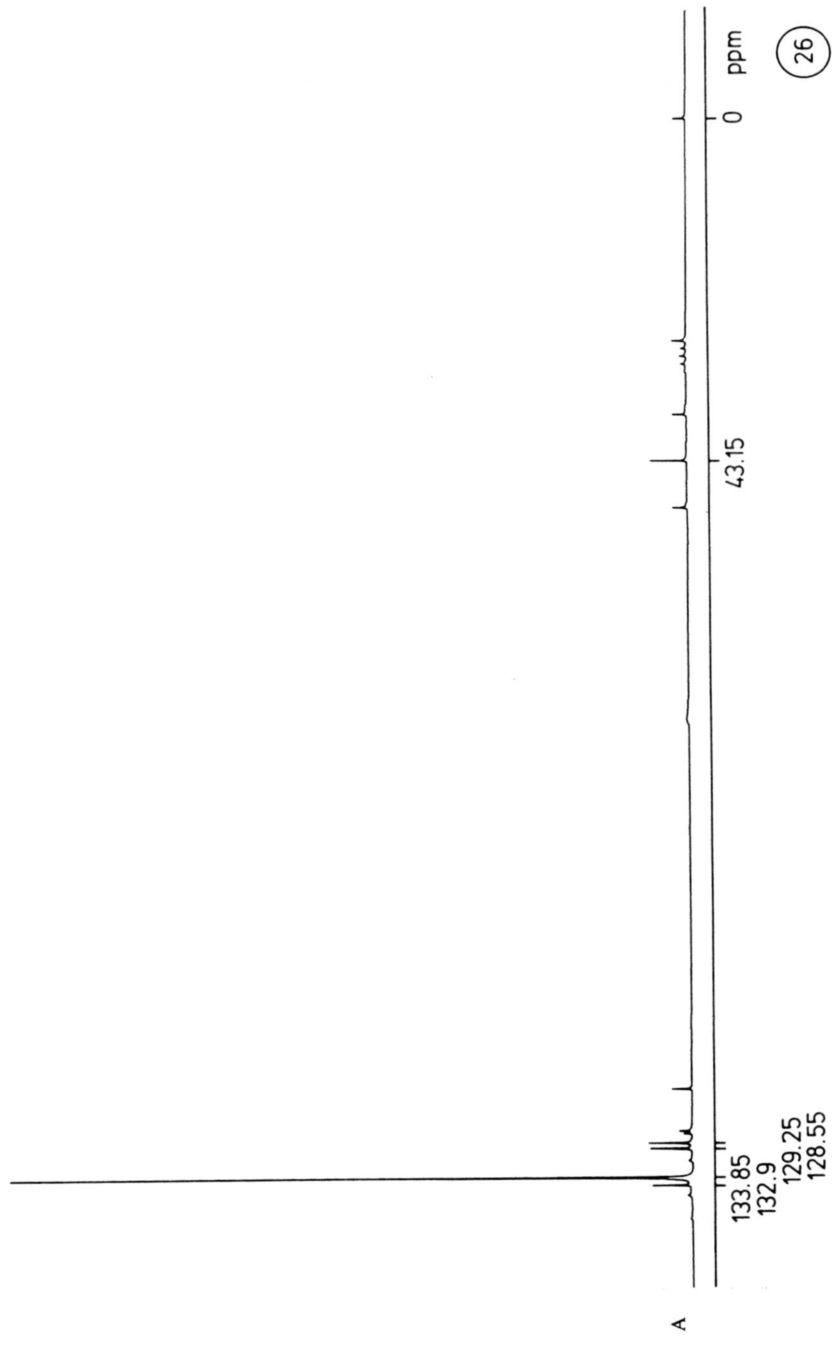

A
133.85
132.9
129.25
128.55
43.15
0
ppm
26

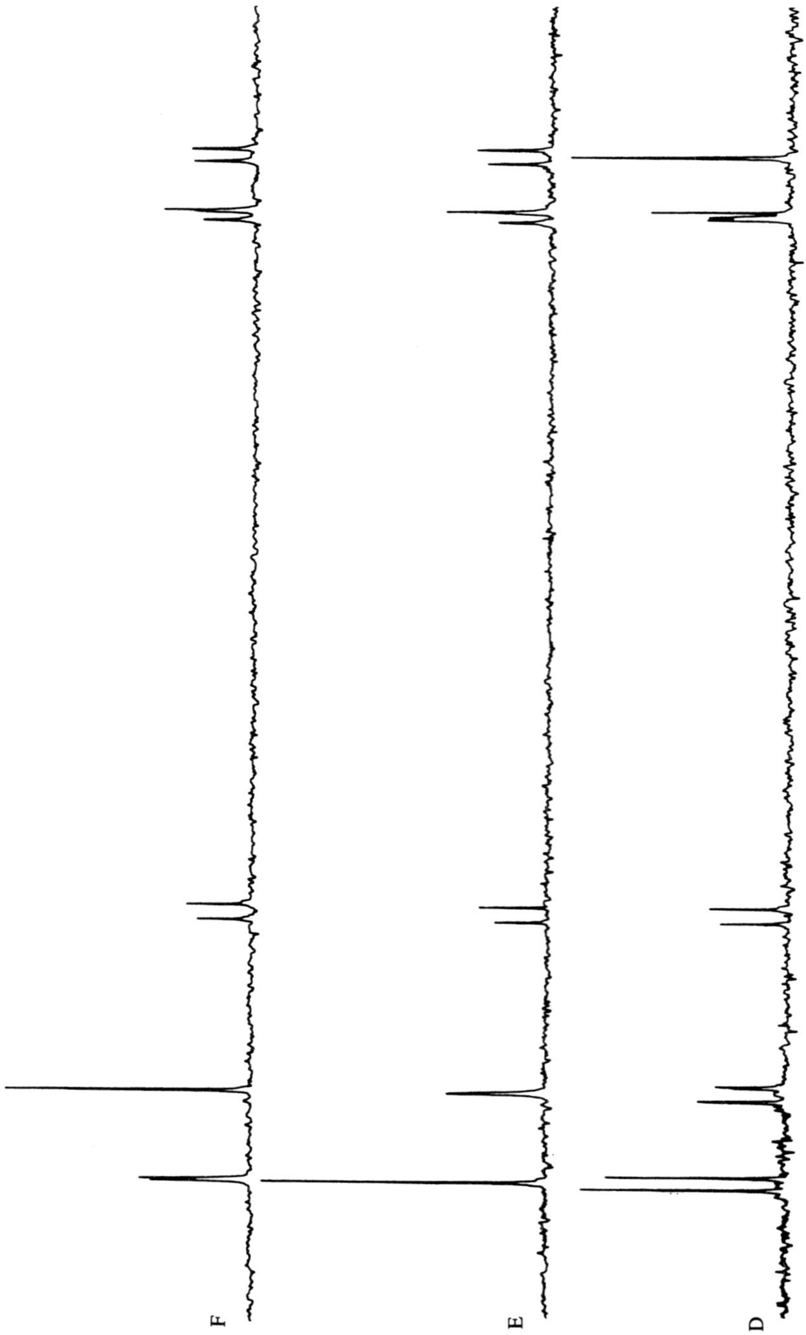
F
E
D

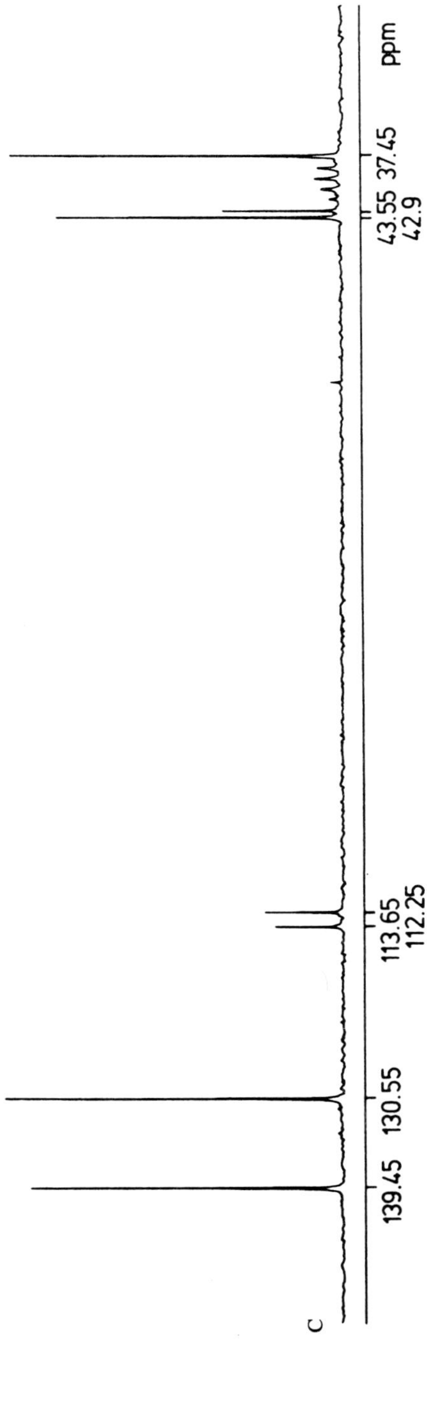
ppm
37.45
43.55
42.9
113.65
112.25
130.55
139.45

C

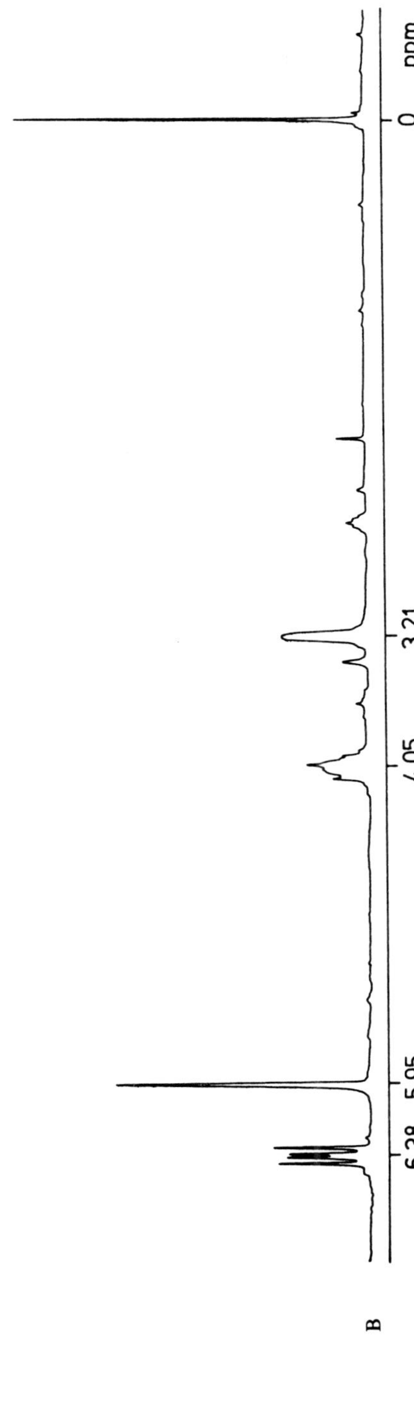
ppm
0
3.21
4.05
5.95
6.38

B

27

For what compound with the molecular formula $C_{16}H_{16}$ was ^{13}C NMR Spectrum 27 recorded?

Recording Conditions

Solvent:	^{12}C-Enriched deuteriochloroform (weak triplet at 77.5 ppm).
Concentration:	10 mg in 1 ml.
Temperature:	25° C.
Operating frequency:	20 MHz.
Decoupling:	Pulsed proton decoupling as in Fig. 3.7A, p. 27 (4,400 scans).
Remarks:	The waiting time between successive pulses was 20 s. Half the width of the 90° pulse was chosen. Under these conditions, the signal intensities (numbers under the signals) can be compared.

$C_{16}H_{16}$

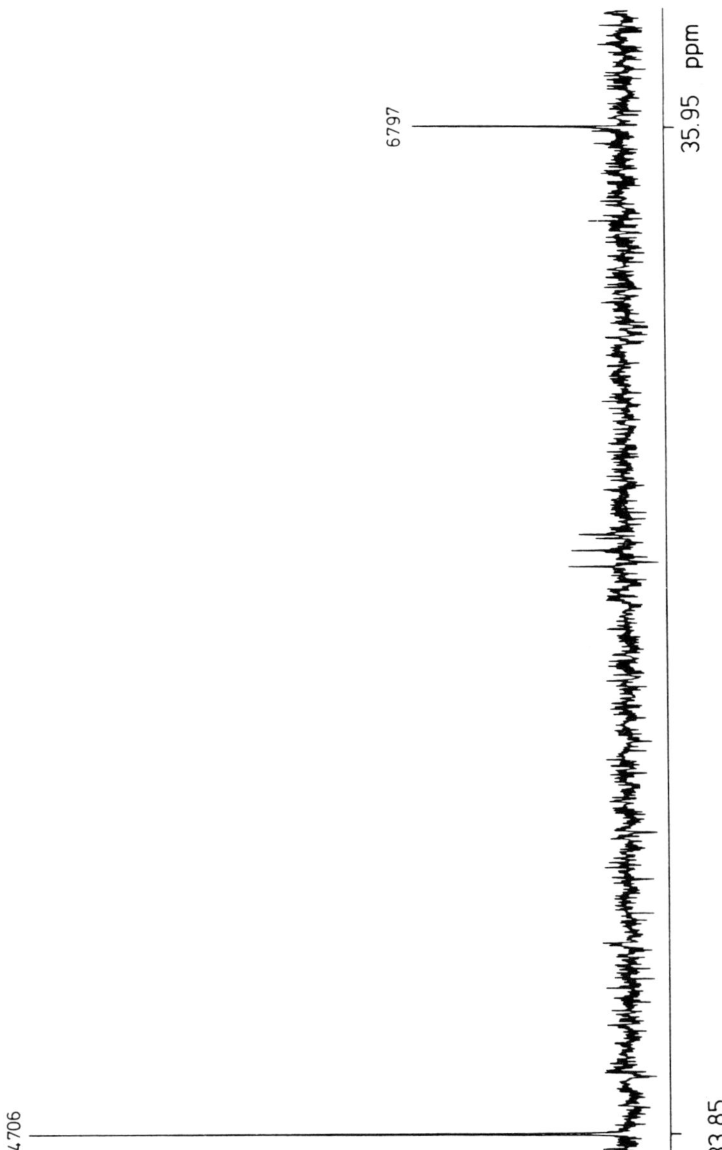

(27)

28

Under different polymerisation conditions, the same monomer affords two polymers with different properties and ^{13}C NMR Spectra 28A and B. Identify the polymers, taking their stereoregularity into account.[1]

Recording Conditions

Solvent:	Deuteriochloroform (triplet at 77.45 ppm).
Concentration:	A. 100 mg in 2 ml. B. 20 mg in 2 ml.
Temperature:	30° C.
Operating frequency:	20 MHz.
Decoupling:	^{1}H Noise-decoupled (A: 3,000 scans; B: 25,000 scans).
Remarks:	The $CDCl_3$ triplet centered at 77.45 ppm was used as an internal reference. The indicated ^{13}C chemical shifts have been recalculated so that they are referred as usual to TMS.

---

[1]General references: F. A. Bovey, *High Resolution Nuclear Magnetic Resonance of Macromolecules,* Academic Press, New York, 1972; and J. Schäfer, *Topics in Carbon-13 Nuclear Magnetic Resonance,* G. C. Levy, ed., Vol. 1, Wiley-Interscience, New York, 1974. Special reference: L. F. Johnson, F. Heatley, F. A. Bovey, *Macromolecules,* **3,** 175 (1970).

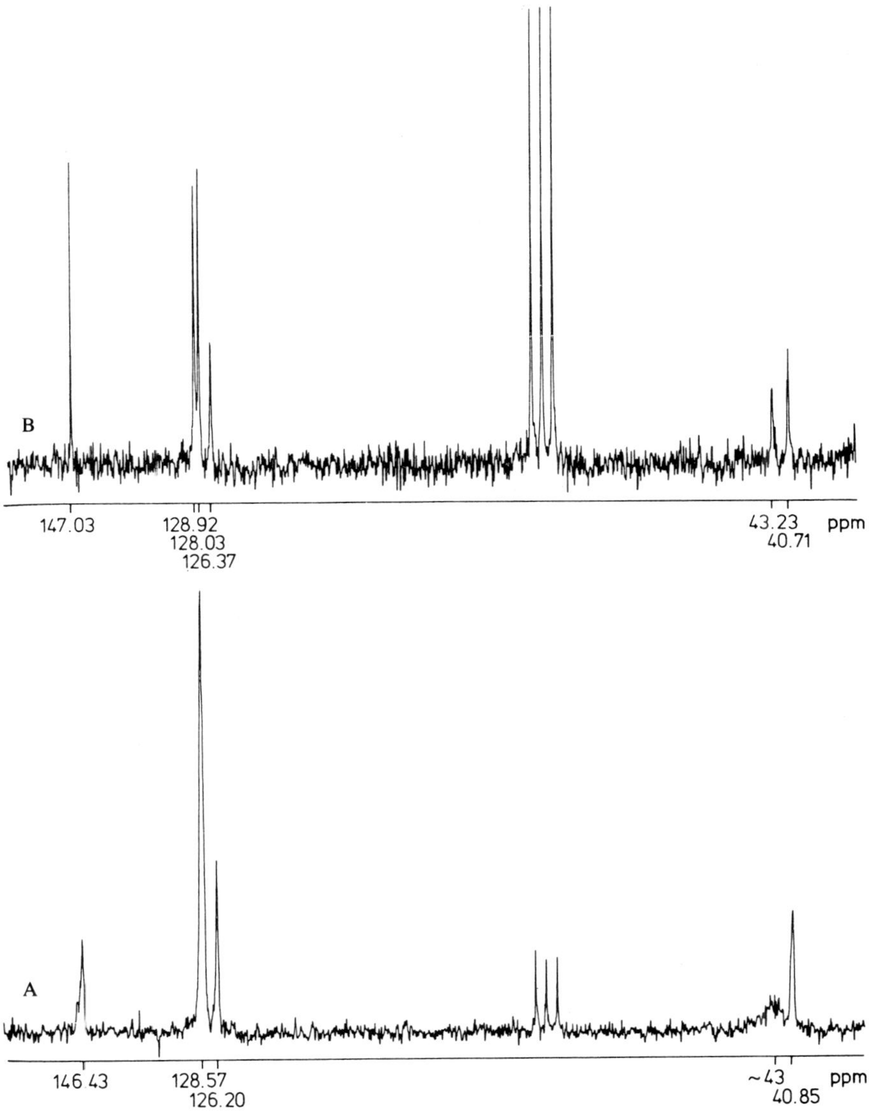
B
147.03
128.92
128.03
126.37
43.23
40.71
ppm
A
146.43
128.57
126.20
~43
40.85
ppm

(28)

29

What hydrocarbon with the molecular formula $C_{11}H_{12}$ fits ^{13}C NMR Spectrum 29? Assign all the signals.

Recording Conditions

Solvent:	Hexadeuterioacetone (septet at 29.8 ppm, singlet at 206.8 ppm not shown)
Concentration:	75% v/v.
Temperature:	28° C.
Operating frequency:	22.63 MHz.
Decoupling:	*Top:* Coupled spectrum (4,000 scans). *Bottom:* ^{1}H noise-decoupled (400 scans).

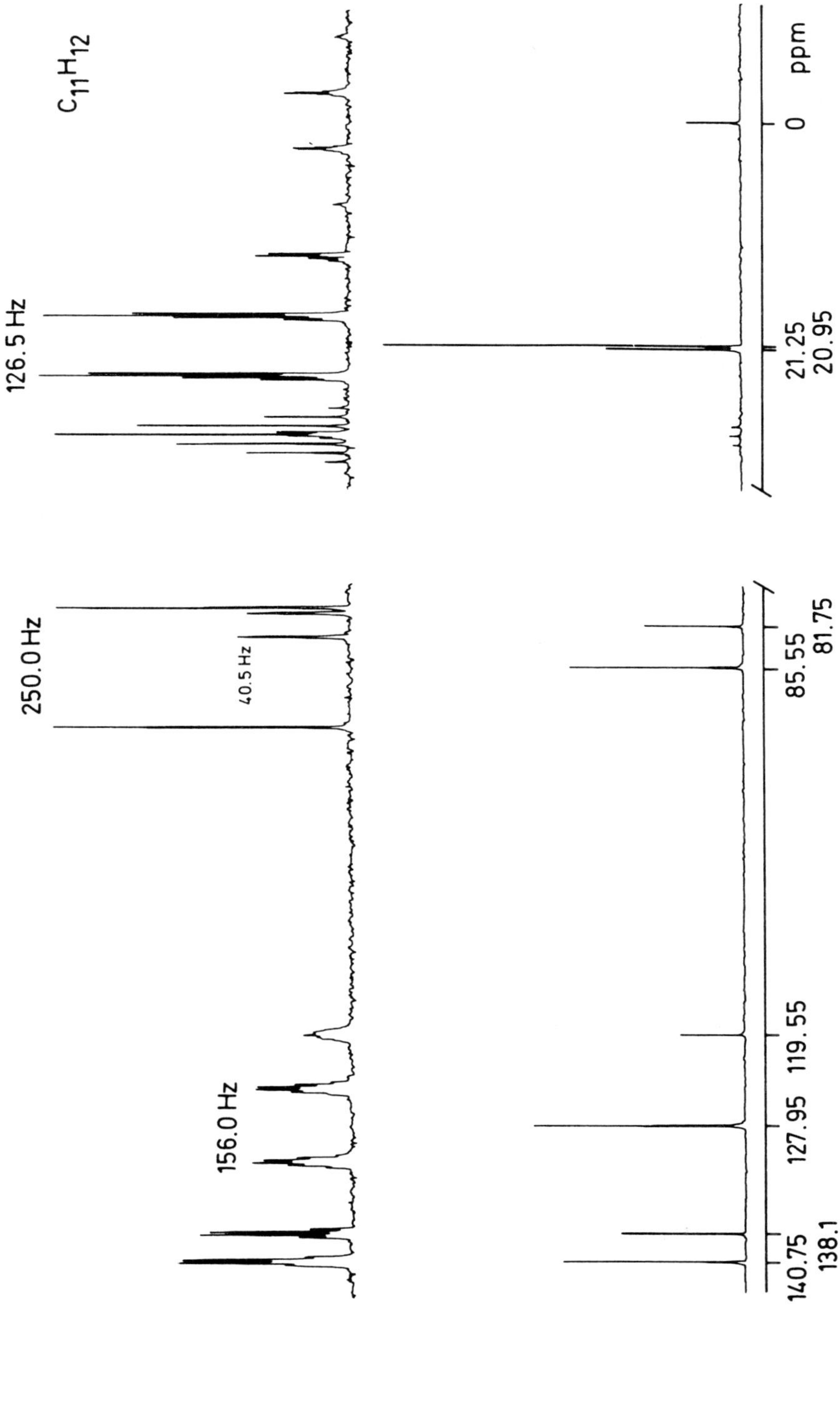
C11H12
126.5 Hz
250.0 Hz
40.5 Hz
156.0 Hz
ppm
0
21.25
20.95
85.55
81.75
119.55
127.95
140.75
138.1

(29)

30

Two mechanisms, A and B, were considered for the formation of benzocyclopropene by base catalyzed dehydrochlorination of 7,7-dichloronorcarene:

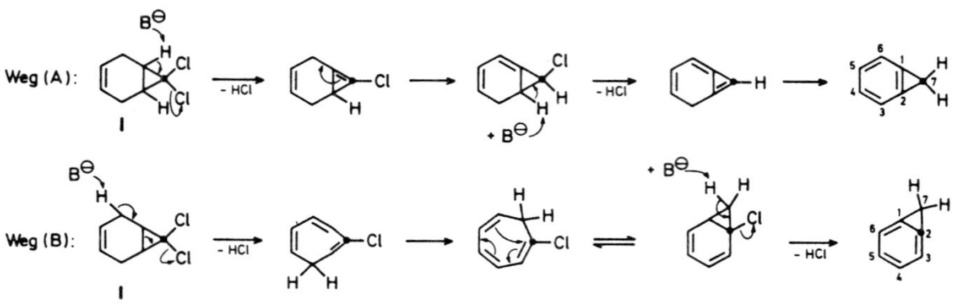

To decide between both routes, 7,7-dichloronorcarene I was prepared first, using ^{12}C-enriched deuteriochloroform (see below), and was then dehydrohalogenated.

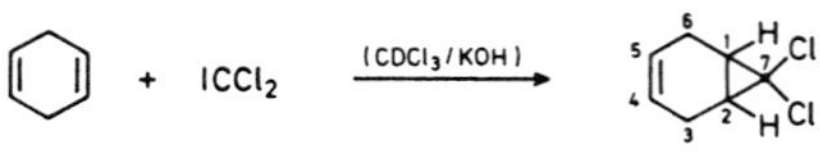

Comparing ^{13}C Spectra 30 for normal benzocyclopropene (A) and for the compound prepared from labeled 7,7-dichloronorcarene (B), decide which is the course of the reaction.[1]

Recording Conditions

Solvent:	Carbon tetrachloride/deuteriochloroform (3.2) (singlet at 97.5 ppm and triplet at 77.25 ppm).
Concentration:	ca. 1.1 mmole/ml.
Decoupling:	A and B, ^{1}H noise-decoupled, 4,096 scans each.

[1]See J. Prestien, H. Günther, *Angew. Chem.* **86,** 278, (1974).

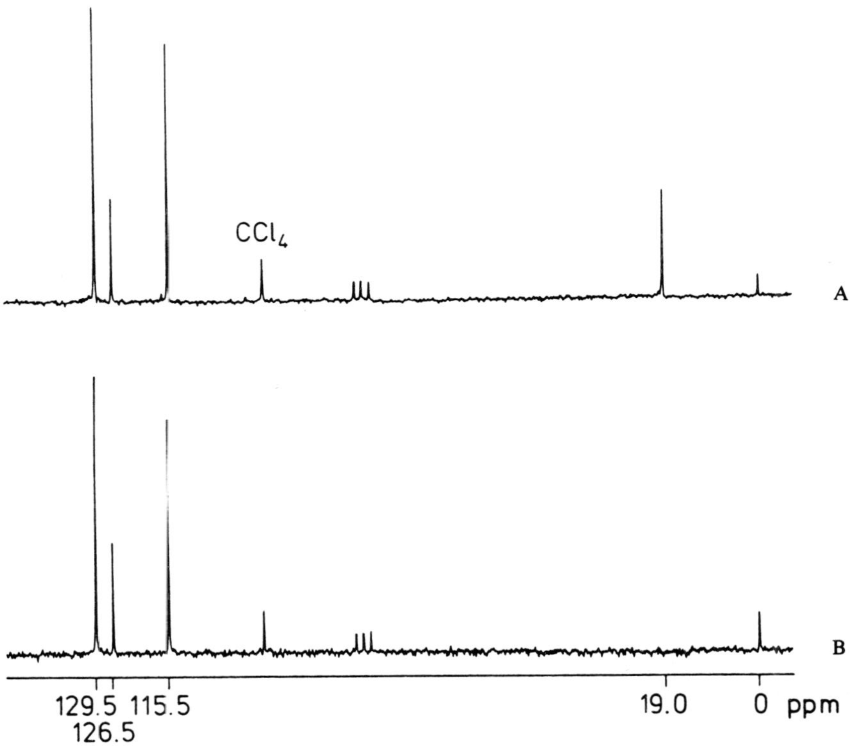

30

31

Identify the compound $C_{15}H_{20}O_4$ on the basis of its ^{13}C NMR Spectrum 31. Consider the signal intensities when making assignments.

Recording Conditions

Solvent:	Hexadeuterioacetone (signals not visible because solution is too concentrated)
Concentration:	90% v/v.
Temperature:	30° C.
Operating frequency:	20 MHz.
Decoupling:	A. Pulsed proton noise-decoupled as in Fig. 3.7A, p. 27 (20 scans); waiting time between 90° pulses = 20 s. B. Off-resonance proton decoupled (200 scans).

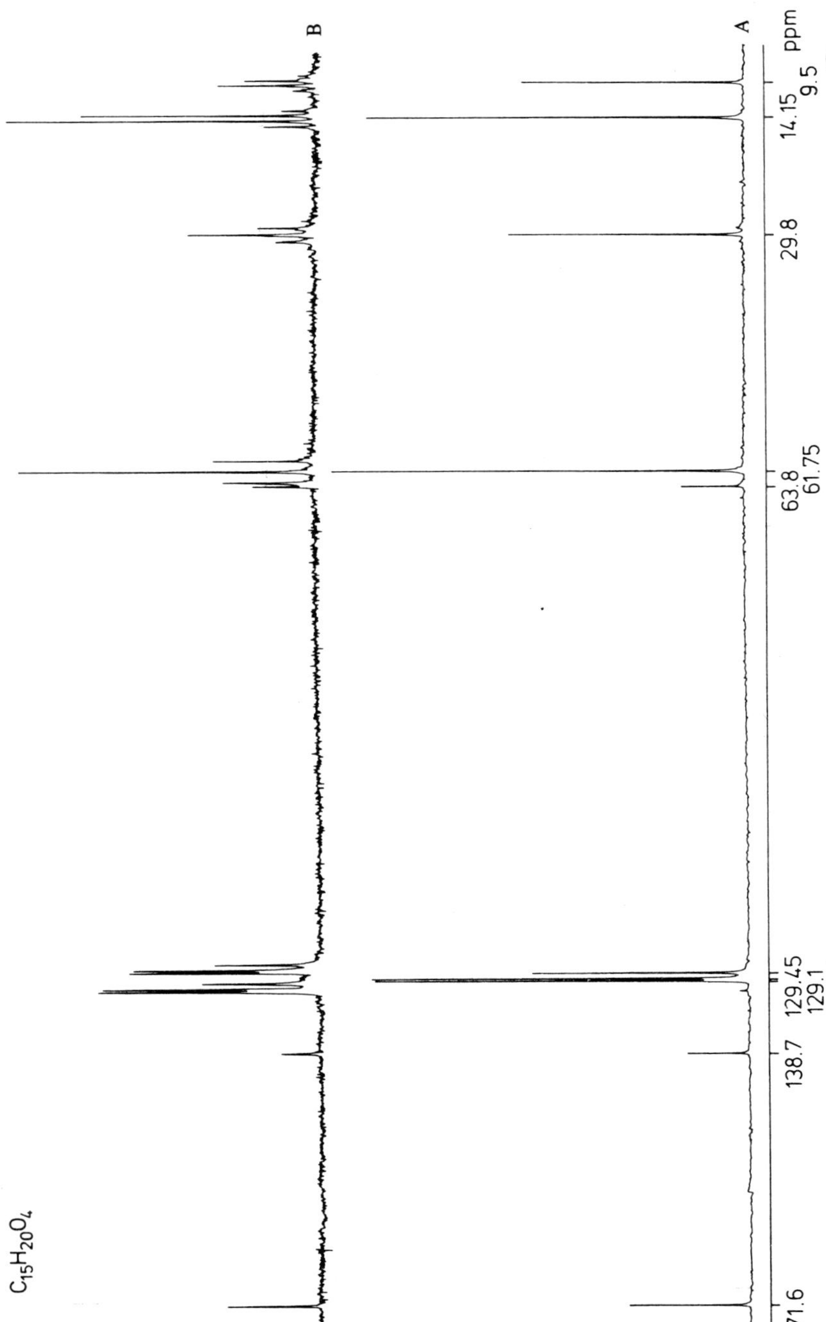

$C_{15}H_{20}O_4$
B
A
ppm
9.5
14.15
29.8
61.75
63.8
128.4
129.1
129.45
138.7
171.6
31

32

Spectrum Series 32 is the result of a T_1 measurement on nitrobenzene using progressive saturation. Determine the T_1 values for the carbon nuclei bonded to hydrogen, and explain the result.

Recording Conditions

Solvent:	Hexadeuteriobenzene (triplet at 128.0 ppm).
Concentration:	90% v/v.
Temperature:	30° C.
Operating frequency:	20 MHz.
Decoupling:	Proton noise-decoupled (30 scans).
Pulse sequence:	90°, τ, 90°, τ, . . . (30 pulses)
Remarks:	The sample was irradiated with four 90° pulses separated by the interval τ before each measurement without accumulating the interferograms, so that the stationary magnetization typical of that interval could be set up. The measurement proceeds under complete computer control and yields a protocol on which the signal integrals needed for the interpretation are recorded.

Interval τ [s]	148.4 ppm Integral	134.8 ppm Integral	129.5 ppm Integral	123.3 ppm Integral
160	6,396.4	16,415.4	35,893.4	34,118.8
80	4,331.3	16,492.2	35,343.5	33,927.5
40	3,286.5	16,246.5	33,867.3	33,557.8
20	1,728.4	15,656.1	29,958.3	29,333.9
19	1,848.9	15,507.9	29,586.0	28,865.6
18	1,375.4	15,452.8	28,253.8	27,884.5
17	1,666.0	15,491.9	28,197.9	27,697.5
16	1,023.4	14,963.8	27,135.2	26,969.7
15	964.9	14,166.1	26,318.4	25,947.8
14	1,245.9	15,001.9	25,817.5	25,542.3
13	754.0	13,801.9	24,568.9	23,987.8
12	680.6	13,570.5	23,143.1	23,290.2
11	226.1	13,245.2	22,289.7	21.793.8
10	436.5	12,212.7	21,144.3	20,601.8
9	297.3	11,557.8	19,664.1	19,384.8
8	476.3	10,722.7	18,048.3	18,007.1
7	164.9	10,285.8	16,741.6	16,834.0
6	420.3	9,550.6	14,630.0	14,987.9
5	177.0	8,314.8	12,918.3	12,724.8
4	—	7,267.3	11,241.5	10,690.8
3	—	5,332.7	8,301.2	8,208.3
450	6,848.1	16,188.2	35,868.5	33,799.4

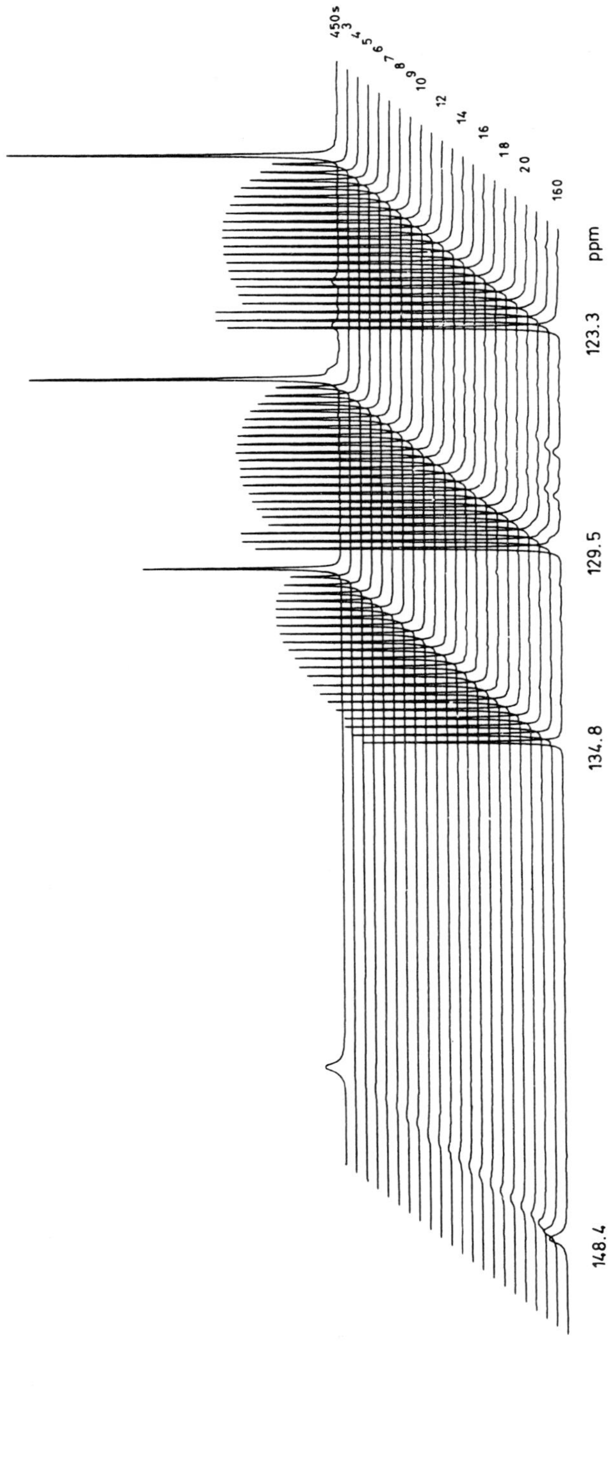

32

33

Assign the signals in the ^{13}C NMR spectrum of picric acid (No. 33).

Recording Conditions

Solvent:	Hexadeuterioacetone (septet at 29.8 ppm and singlet at 206.7 ppm, neither of them shown).
Concentration:	Saturated solution.
Temperature:	28° C.
Operating frequency:	22.63 MHz.
Decoupling:	Coupled spectrum (2,500 scans).

$C_6H_3N_3O_7$

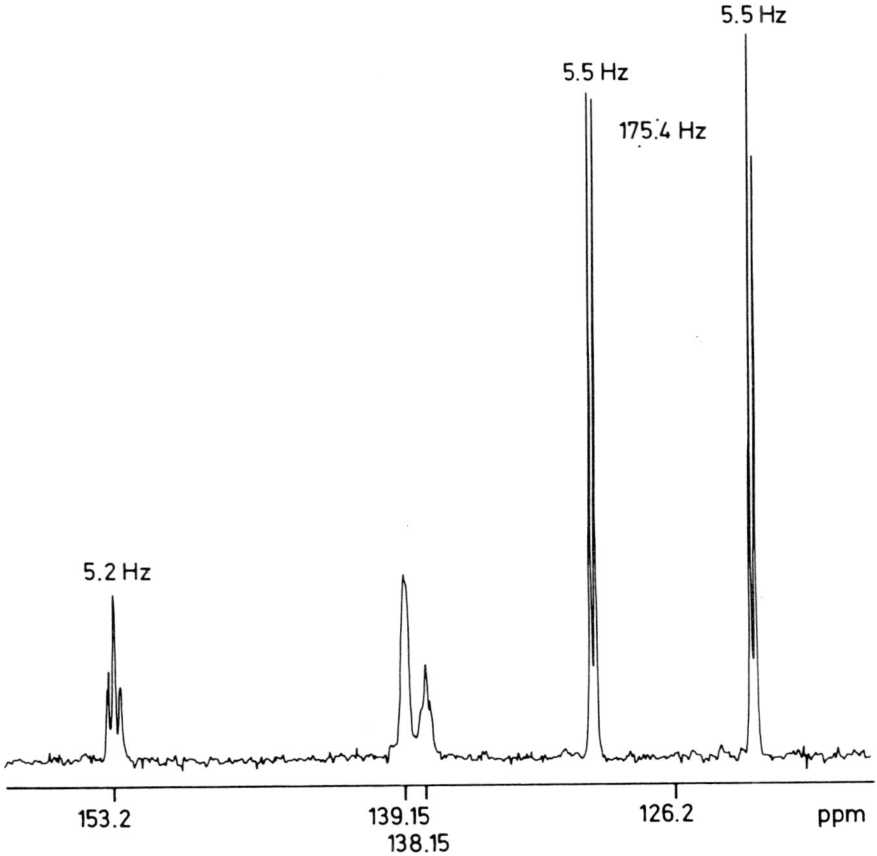

33

34

2,4,6-Trinitrochlorobenzene and acetone react, giving a compound with the molecular formula $C_9H_7N_3O_7$ and ^{13}C NMR Spectrum 34. What is the structure of the reaction product? Compare the ^{13}C chemical shifts of the phenyl carbon atoms in spectra 33 and 34.

Recording Conditions

Solvent:	Hexadeuterioacetone (septet at 29.75 ppm with J_{C-D} = 20.6 Hz and singlet at 206.8 ppm.
Concentration:	Saturated solution.
Temperature:	28° C.
Operating frequency:	22.63 MHz.
Decoupling:	Coupled spectrum (12,500 scans).

$C_9H_7N_3O_7$

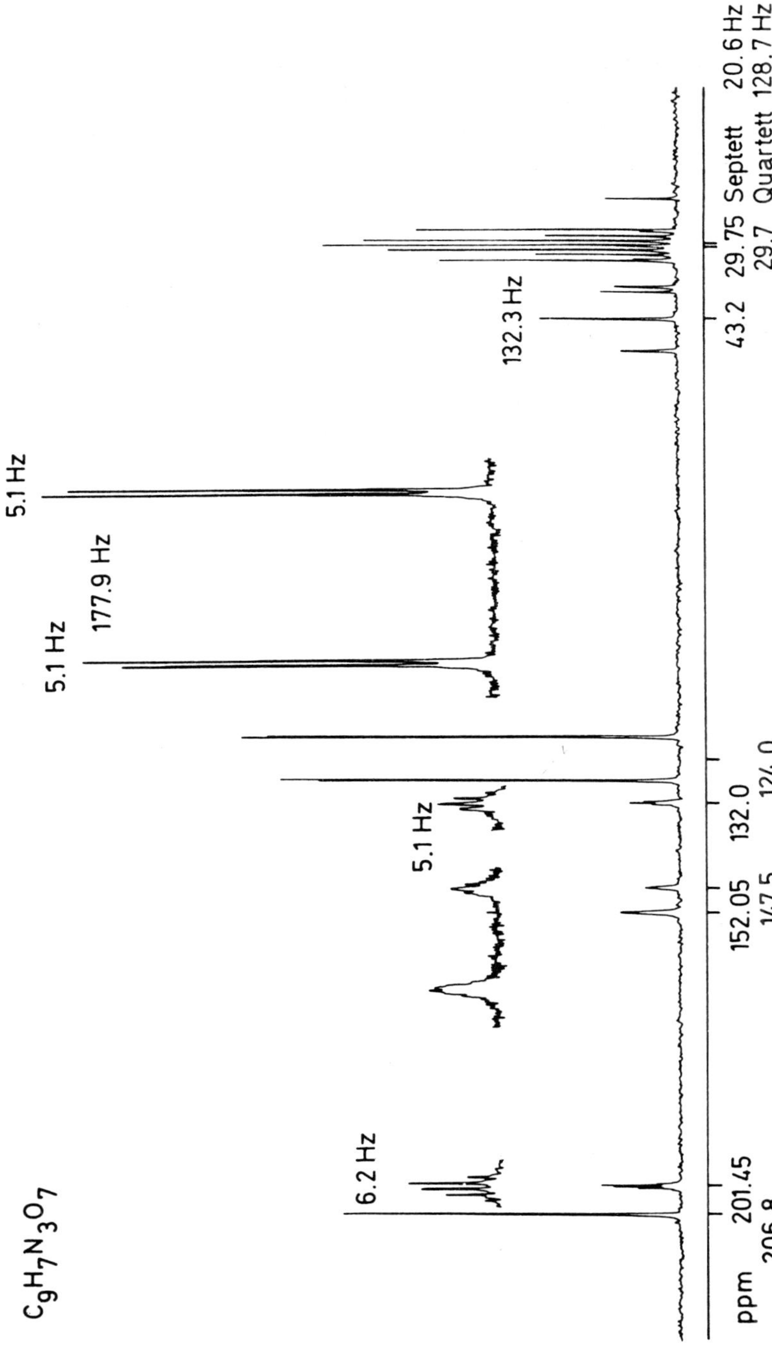

5.1 Hz
5.1 Hz
177.9 Hz
5.1 Hz
6.2 Hz
132.3 Hz
ppm
201.45
206.8
152.05
147.5
132.0
124.0
43.2
29.75 Septett 20.6 Hz
29.7 Quartett 128.7 Hz

34

35

1,3,5-Trinitrobenzene and acetone react in the presence of methanolic KOH, giving a blue-violet Meisenheimer potassium salt with ^{13}C NMR Spectrum 35. Analyze the spectrum and give the most important spectral arguments supporting the structure of the Meisenheimer salt.

Recording Conditions

Solvent:	Hexadeuteriodimethylsulfoxide (septet at 39.7 ppm with J_{C-D} = 20.5 Hz).
Concentration:	Saturated solution.
Temperature:	28° C.
Operating frequency:	22.63 MHz.
Decoupling:	*Top:* Coupled spectrum (25,000 scans) with parts expanded fourfold. *Bottom:* Proton noise-decoupled (5,000 scans).

2,4,6-Trinitrophenylacetone,
Meisenheimer salt with $KOCH_3$

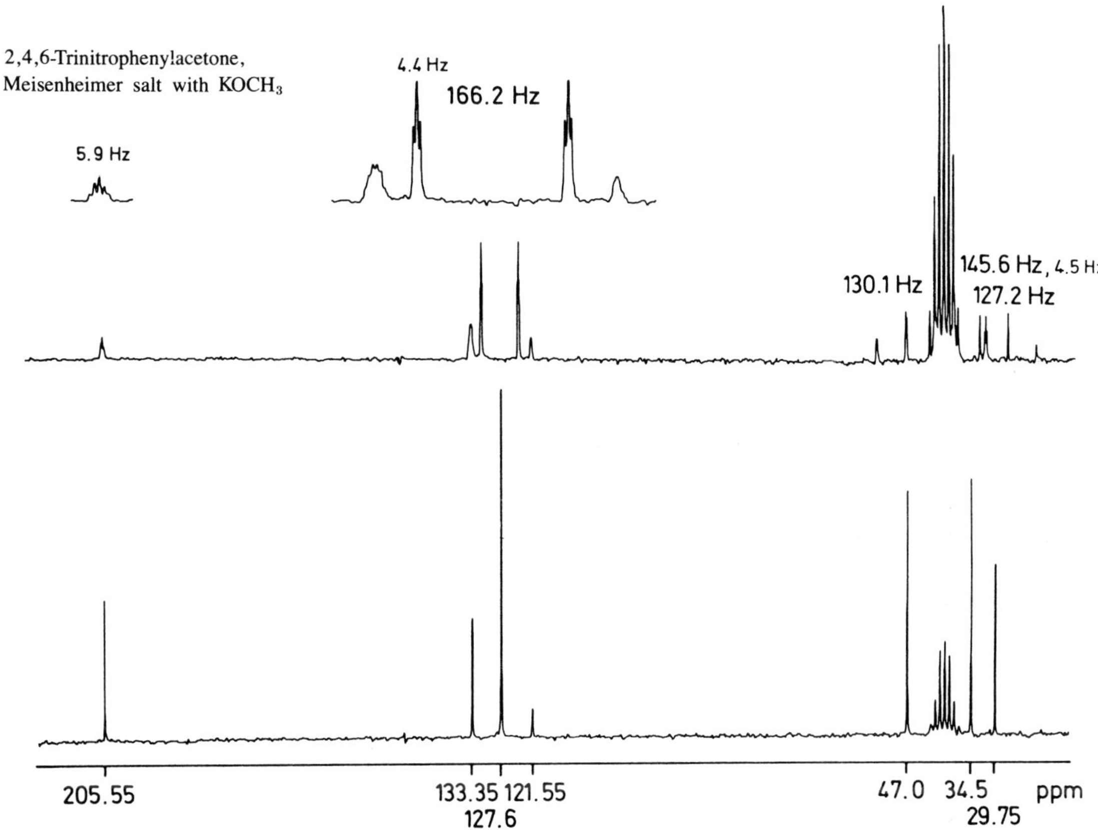

35

36

Spectra 36A and B correspond to the isomeric phenols thymol and carvacrol. Using the chemical shifts of toluene and the OH and $CH(CH_3)_2$ substituent increments for the benzene ring carbon atoms (Table 7.10, p. 61), decide which spectrum belongs to thymol and which to carvacrol.

^{13}C chemical shifts of toluene

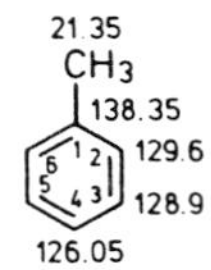

Recording Conditions

Solvent:	Deuteriochloroform (triplet at 77.1 ppm).
Concentration:	About 50% w/w.
Temperature:	28° C.
Operating frequency:	22.63 MHz.
Decoupling:	*Top:* Off-resonance proton decoupled (500 scans). *Bottom:* Proton noise-decoupled (200 scans).

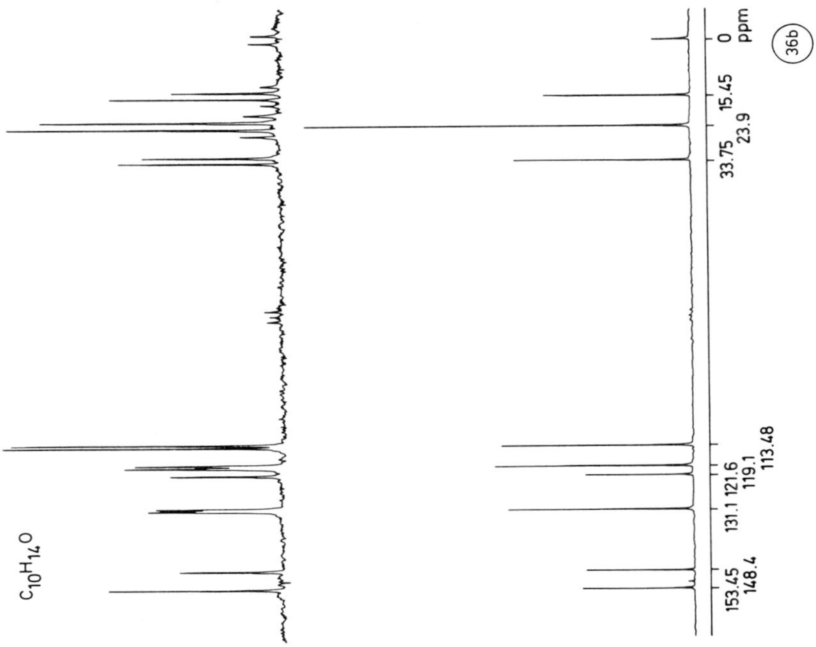

$C_{10}H_{14}O$
153.45 148.4 131.1 121.6 119.1 113.48 33.75 23.9 15.45 0 ppm
36b

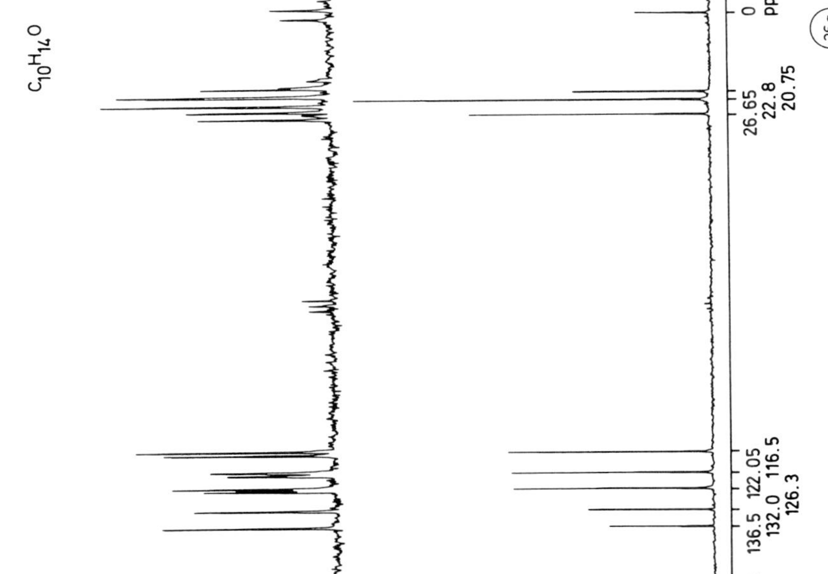

$C_{10}H_{14}O$
152.4 136.5 132.0 126.3 122.05 116.5 26.65 22.8 20.75 0 ppm
36a

37

Which is the structure of the compound with the molecular formula $C_8H_{11}N$ and ^{13}C NMR Spectrum 37? Assign the signals using appropriate chemical shift increments.

Recording Conditions

Solvent:	Deuteriochloroform (triplet at 77.25 ppm in the upper spectrum).
Concentration:	75% v/v.
Temperature:	28° C.
Operating frequency:	22.63 MHz.
Decoupling:	*Top:* Off-resonance proton decoupled (500 scans). *Bottom:* Proton noise-decoupled (100 scans).

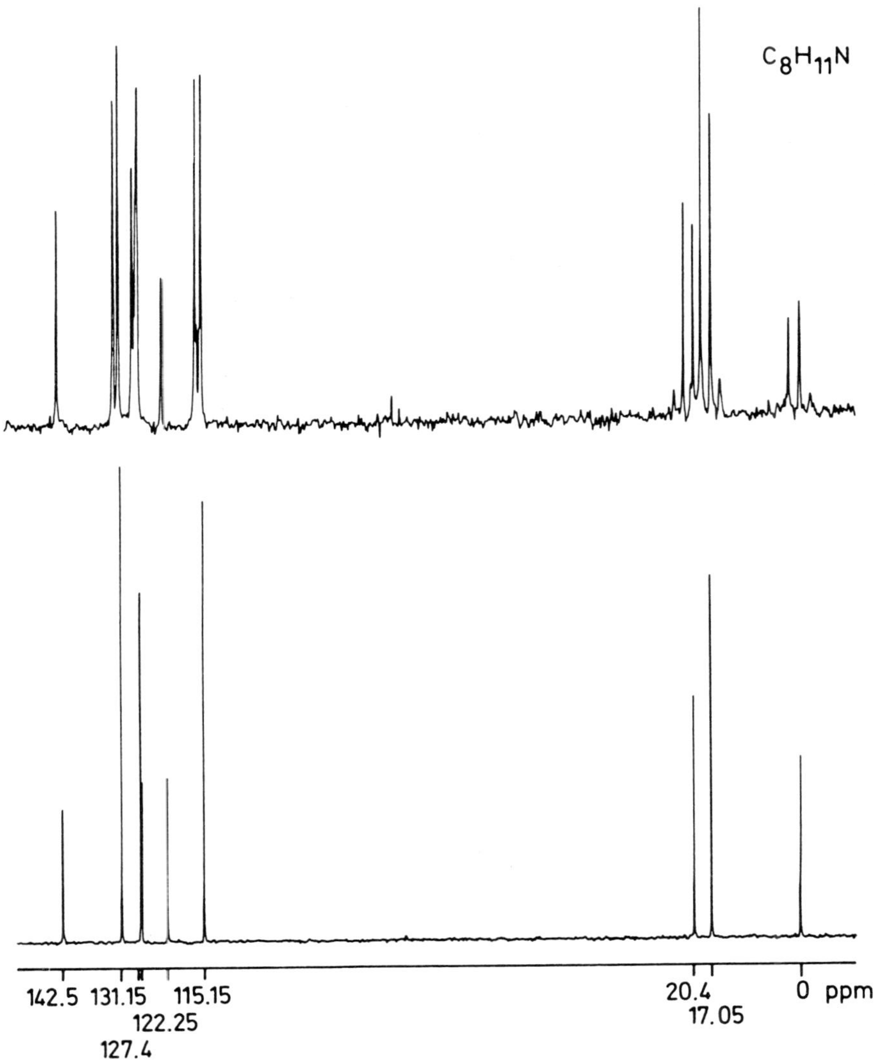

37

38

Upon the alkylation of o-aminothiophenol with 1,3-dibromopropane, a product was obtained for which ^{13}C NMR Spectrum 38 was recorded. What has occurred: an N- or an S-alkylation? Assign all the signals.

Recording Conditions

Solvent:	Deuteriochloroform (triplet at 77.6 ppm).
Concentration:	100 mg in 1 ml.
Temperature:	25° C.
Operating frequency:	20 MHz.
Decoupling:	*Top:* Off-resonance proton decoupled (2,000 scans). *Bottom:* Proton noise-decoupled (600 scans).

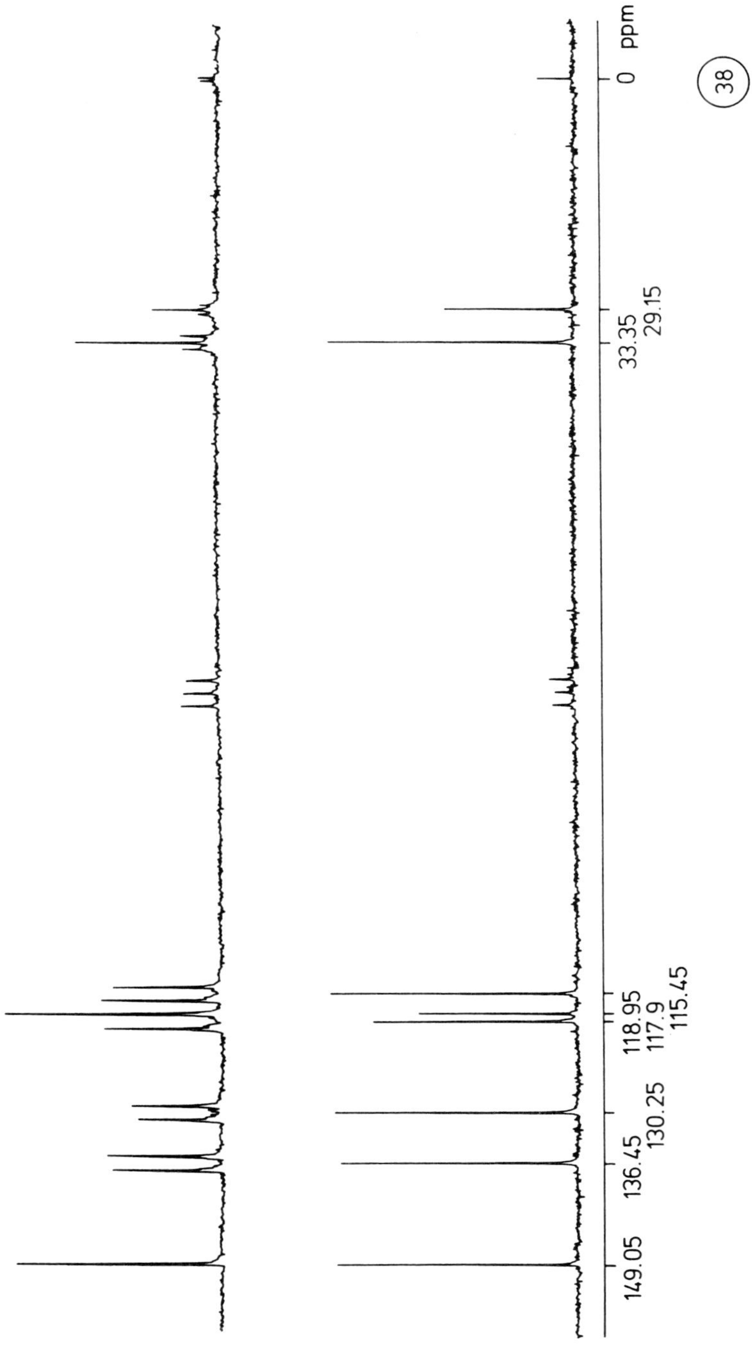
ppm
0
29.15
33.35
115.45
117.9
118.95
130.25
136.45
149.05
38

39

Assign the signals in ^{13}C NMR Spectrum 39, corresponding to phenylthiourea. What is the state of the compound in solution?

Recording Conditions

Solvent:	Hexadeuteriodimethylsulfoxide (septet at 39.5 ppm, not shown).
Concentration:	Saturated solution at 30° C.
Standard:	Hexadeuteriodimethylsulfoxide (see above).
Temperature:	30° C.
Operating frequency:	20.0 MHz.
Decoupling:	A. Pulsed proton decoupling according to Fig. 3.7A, p. 27 (360 scans). B. Pulsed proton decoupling according to Fig. 3.6A, p. 26 (925 scans).
Remarks:	The waiting time between pulses was 10 s in A and B. Half the value required for a 90° pulse was chosen as pulse width.

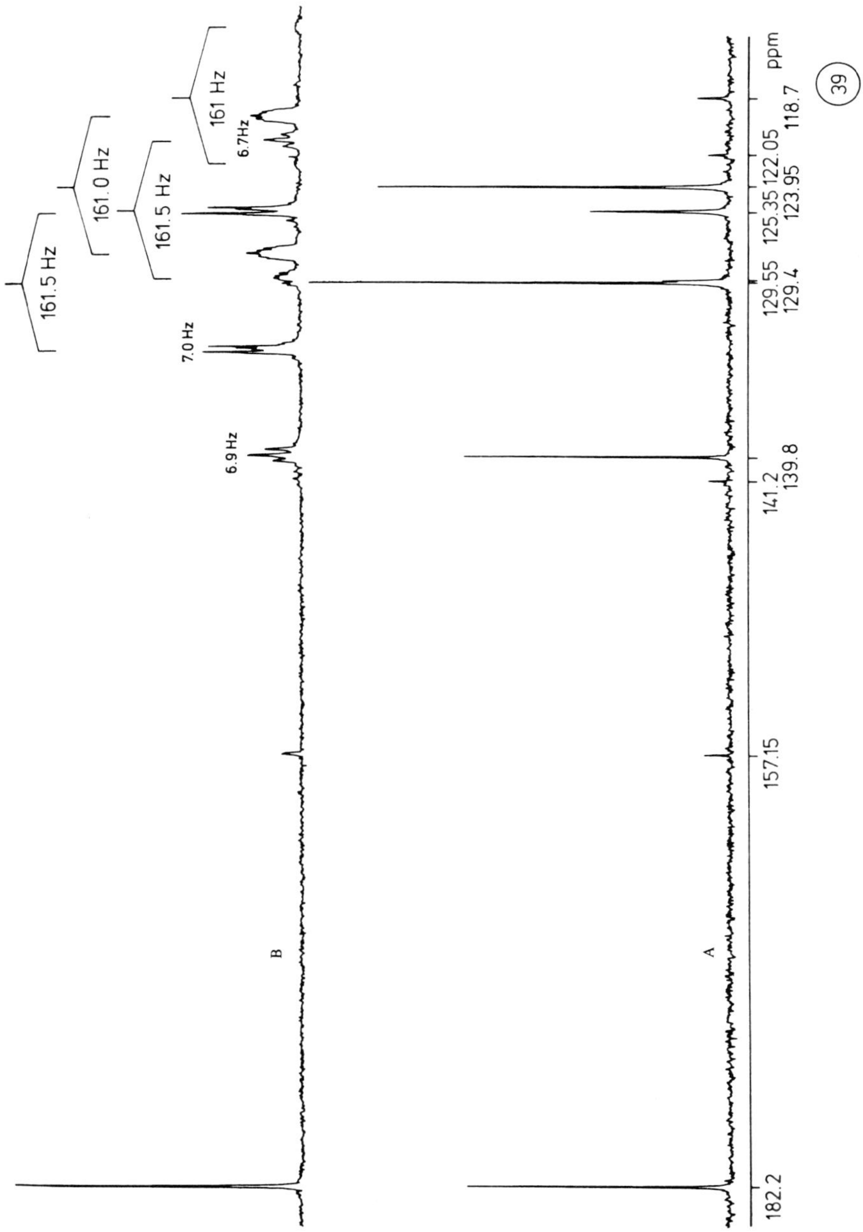

161.5 Hz
161.0 Hz
161.5 Hz
161 Hz
6.7Hz
7.0 Hz
6.9 Hz
B
A
ppm
118.7
122.05
123.95
125.35
129.55
129.4
139.8
141.2
157.15
182.2
39

40

3-Amino-2-methylacrolein reacts with phenyl isocyanate giving a product with ^{13}C NMR Spectrum 40. What is this product? Assign all the signals in the spectrum, using the ^{13}C chemical shifts of 3-amino-2-methylacrolein (191.3, 152.15, 117.6, and 7.7 ppm).

Recording Conditions

Solvent:	Hexadeuteriodimethylsulfoxide (septet at 39.7 ppm).
Concentration:	Saturated solution.
Temperature:	30° C.
Operating frequency:	15.08 MHz.
Decoupling:	*Top:* Off-resonance proton decoupled (2,000 scans, decoupling frequency near the aldehyde proton resonance). *Middle:* Off-resonance proton decoupled (2,000 scans, decoupling frequency near the methyl proton resonance). *Bottom:* Proton noise-decoupled (1,000 scans).
Remarks:	Two off-resonance experiments were recorded so that the multiplicity of the signal at 116.9 ppm could be seen clearly.

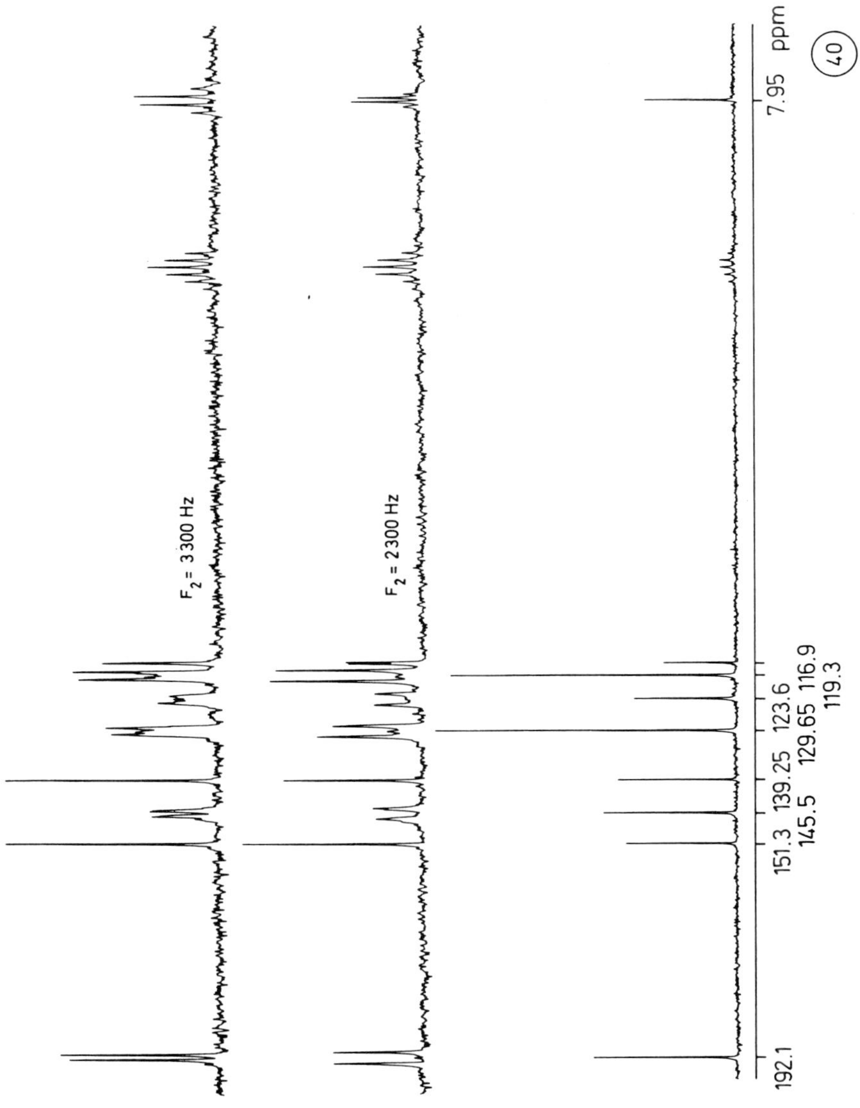
F2 = 3300 Hz
F2 = 2300 Hz
192.1
151.3
145.5
139.25
129.65
123.6
119.3
116.9
7.95
ppm
40

41

An active component of many drugs has the elementary composition $C_{10}H_{13}NO_2$ and ^{13}C NMR Spectrum 41. What compound is it? Assign all the ^{13}C chemical shifts.

Recording Conditions

Solvent:	Hexadeuteriodimethylsulfoxide (septet at 39.7 ppm).
Concentration:	100 mg in 1 ml.
Temperature:	28° C.
Operating frequency:	15.08 MHz.
Decoupling:	*Top:* Proton off-resonance decoupled (2,000 scans). *Bottom:* Proton noise-decoupled (500 scans).

$C_{10}H_{13}NO_2$

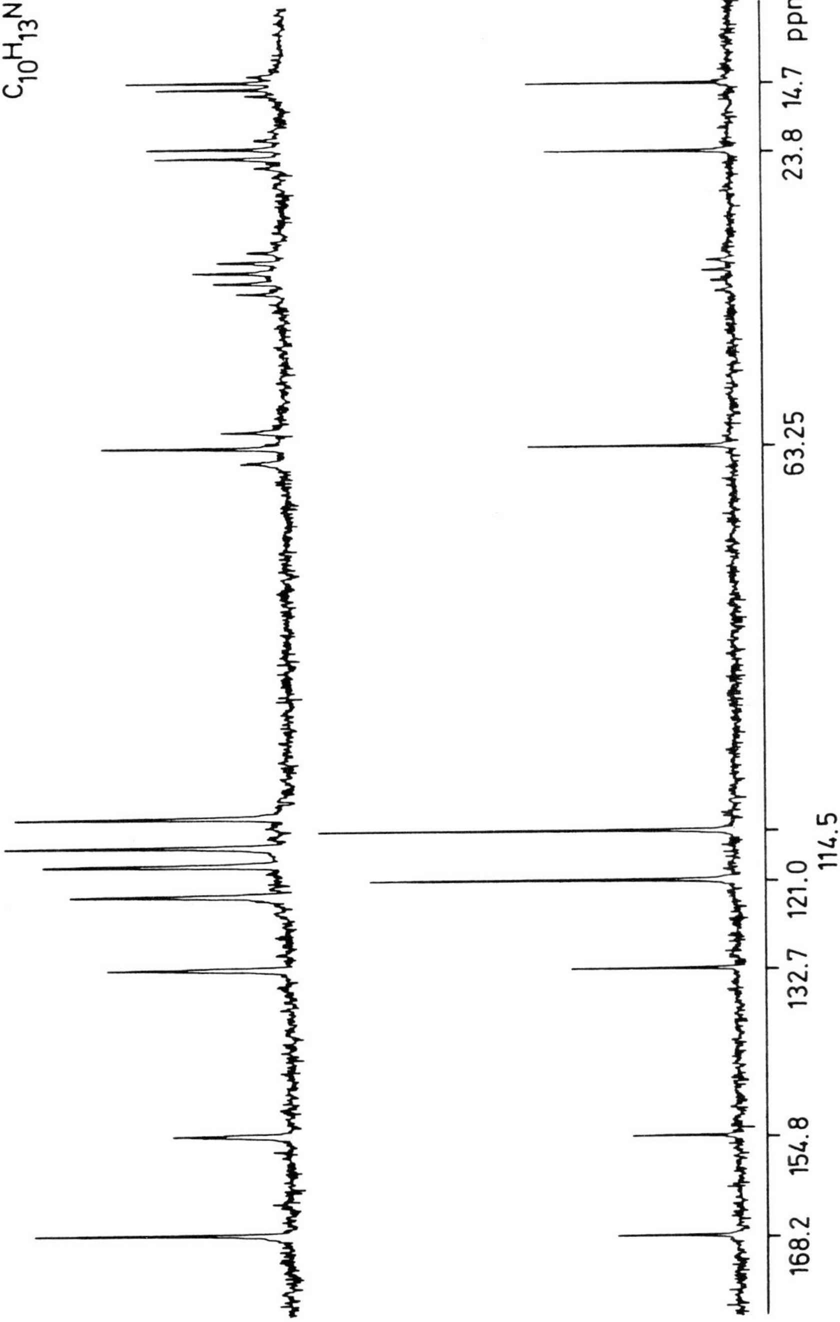

41

42

Successive Friedel-Crafts alkylation and nitration of acetanilide afford a product with molecular formula $C_{12}H_{16}N_2O_3$ and ^{13}C NMR Spectra 42 B-E. What product was obtained? Assign the spectra in detail.

Recording Conditions

The same sample solution was used for all spectra.

Solvent:	Deuteriochloroform (triplet at 77.25 ppm in spectra B-E).
Concentration:	250 mg/ml.
Temperature:	30° C.
Operating frequency:	80 MHz (^{1}H). 20 MHz (^{13}C).
Decoupling:	A. Undecoupled (single scan). B. Proton noise-decoupled (200 scans, 5 W decoupler power). C. Proton selectively decoupled (400 scans, 0.05 W, decoupling frequency F_2 = 6001 D. Proton selectively decoupled (400 scans, 0.05 W, F_2 = 6,041 Hz). E. Proton selectively decoupled (400 scans, 0.05 W, F_2 = 6,076 Hz).
Remarks:	The instrument-specific frequencies given in Spectrum A are the decoupling frequencies F_2 used in Spectra C-D.

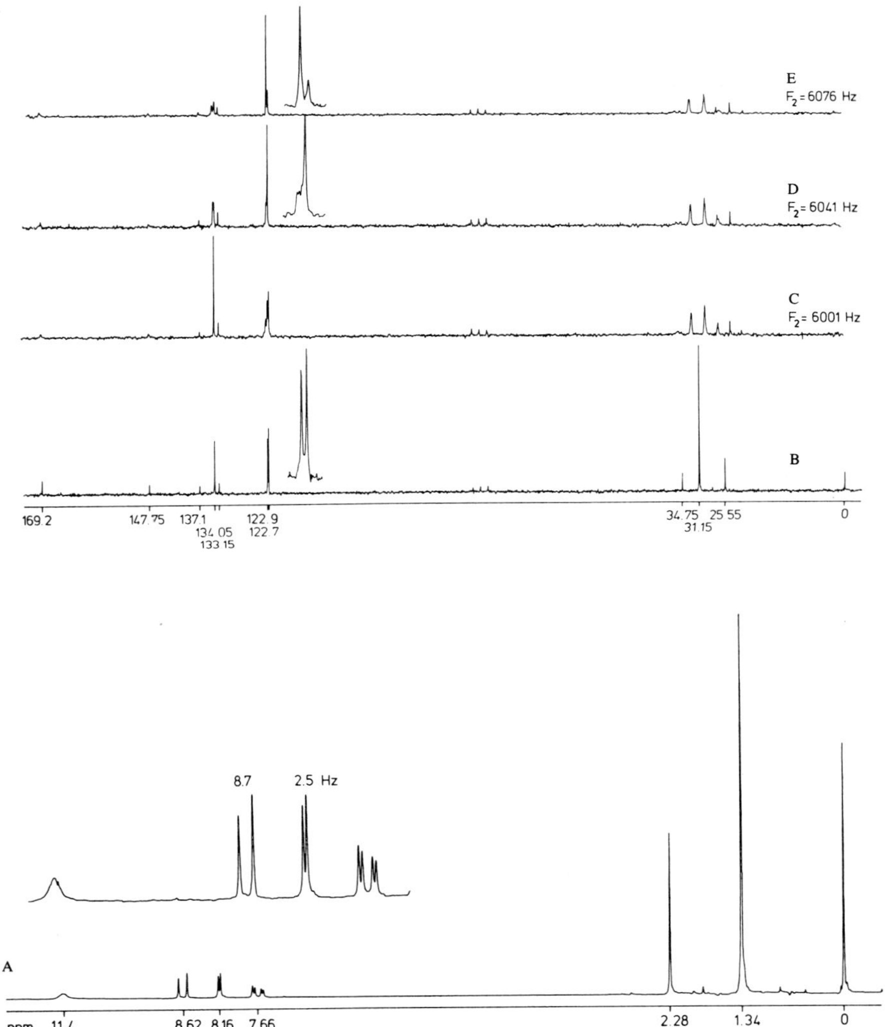
E
F2 = 6076 Hz
D
F2 = 6041 Hz
C
F2 = 6001 Hz
B
169.2
147.75
137.1
134.05
133.15
122.9
122.7
34.75
25.55
31.15
0
8.7
2.5 Hz
A
ppm
11.4
8.62
8.16
7.66
6076
6041
6001
Hz
2.28
1.34
0

42

43

^{13}C NMR Spectrum 43 was recorded for an aroma component with the elementary composition $C_8H_8O_3$. What compound is it? Assign the spectrum.

Recording Conditions

Solvent:	Deuteriochloroform (triplet at 77.2 ppm in the top spectrum).
Concentration:	150 mg in 1 ml.
Temperature:	28° C.
Operating frequency:	22.63 MHz.
Decoupling:	A. Proton noise-decoupled (200 scans). B. Gated proton decoupling as in Fig. 3.6A, p. 26, with C-H direct couplings (1,600 scans). C. Expanded portions of Spectrum B, with C-H long-range couplings.

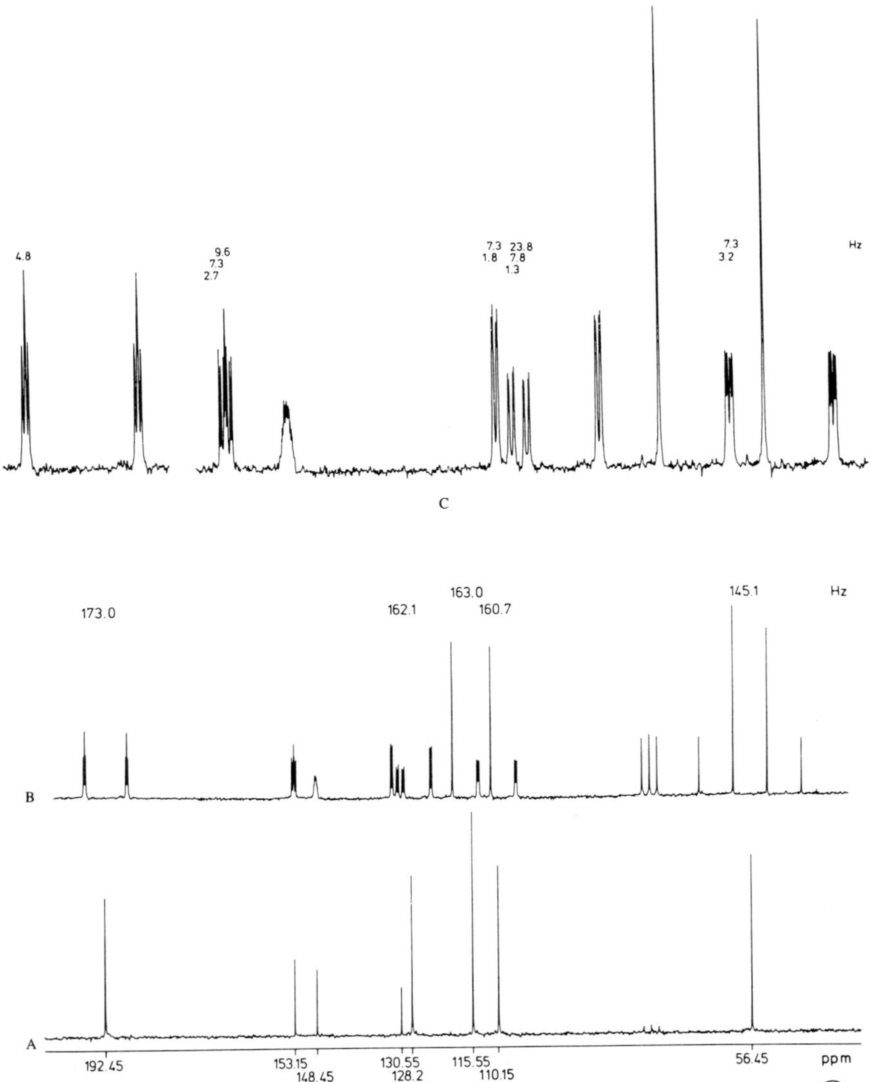

4.8
9.6
7.3
2.7
7.3 23.8
1.8 7.8
1.3
7.3
3.2
Hz
C
163.0
173.0
162.1
160.7
145.1
Hz
B
A
192.45
153.15
148.45
130.55
128.2
115.55
110.15
56.45
ppm

43

44

Analyze ^{13}C NMR Spectra 44 A and B of the compound $C_{16}H_{14}O_4$, considering the signal intensities. Determine the structure of the compound from its formula and spectra.

Recording Conditions

Solvent:	Deuteriochloroform (triplet at 77.2 ppm).
Concentration:	250 mg/ml.
Temperature:	30° C.
Operating frequency:	20 MHz.
Decoupling:	A. Proton noise-decoupled (500 scans), no waiting time between pulses. B. Pulsed proton noise-decoupling as in Fig. 3.7A, p. 27 (500 scans), waiting time between pulses: 30 s.

44

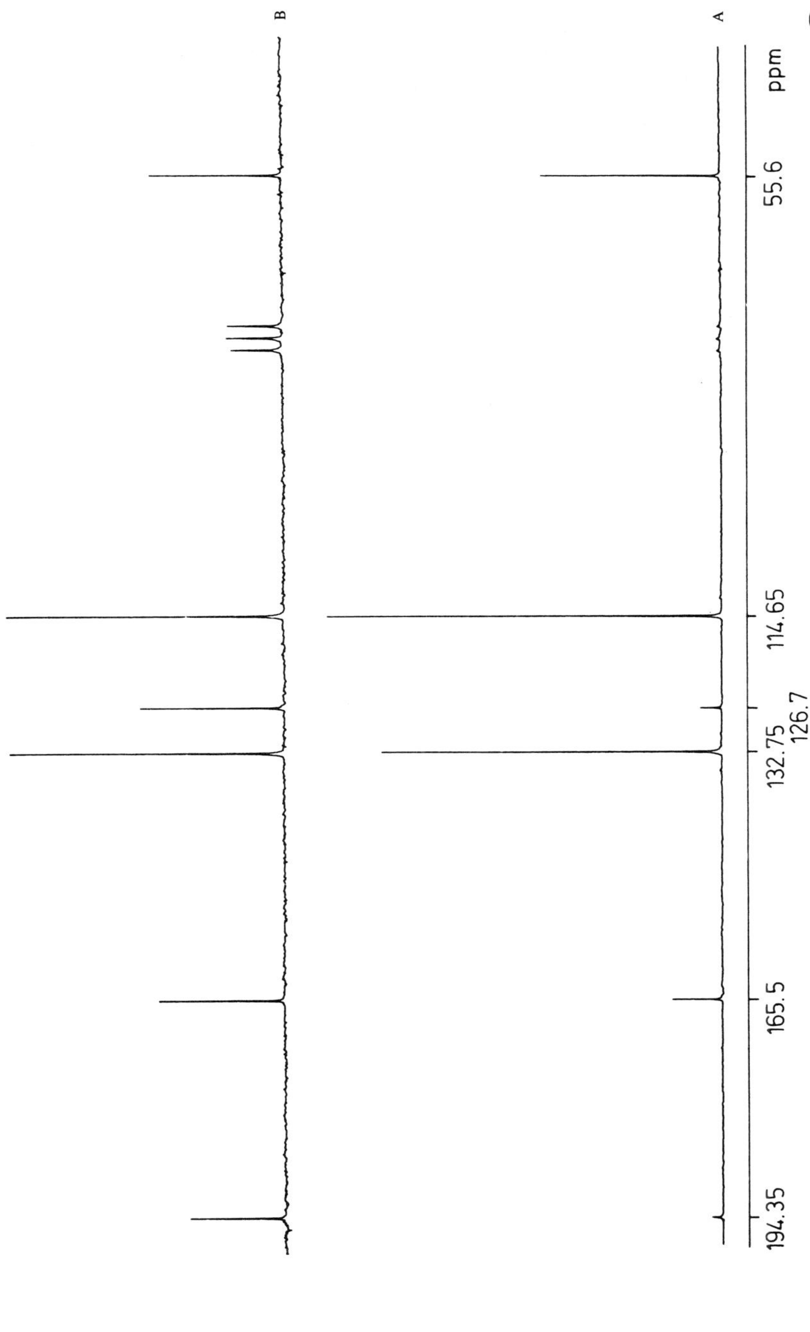

45

On attempting to distill ethyl 3-dimethylamino-3-phenylpropionate, a product was obtained giving ^{13}C NMR Spectrum 45. On the basis of the ^{13}C NMR spectrum, what happened during the distillation? Assign the spectrum.

Recording Conditions

Solvent:	Deuteriochloroform (triplet at 78.05 ppm).
Concentration:	50% v/v.
Temperature:	25° C.
Operating frequency:	20 MHz.
Decoupling:	Proton noise-decoupled (500 scans).

P4224-7009 Appendix

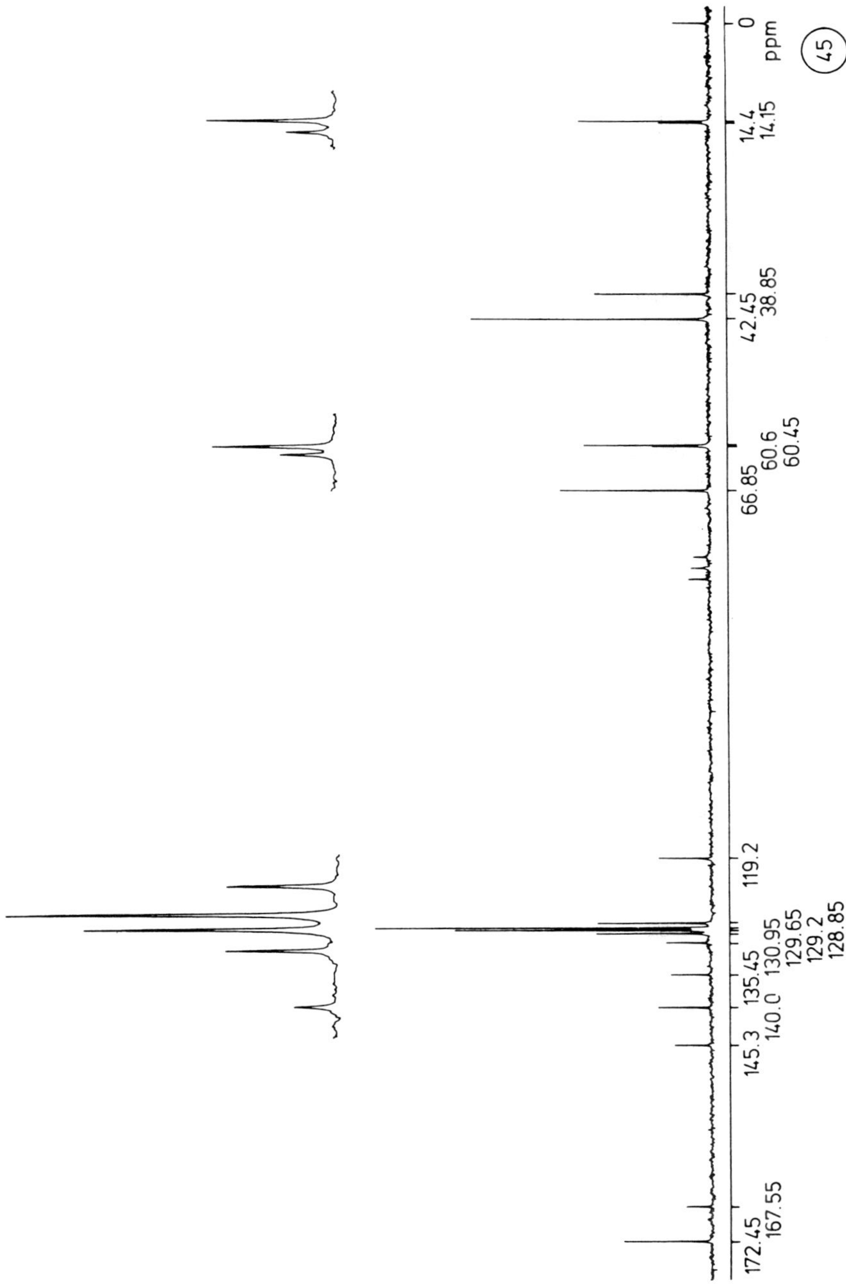
0 ppm
45
14.4
14.15
42.45
38.85
66.85
60.6
60.45
119.2
145.3
140.0
135.45
130.95
129.65
129.2
128.85
128.15
172.45
167.55

46

What derivative of azobenzene with the formula $C_{15}H_{14}O_2N_2$, gave ^{13}C NMR Spectrum 46?

^{13}C chemical shifts of *trans*-azobenzene in hexadeuteriodimethylsulfoxide at 25° C

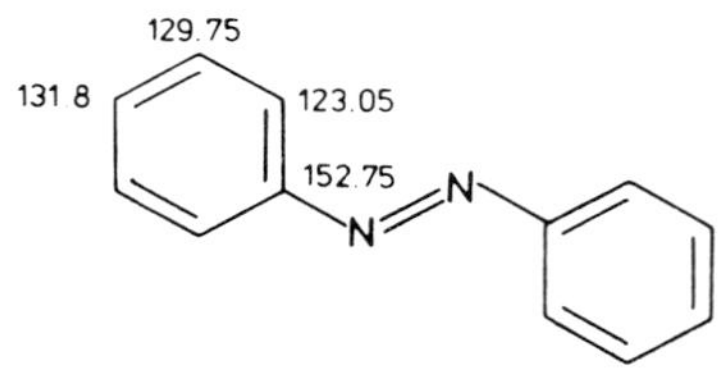

Recording Conditions

Solvent:	Hexadeuteriodimethylsulfoxide (septet at 39.8 ppm).
Concentration:	Saturated solution.
Temperature:	25° C.
Operating frequency:	20 MHz.
Decoupling:	*Top:* Off-resonance proton-decoupled (1,500 scans). *Bottom:* Proton noise-decoupled (300 scans) with expanded partial spectrum.

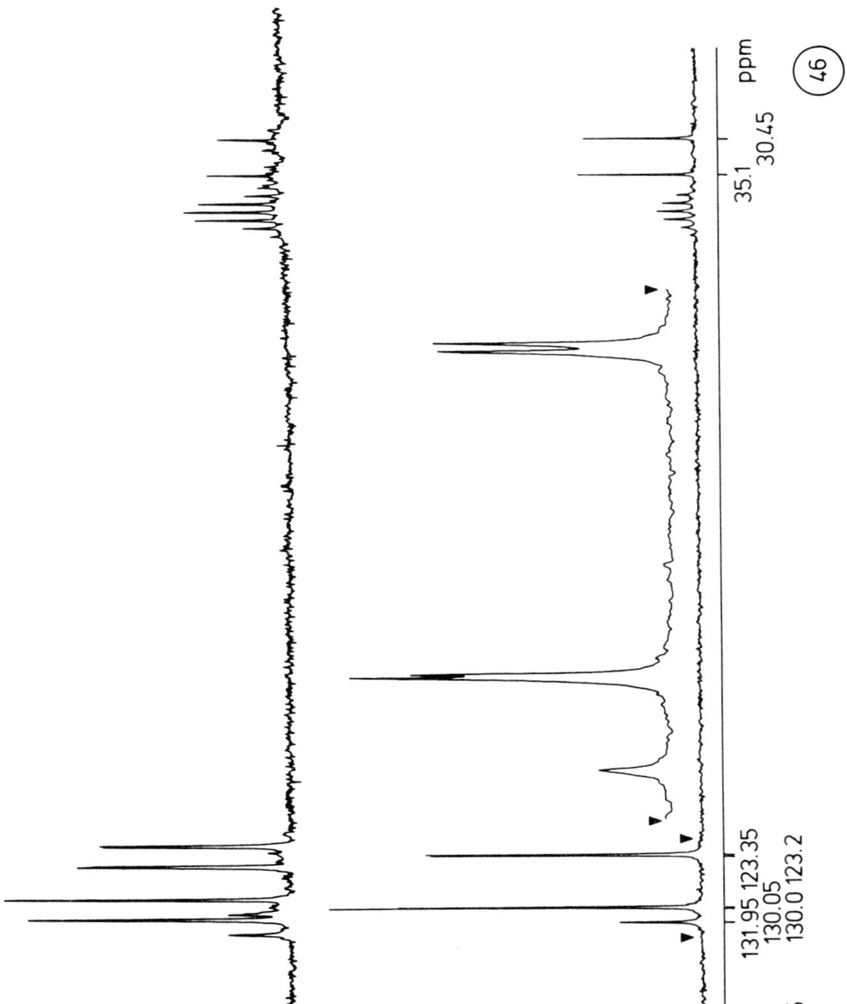
174.55
152.95
151.45
145.75
131.95
130.05
130.0
123.35
123.2
35.1
30.45
ppm
46

47

A compound with the elementary composition $C_{20}H_{16}O_2$ and ^{13}C NMR Spectrum 47A is formed from fumaric dialdehyde (trans-2-butendial) and phenylacetylene. The oxidation of this reaction product yields $C_{20}H_{12}O_2$ with ^{13}C NMR Spectrum 47B. What are the structures of the reaction products?

Recording Conditions for Both Spectra

Solvent:	Deuteriochloroform (triplet at 77.25 ppm).
Concentration:	150 mg in 1 ml.
Temperature:	28° C.
Operating frequency:	22.63 MHz.
Decoupling:	*Top:* Coupled (25,000 scans). *Bottom:* Proton noise-decoupled (4,000 scans).

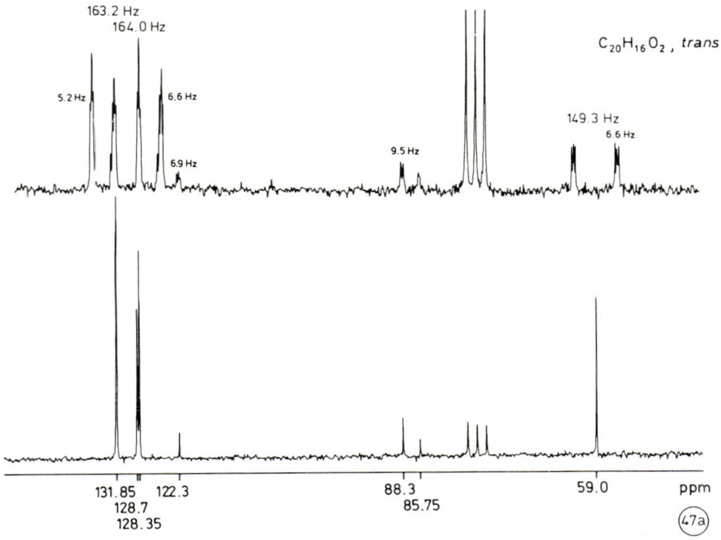
163.2 Hz
164.0 Hz
5.2 Hz
6.6 Hz
6.9 Hz
9.5 Hz
$C_{20}H_{16}O_2$, *trans*
149.3 Hz
6.6 Hz
131.85
128.7
128.35
122.3
88.3
85.75
59.0
ppm

47a

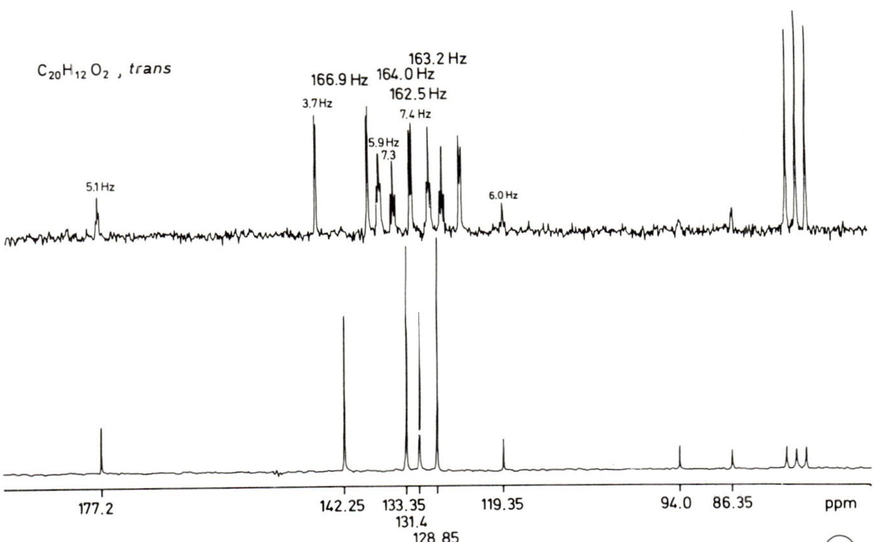
$C_{20}H_{12}O_2$, *trans*
5.1 Hz
166.9 Hz
3.7 Hz
163.2 Hz
164.0 Hz
162.5 Hz
7.4 Hz
5.9 Hz
7.3
6.0 Hz
177.2
142.25
133.35
131.4
128.85
119.35
94.0
86.35
ppm

47b

48

Assign ^{13}C NMR Spectrum 48 of 1,8-diiodonaphthalene. On the basis of your assignment, determine the influence of the iodine substituents on the ^{13}C chemical shifts of naphthalene (Table 7.13, p. 66) and compare your result with the iodine increments for the benzene chemical shifts (Table 7.10, p. 61).

Recording Conditions

Solvent:	Deuteriochloroform (triplet at 77.45 ppm, not shown).
Concentration:	Saturated solution.
Temperature:	25° C.
Operating frequency:	20 MHz.
Decoupling:	A. Pulsed proton decoupling as in Fig. 3.7A (200 scans). B. Pulsed proton decoupling as in Fig. 3.6A (675 scans).
Remarks:	The waiting time between pulses was 5 s in both spectra. The pulse width was chosen as one-third of the value required for a 90° pulse.

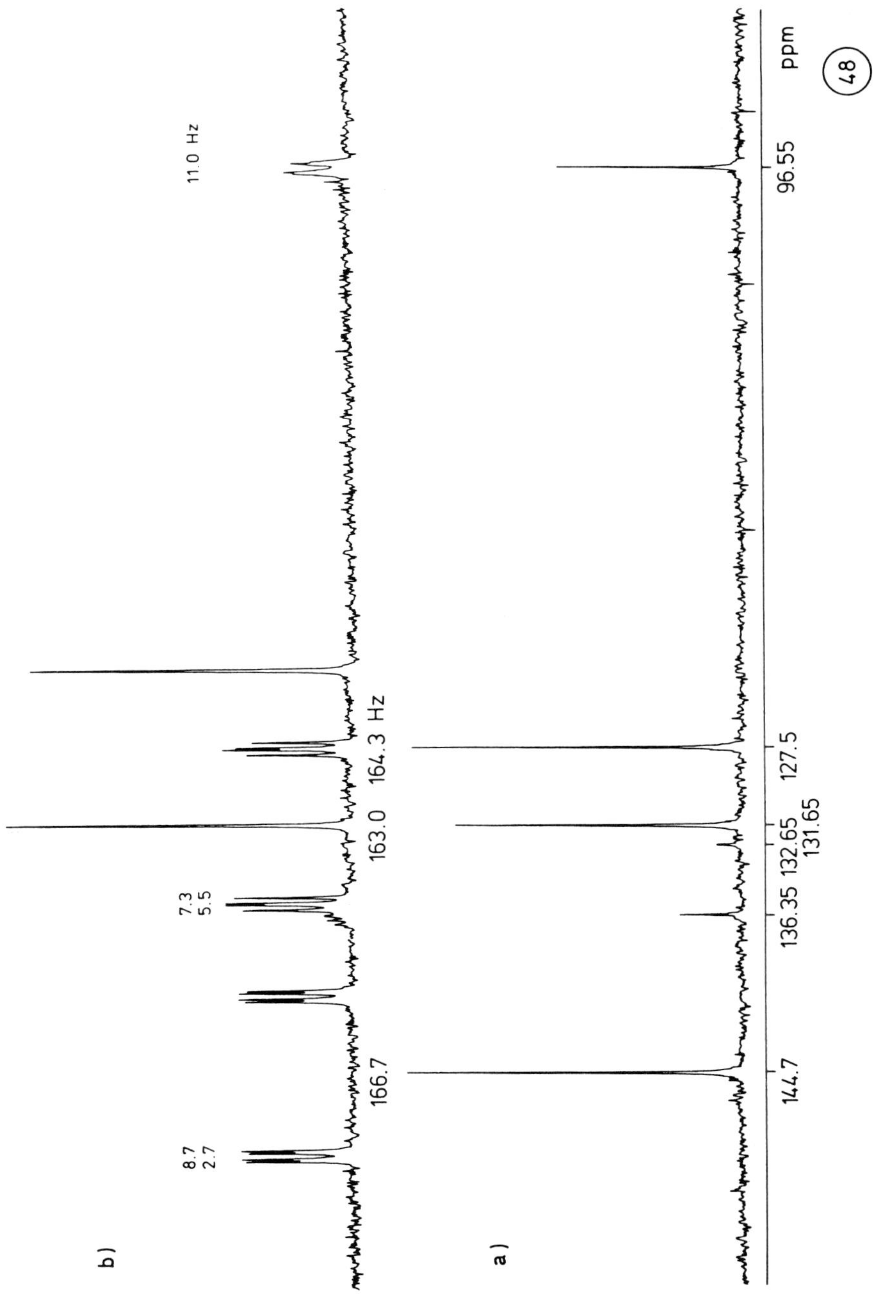
b)
8.7
2.7
7.3
5.5
11.0 Hz
166.7
163.0
164.3 Hz
a)
144.7
136.35
132.65
131.65
127.5
96.55
ppm
48

49

The assigned ^{13}C chemical shifts of tetralin and 1,4-naphthoquinone and the ^{13}C NMR spectra of α-tetralone (49A and B) and 6-methoxy-α-tetralone (49C and D) are given. Assign the spectra as completely as possible, using meaningful chemical shift comparisons, the pseudocontact shifts from Spectrum B, the off-resonance multiplicities from Spectrum D, and the OCH_3 shift increments from Table 7.10.

^{13}C chemical shifts of tetralin[1] and 1,4-naphthoquinone[2]

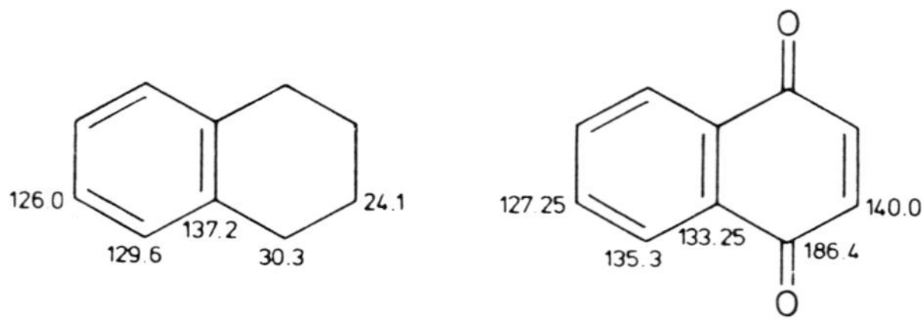

Recording Conditions

Solvent:	Deuteriochloroform (triplet at 77.6 ppm in A and B, at 78.0 ppm in C and D).
Concentration:	A. 0.1 ml in 1 ml.
	B. 0.1 ml in 1 ml, with addition of 200 mg of tris(1,1,1,2,2,3,3-heptafluoro-7,7-dimethyl-4,6-octandionato) europium (III).
	C-D. 150 mg in 1 ml.
Temperature:	25° C.
Operating frequency:	20 MHz.
Decoupling:	A-B. Proton noise-decoupled (1,500 scans each).
	C. Proton noise-decoupled (2,000 scans).
	D. Proton off-resonance decoupled (3,000) scans).

[1] J. B. Stothers, *Carbon-13 NMR Spectroscopy*, Academic Press, New York, 1972, p. 99.
[2] Unpublished result; saturated solution in hexadeuterioacetone.

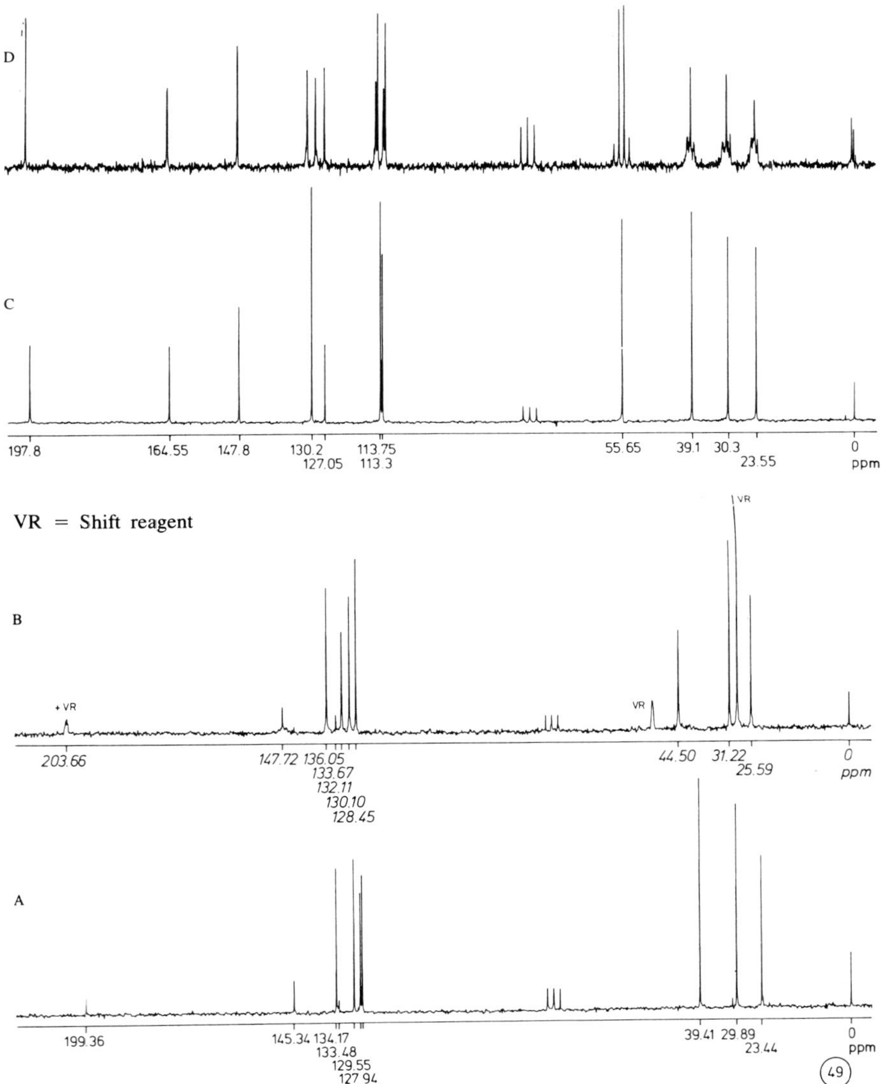
D
C
197.8
164.55
147.8
130.2
127.05
113.75
113.3
55.65
39.1
30.3
23.55
0
ppm
VR = Shift reagent
B
+VR
VR
VR
203.66
147.72
136.05
133.67
132.11
130.10
128.45
44.50
31.22
25.59
0
ppm
A
199.36
145.34
134.17
133.48
129.55
127.94
127.40
39.41
29.89
23.44
0
ppm
49

50

^{13}C NMR Spectrum 50A was recorded for the reaction product $C_{24}H_{22}N_4O_2$ of o-acetylaminobenzaldehyde and o-phenylenediamine. What is the compound's structure?

The reaction of o,o′-diformyloxaldianilide with o-phenylenediamine afforded a product with a mass-spectrometrically determined molecular mass of 368 and ^{13}C NMR Spectrum 50B. What product was formed? Assign both spectra as completely as possible.[1]

Recording Conditions

Solvent:	Deuteriochloroform (triplet at 77.5 ppm (A) and 77.45 ppm (B), respectively).
Concentration:	Saturated solutions.
Temperature:	30° C.
Operating frequency:	20 MHz.
Decoupling:	Proton noise-decoupled (3,000 scans each).
Remarks:	The aromatic region is shown expanded above the complete spectra.

[1]Suggested reading: E. Lorch, E. Breitmaier, *Chemiker Ztg.*, **99**, 87 (1975), similar compounds.

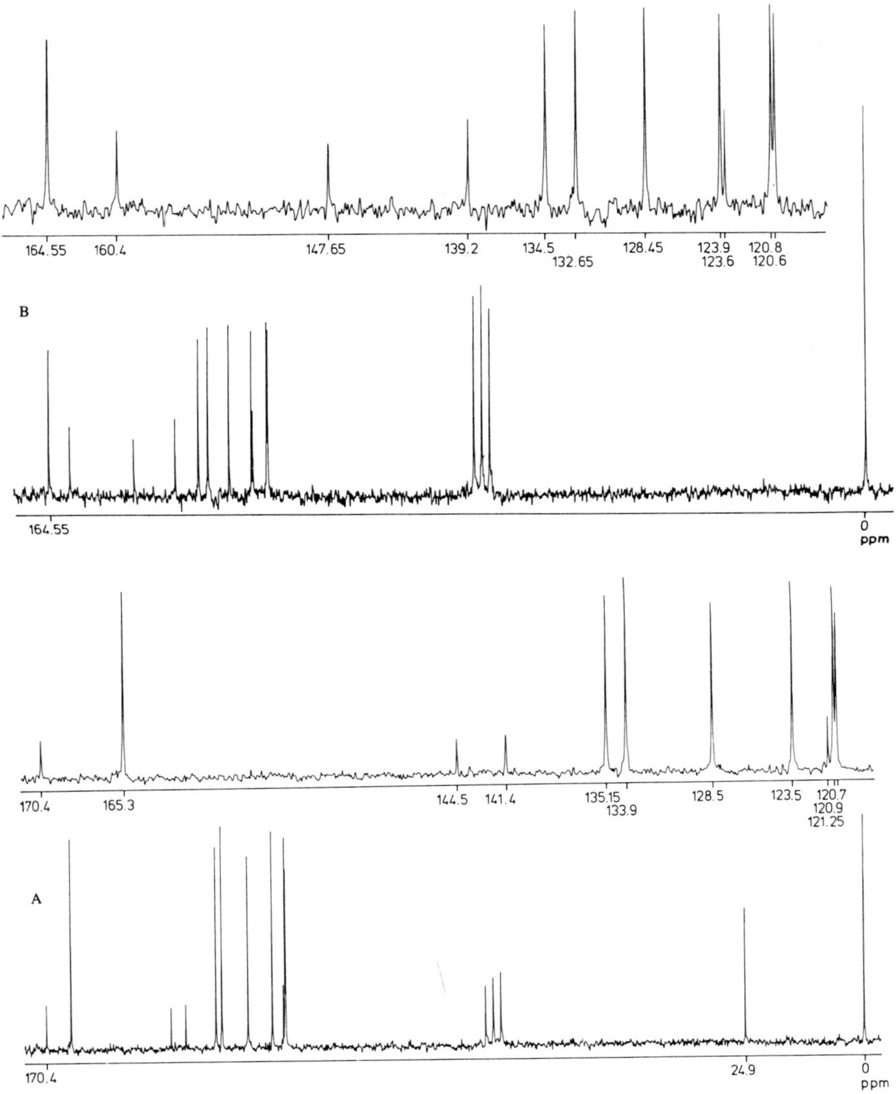
164.55 160.4 147.65 139.2 134.5 132.65 128.45 123.9 123.6 120.8 120.6
B
164.55
0
ppm
170.4 165.3 144.5 141.4 135.15 133.9 128.5 123.5 120.7 120.9 121.25
A
170.4
24.9
0
ppm

(50)

51

^{1}H NMR Spectrum 51A and ^{13}C NMR Spectra 51B-F were recorded for 2-acetyl-1-methylpyrrole. Four of the five proton types of the pyrrole derivative were decoupled separately while recording ^{13}C NMR spectra C-F. Assign ^{13}C NMR Spectrum 51B on the basis of decoupling experiments 51C-F.

Recording Conditions

Solvent:	Hexadeuterioacetone (weak septet at 29.8 ppm).
Concentration:	50% v/v.
Temperature:	30° C.
Operating frequencies:	^{1}H: 80 MHz. ^{13}C: 20 MHz (the same spectrometer).
Decoupling:	A. Coupled with expanded partial spectrum (6.06 — 6.91 ppm). B. Proton noise-decoupled (decoupler power 5 W; 50 scans). C-F. Decoupling of the proton resonances (decoupler power 0.05 W) at 2.31 ppm (C), 3.84 ppm (D), 6.06 ppm (E), and 6.91 ppm (F).
Remarks:	The (instrument-specific) decoupling frequencies (F_2) are written above the corresponding ^{1}H signals and on the specifically decoupled ^{13}C NMR spectra ^{13}C:

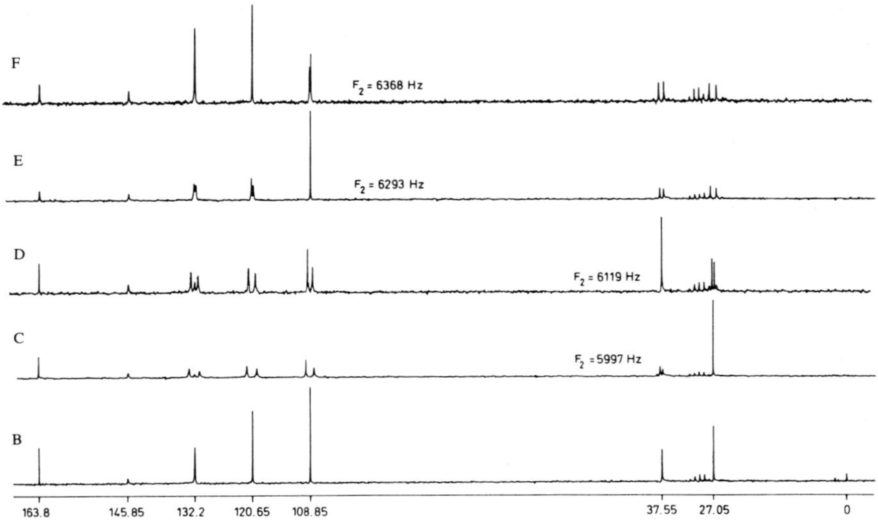
F
F_2 = 6368 Hz
E
F_2 = 6293 Hz
D
F_2 = 6119 Hz
C
F_2 = 5997 Hz
B
163.8
145.85
132.2
120.65
108.85
37.55
27.05
0

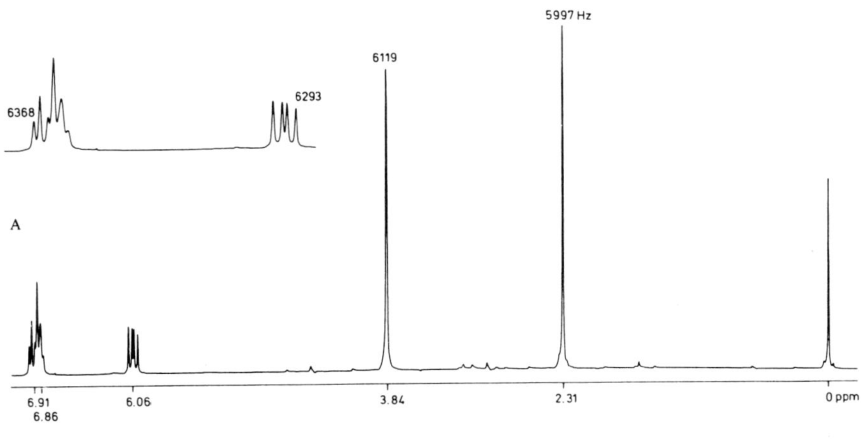
5997 Hz
6119
6293
6368
A
6.91
6.86
6.06
3.84
2.31
0 ppm

52

What furan derivative with formula $C_5H_4O_3$ fits ^{13}C NMR Spectrum 52? Assign all chemical shifts and couplings in the spectrum, using the data known for furan.[1]

Chemical shifts and C-H coupling constants of furan

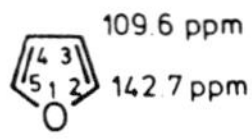

$J_{C-2-H-2}$ = 201.0 Hz

$J_{C-2-H-3}$ = 7.0 Hz

$J_{C-2-H-4}$ = 10.8 Hz

$J_{C-2-H-5}$ = 7.0 Hz

$J_{C-3-H-3}$ = 175.0 Hz

$J_{C-3-H-2}$ = 14.0 Hz

$J_{C-3-H-4}$ = 4.0 Hz

$J_{C-3-H-5}$ = 5.8 Hz

Recording Conditions

Solvent:	Hexadeuterioacetone (not shown).
Concentration:	Saturated solution.
Temperature:	28° C.
Operating frequency:	22.63 MHz.
Decoupling:	Coupled, 10,000 scans.

---

[1]See F. J. Weigert, J. D. Roberts, *J. Am Chem. Soc.*, **90,** 3543 (1968) and **91,** 4940 (1969); and J. B. Stothers, *Carbon-13 NMR Spectroscopy*, Academic Press, New York, 1972, p. 343.

$C_5H_4O_3$

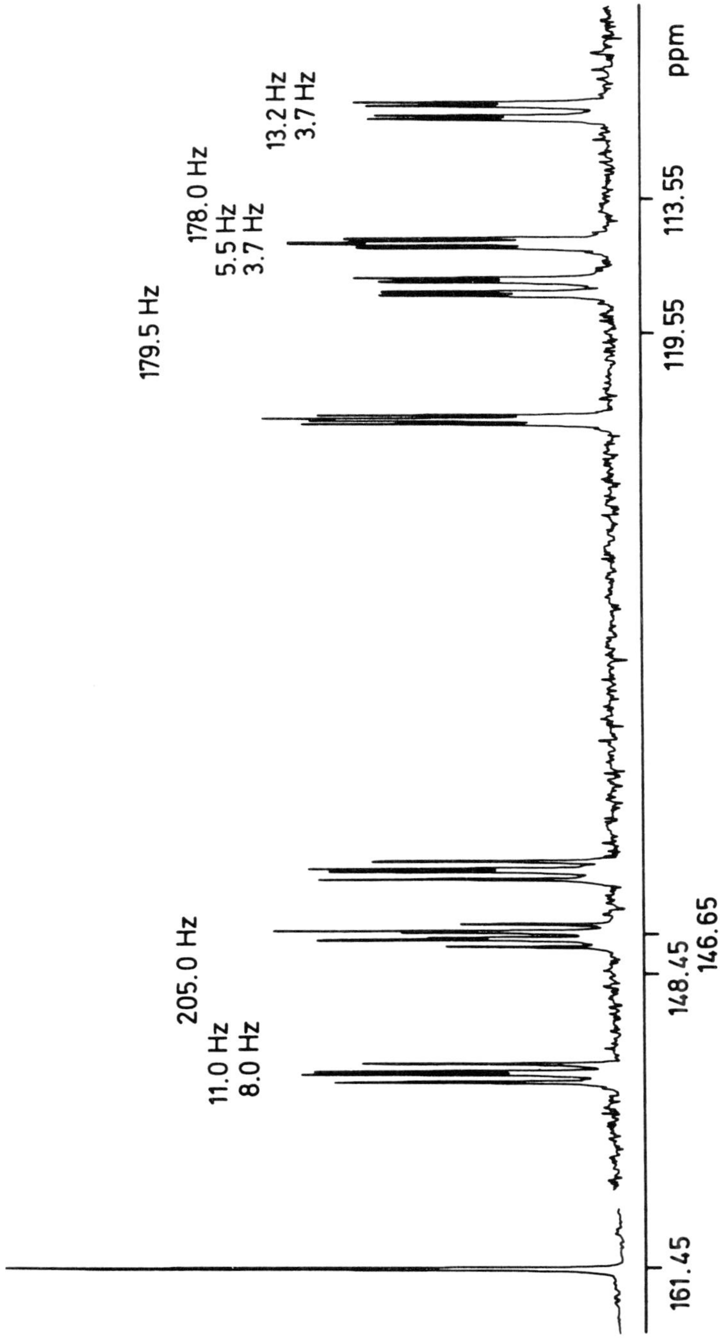
205.0 Hz
11.0 Hz
8.0 Hz
179.5 Hz
178.0 Hz
5.5 Hz
3.7 Hz
13.2 Hz
3.7 Hz
161.45
148.45
146.65
119.55
113.55
ppm

52

53

What tetrahydrofuran derivative with formula $C_8H_{16}O_3$ fits ^{13}C NMR Spectrum 53?

Recording Conditions

Solvent:	Deuteriochloroform (not visible because the solution is too concentrated).
Concentration:	80% v/v.
Temperature:	30° C.
Operating frequency:	15.08 MHz.
Decoupling:	*Top:* Proton off-resonance-decoupled (500 scans). *Bottom:* Proton noise-decoupled (100 scans).

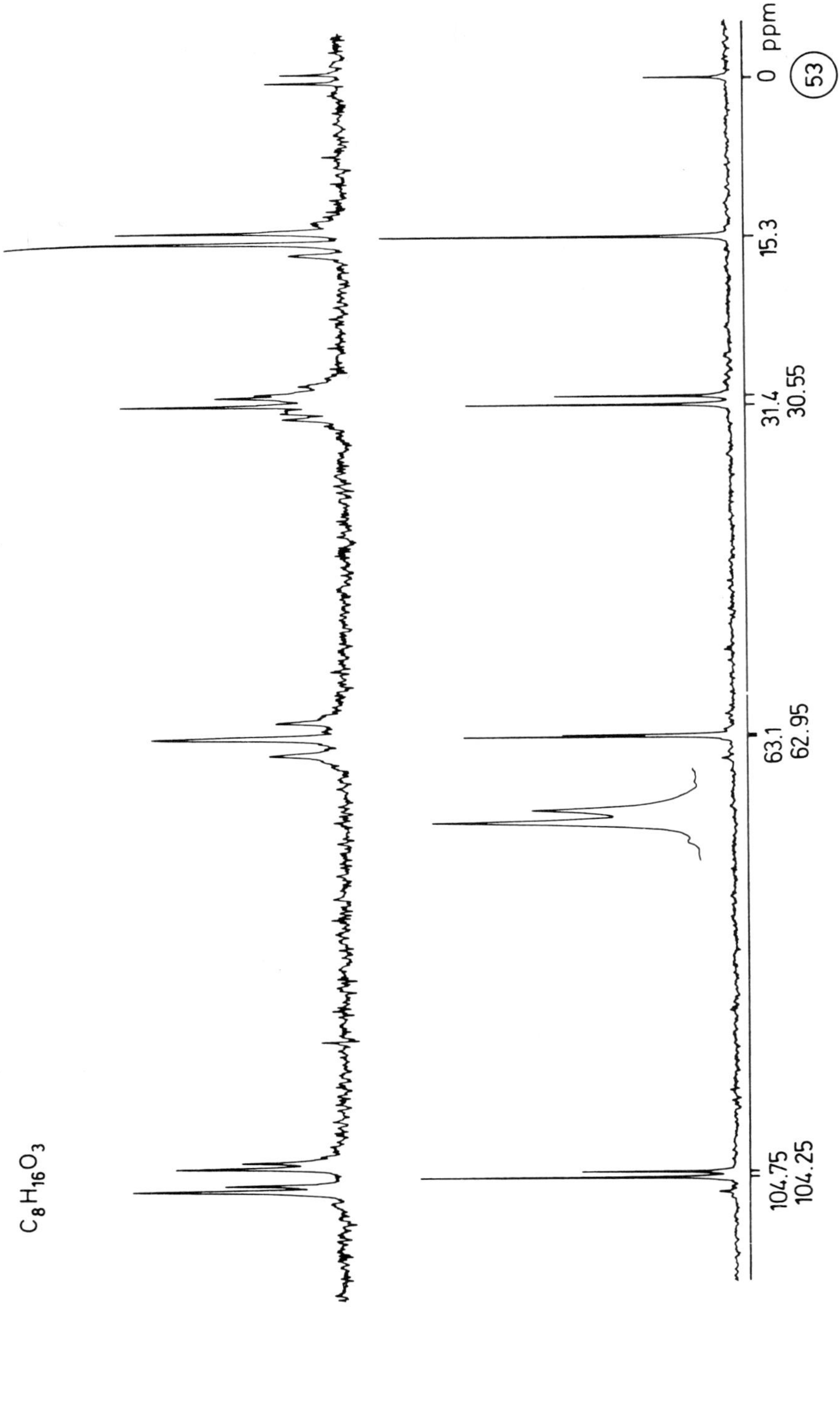

$C_8H_{16}O_3$
104.75
104.25
63.1
62.95
31.4
30.55
15.3
0 ppm
53

54

Assign all the chemical shifts and couplings in ^{13}C NMR Spectrum 54, of 1-(2-thienoyl)-3,3,3-trifluoroacetone. Use the known thiophene chemical shifts and coupling constants. In what form is the compound present?

Chemical shifts and coupling constants of thiophene[1]

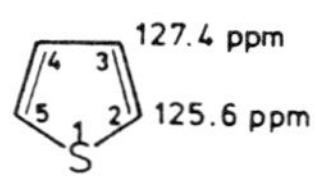

$J_{C-2-H-2}$ = 185.0 Hz
$J_{C-2-H-3}$ = 7.35 Hz
$J_{C-2-H-4}$ = 10.0 Hz
$J_{C-2-H-5}$ = 5.15 Hz
$J_{C-3-H-3}$ = 168.0 Hz
$J_{C-3-H-2}$ = 4.7 Hz
$J_{C-3-H-4}$ = 5.9 Hz
$J_{C-3-H-5}$ = 9.5 Hz

Recording Conditions

Solvent:	Deuteriochloroform (triplet at 77.25 ppm).
Concentration:	500 mg in 1 ml.
Temperature:	30° C.
Operating frequency:	20.0 MHz.
Decoupling:	*Top:* Coupled (2,705 scans). *Bottom:* Proton noise-decoupled (500 scans).
Remarks:	The coupled spectrum was expanded fourfold (upper frequency scale) in the regions indicated by Δ, ▼, and ▲. The "Frequency" column of the computer printout of the coupled spectrum can be used to determine the C-H and C-F coupling constants.

---

[1]See J. B. Stothers, *Carbon-13 NMR Spectroscopy*, Academic Press, New York, 1972, p. 253 and 343; and E. Breitmaier, W. Voelter, *^{13}C NMR Spectroscopy: Methods and Applications*, Monographs in Modern Chemistry," Vol. 5, E. H. Ebel, ed., Verlag Chemie GmbH, Weinheim, 1974, p. 98, and references therein.

Printout of Spectrum 54 of 1-(2-thienoyl)-3,3,3-trifluoroacetone

Address	Frequency	ppm	Integral	Intensity
210	3,671.770	183.588	21,438.400	1,581.184
220	3,667.192	183.359	1,227.264	685.184
611	3,488.206	174.410	1,617.728	445.248
685	3,454.332	172.716	4,601.472	1,180.736
690	3,452.043	172.602	3,675.328	1,093.696
765	3,417.711	170.885	4,809.280	1,256.512
770	3,415.422	170.771	3,455.552	1,195.264
845	3,881.090	169.054	907.072	330.944
850	3,378.801	168.940	1,298.432	472.576
2,074	2,818.498	140.924	16,978.688	4,527.296
2,090	2,811.174	140.558	19,097.152	5,227.264
2,099	2,807.054	140.352	17,569.088	4,996.352
2,114	2,800.188	140.009	24,667.200	6,013.248
2,126	2,794.694	139.734	3,537.472	648.704
2,135	2,790.574	139.528	3,961.344	960.576
2,147	2,785.081	139.254	2,492.928	581.696
2,150	2,783.708	139.185	2,312.448	981.248
2,155	2,781.419	139.070	2,221.376	825.344
2,207	2,757.615	137.880	17,079.552	5,108.672
2,219	2,752.122	137.606	15,000.256	4,980.480
2,227	2,748.460	137.423	15,882.048	5,580.672
2,239	2,742.967	137.148	14,422.336	5,328.896
2,387	2,675.218	133.760	16,730.048	5,572.800
2,396	2,671.098	133.554	31,682.176	7,481.088
2,405	2,666.978	133.348	15,937.152	6,069.312
2,483	2,631.273	131.563	20,329.344	5,116.480
2,492	2,627.153	131.357	1,440.192	303.168
2,499	2,623.948	131.197	17,438.400	4,518.016
2,507	2,620.286	131.014	19,258.432	5,107.584
2,523	2,612.962	130.648	16,244.992	4,694.848
2,577	2,588.243	129.412	18,593.792	5,372.544
2,589	2,582.750	129.137	15,097.344	5,319.104
2,597	2,597.088	128.954	16,737.536	5,302.080
2,609	2,573.594	128.679	14,781.824	5,359.424
2,763	2,503.099	125.154	20,678.848	6,442.304
2,767	2,501.268	125.063	8,204.800	2,604.096
2,771	2,499.437	124.971	30,719.360	8,611.328
2,780	2,495.317	124.765	15,029.696	5,566.912
3,375	2,222.948	111.147	6,124.544	1,661.568
3,379	2,221.116	111.055	4,688.960	1,988.608
3,947	1,961.107	98.055	11,782.784	2,264.320
3,953	1,958.360	97.918	21,604.224	5,974.336
3,959	1,955.614	97.780	21,455.808	5,837.760
3,964	1,953.325	97.666	6,131.520	2,076.160
3,975	1,948.289	97.414	796.928	275.200
3,987	1,942.796	97.139	2,673.920	705.984
3,991	1,940.965	97.048	1,405.696	710.336
4,038	1,919.450	95.972	816.960	281.472
4,312	1,794.023	89.701	5,456.000	439.552

Printout of Spectrum 54 of 1-(2-thienoyl)-3,3,3-trifluoroacetone (continued)

Address	Frequency	ppm	Integral	Intensity
4,317	1,791.734	89.586	8,583.360	2,345.600
4,323	1,788.988	89.449	21,299.904	6,136.512
4,329	1,786.241	89.312	21,252.736	6,019.008
4,335	1,783.494	89.174	7,377.408	2,169.792
4,786	1,577.043	78.852	9,847.744	1,086.336
4,856	1,545.000	77.250	6,690.048	1,128.832
4,926	1,512.956	75.647	5,262.592	1,125.888
6,526	780.534	39.026	9,805.376	316.544
8,055	80.614	4.030	865.600	348.544
8,198	15.153	0.757	1,113.792	295.808
8,222	4.167	0.208	2,444.544	266.496

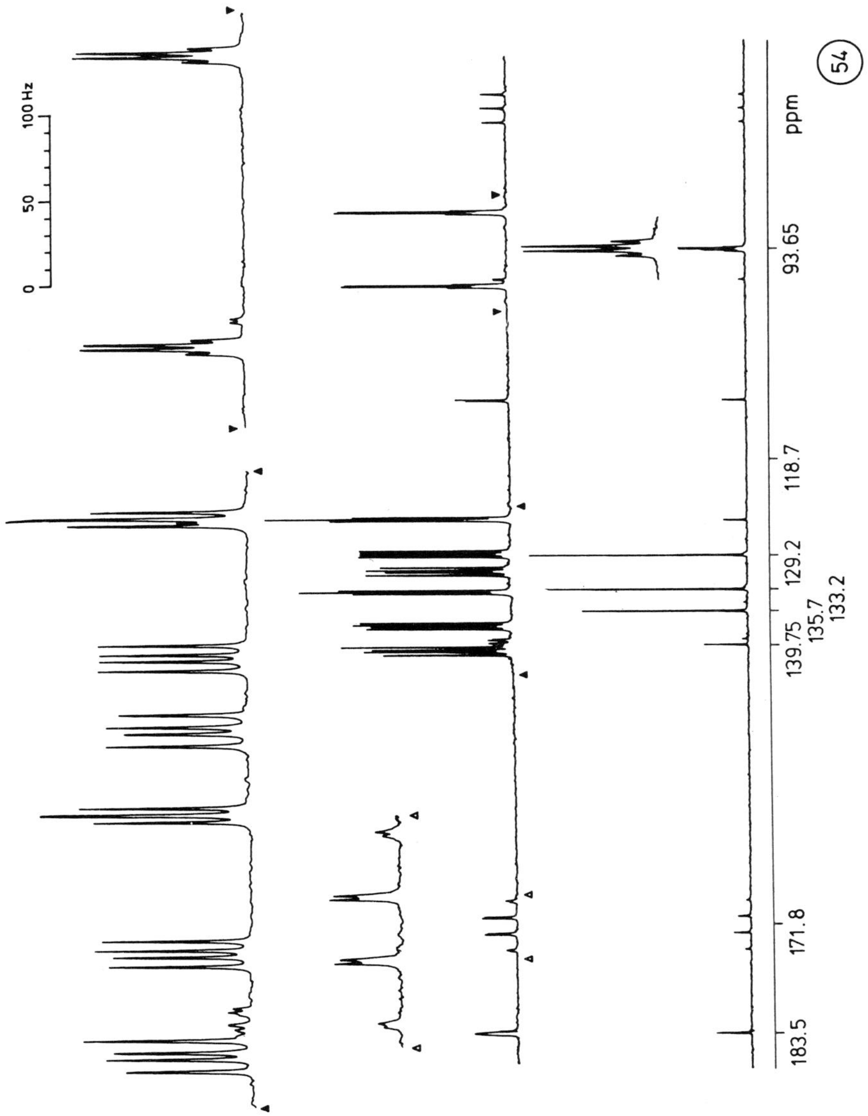

0
50
100 Hz
ppm
93.65
118.7
129.2
133.2
135.7
139.75
171.8
183.5
54

55

^{13}C NMR Spectrum 55 was recorded for 3-methyl-5-oxo-1-phenyl-4,5-dihydropyrazole in dimethylsulfoxide. Try to assign all the signals and couplings in this spectrum.

Recording Conditions

Solvent:	Hexadeuteriodimethylsulfoxide (septet at 39.7 ppm).
Concentration:	Saturated solution.
Temperature:	30° C.
Decoupling:	A. Decoupling and measurement frequencies pulsed alternately (gated decoupling, 5,000 scans). A′. Expanded portion of Spectrum A: sp^2 carbon region. B. Proton noise-decoupled only during the recording of the pulse interferograms (inverse gated decoupling, 3,000 scans).
Remarks:	In both experiments, a reduced pulse-width (5μs instead of 8.5 μs for the 90° pulse) and a waiting time of 5 s between individual FIDs were used to permit a better comparison of the signal intensities.

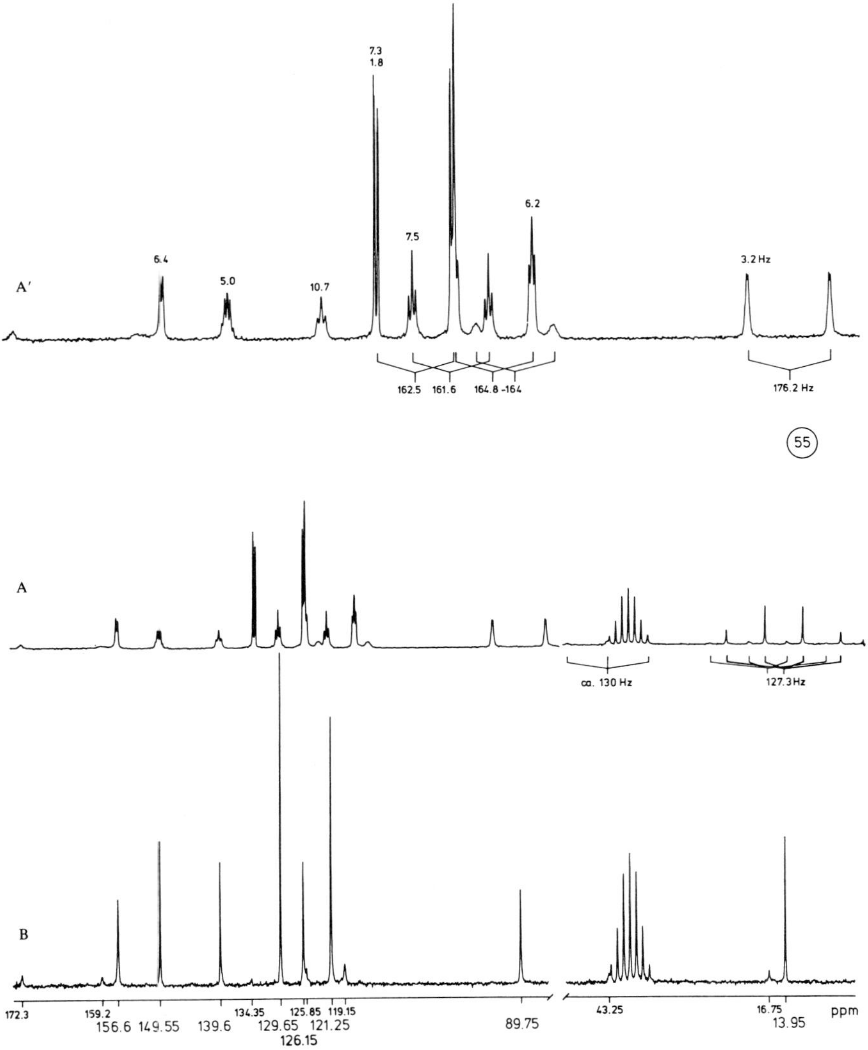

A′
6.4
5.0
10.7
7.3
1.8
7.5
6.2
3.2 Hz
162.5 161.6 164.8 -164
176.2 Hz
A
ca. 130 Hz
127.3 Hz
B
172.3
159.2
156.6 149.55
139.6
134.35
129.65
126.15
125.85
121.25
119.15
89.75
43.25
16.75
13.95
ppm

55

56

A crystalline product with ^{13}C NMR Spectrum 56 is formed when 5-amino-3-methyl-1-phenylpyrazole reacts with dimethylformamide and phosphorus oxychloride. What is the product's structure?

Recording Conditions

Solvent:	Tetradeuteriomethanol (septet at 49.0 ppm).
Concentration:	Saturated solution.
Temperature:	About 30° C.
Operating frequency:	15.08 MHz.
Decoupling:	*Top:* Proton off-resonance-decoupled (2,000). *Bottom:* Proton noise-decoupled (500 scans).

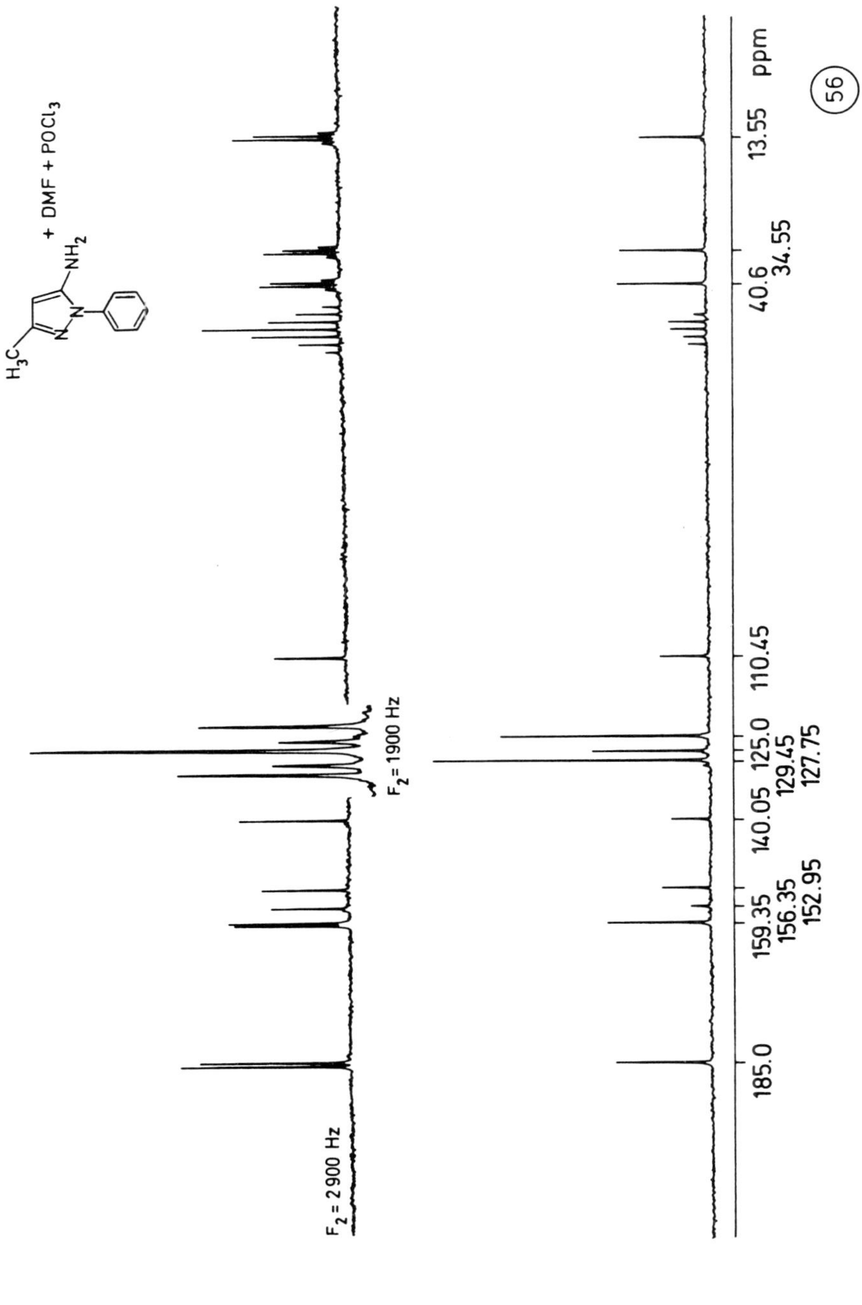
H3C
NH2
+ DMF + POCl3
F2 = 2900 Hz
F2 = 1900 Hz
185.0
159.35
156.35
152.95
140.05
129.45
127.75
125.0
110.45
40.6
34.55
13.55
ppm
56

57

What reaction product fits Spectrum 57? It is formed by reaction of 2-amino-2-methylpropanoic acid with trifluoroacetic anhydride, and has the formula $C_6H_6O_2NF_3$.

Recording Conditions

Solvent:	Deuteriochloroform (triplet at 77.25 ppm).
Concentration:	50% v/v.
Temperature:	28° C.
Operating frequency:	22.63 MHz.
Decoupling:	Proton noise-decoupled (500 scans).

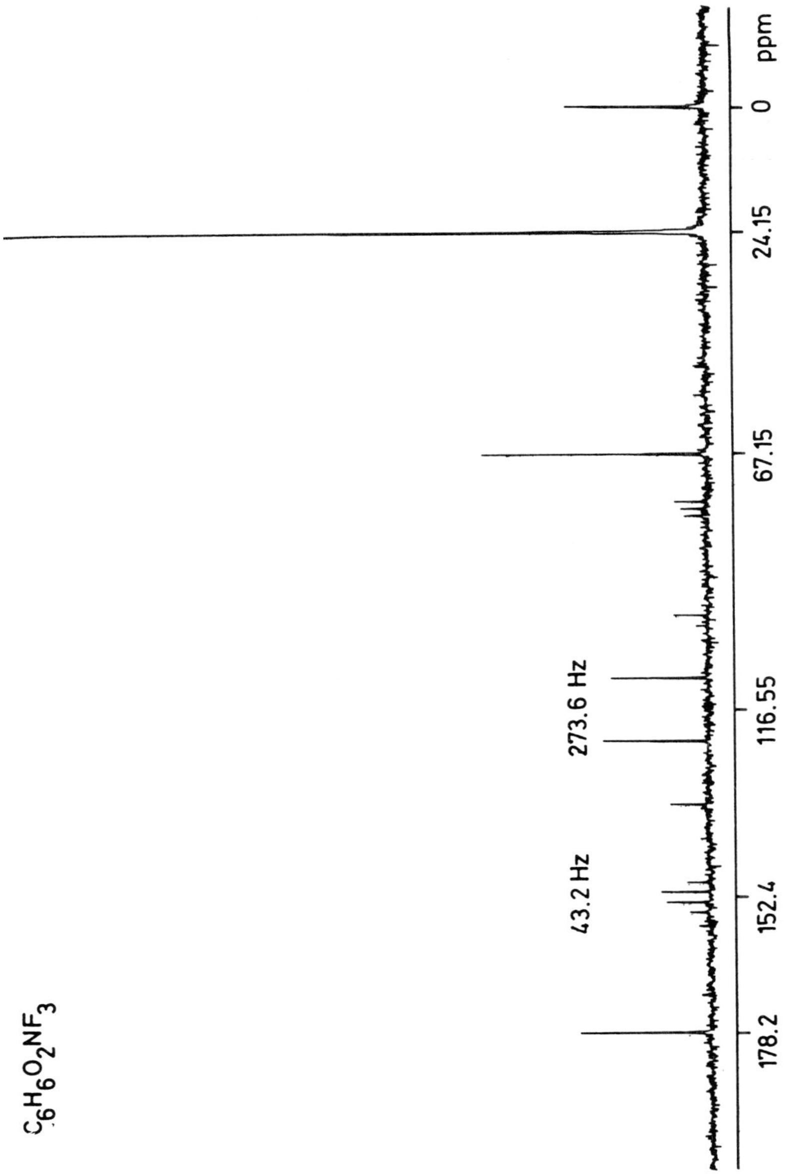

57

58

2-Aminothiophenol and 4,4′-dimethoxybenzil (^{13}C chemical shifts: 194.35, 165.5, 132.75, 114.65, and 55.6 ppm) react in ethanol solution affording a product with formula $C_{22}H_{19}NO_3S$. What is this product's structure, according to ^{13}C NMR Spectrum 58?

Recording Conditions

Solvent:	Deuteriochloroform (triplet at 77.25 ppm).
Concentration:	Saturated solution.
Temperature:	30° C.
Operating frequency:	20 MHz.
Decoupling:	Proton noise-decoupled (2,000 scans). *Top:* With low power (0.05 W, "low-power noise decoupling"). *Bottom:* With high power (5.0 W).

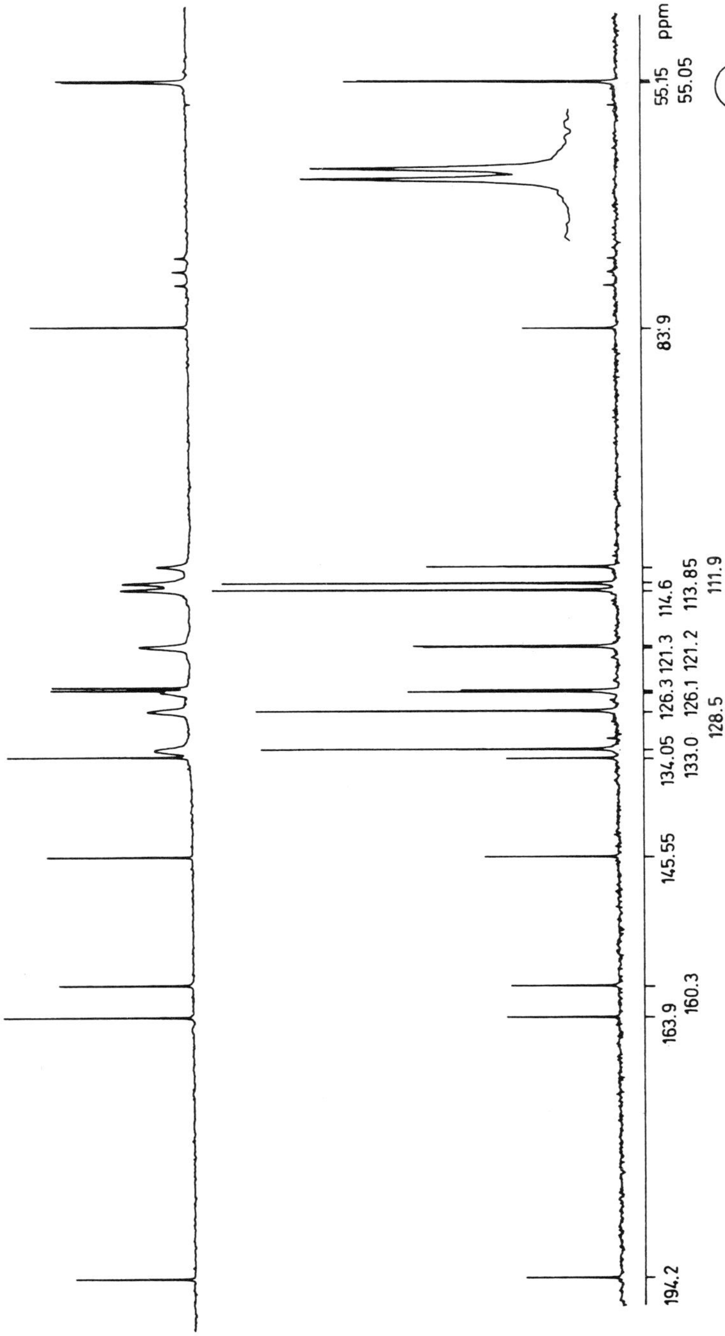

ppm
55.15
55.05
83.9
114.6
113.85
111.9
121.3
121.2
126.3
126.1
128.5
134.05
133.0
145.55
160.3
163.9
194.2

58

59

A T_1 measurement on pyridine in deuterium oxide, using the inversion recovery method, gave Spectral Series 59. Determine the T_1 values, and try to explain the result, considering the interaction between pyridine and water. In this connection, comment briefly on the T_1 measurement depicted on the cover of this book.

Recording Conditions for Spectral Series 59

Solvent:	Deuterium oxide.
Concentration:	75% v/v.
Temperature:	30° C.
Operating frequency:	15.08 MHz.
Decoupling:	Proton noise-decoupled (10 scans) T_1 experiment with 180°, τ, 90° pulse sequence. Spectrum series for τ = 1.0, 1.5, 2.0, 2.5, 3.0, 3.5, 4.0, 4.5, 5.0, 5.5, 6.0, 6.5, 7.0, 7.5, 8.0, 8.5, 9.0, 9.5, 10.0, 10.5, 11.0, 11.5, 12.0, and 50 s (for approximately complete relaxation). The waiting time between successive 180°, τ, 90° sequences was 35 s.

Recording Conditions for the Spectral Series Depicted on the Cover

Solvent:	Hexadeuteriobenzene (triplet at 128.7 ppm).
Concentration:	400 mg in 1 ml.
Temperature:	30° C.
Operating frequency:	15.08 MHz.
Decoupling:	Proton noise-decoupled (16 scans for each individual measurement). The waiting time between the 180°, τ, 90° pulse sequences, because of the slowly relaxing carbon atoms C-2 and C-2′, was 200 s. The measurement series was programmed for the following intervals: τ = 0.2, 0.4, 0.6, 0.8, 1.0, 1.2, 1.4, 1.6, 1.8, 2.0, 2.4, 2.8, 3.2, 3.6, 4.0, 4.5, 5.0, 5.5, 6.0, 6.5, 7.0, 7.5, 10.0, 12.5, 15.0, 17.5, 20.0, 22.5, 25.0, 27.5, 30.0, 32.5, and 35.0 s.

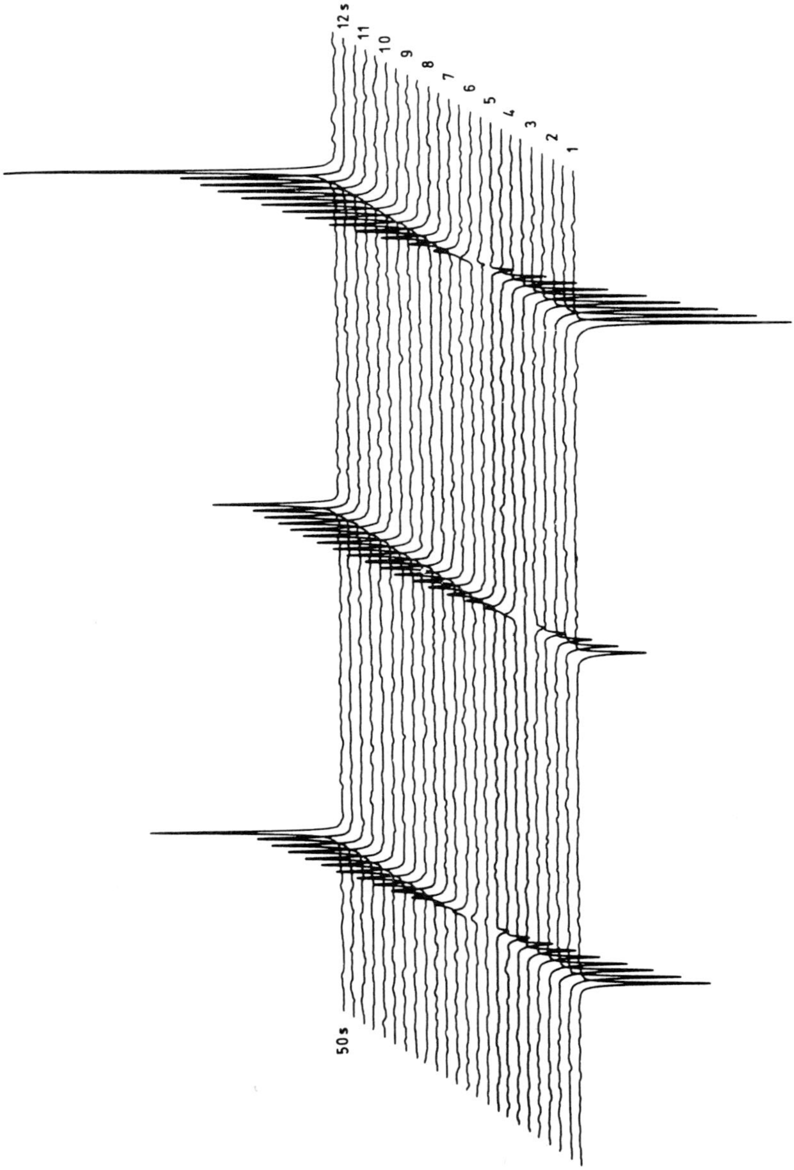

59

60

^{13}C NMR Spectrum 60A was recorded for a compound with formula C_6H_7N. What heterocycle is this? What derivative with formula C_6H_7NO corresponds to Spectrum 60B? Assign both spectra and explain the chemical shift differences considering resonance formulae.

Recording Conditions for All Spectra

Solvent:	Deuteriochloroform (triplet at 77.25 ppm).
Concentration:	50%.
Temperature:	28° C.
Operating frequency:	22.63 MHz.
Decoupling:	*Top:* Proton off-resonance-decoupled (1,200 scans). *Bottom:* Proton noise-decoupled (400 scans).

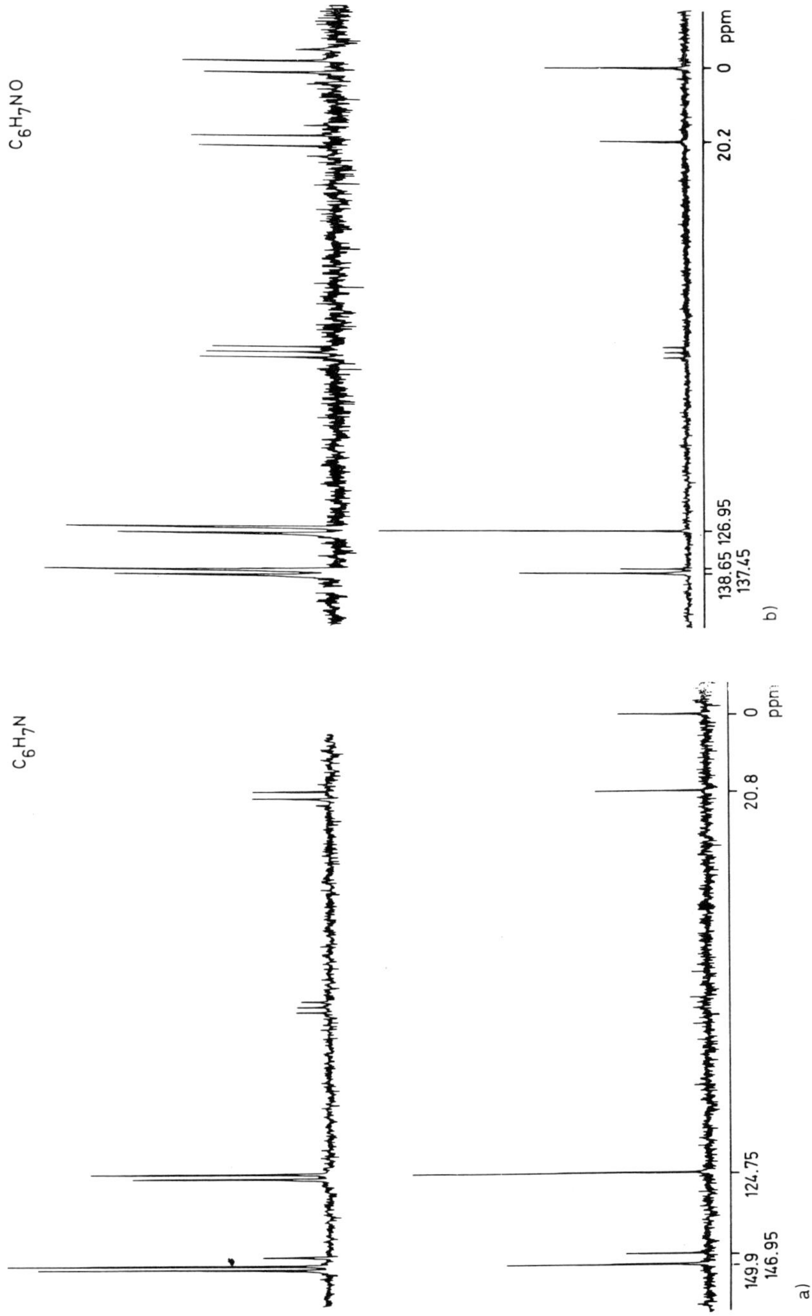

C6H7NO
138.65 126.95
137.45
20.2
0
ppm
b)
C6H7N
149.9
146.95
124.75
20.8
0
ppm
a)

(60)

61

^{13}C NMR Spectrum 61 was recorded for a disubstituted pyridine with formula $C_7H_5NO_4$. What pyridine derivative is this? Assign the spectrum. Use the coupling pattern of pyridine as a starting point.

C-H couplings in pyridine

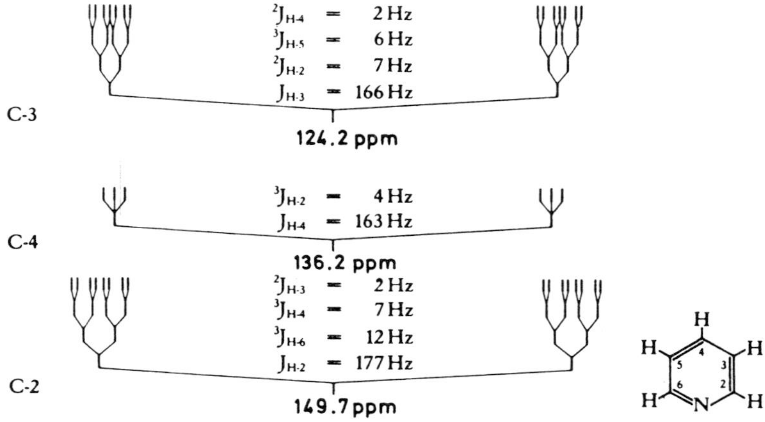

Recording Conditions

Solvent:	Hexadeuteriodimethylsulfoxide (septet at 39.7 ppm with J_{C-D} = 20.5 Hz, not shown).
Concentration:	Saturated solution.
Temperature:	28° C.
Operating frequency:	22.63 MHz.
Decoupling:	Coupled spectrum (all the compound's resonances are shown).

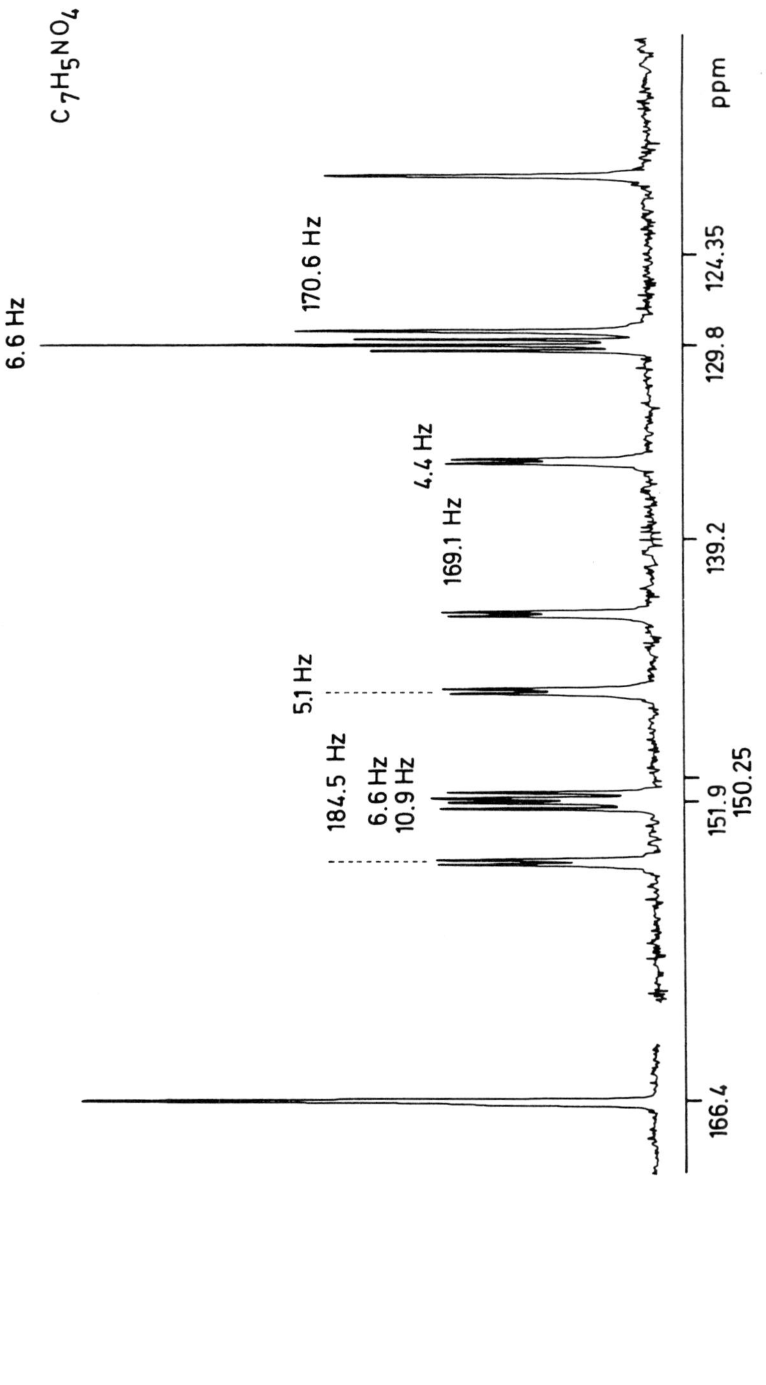
$C_7H_5NO_4$
6.6 Hz
170.6 Hz
4.4 Hz
169.1 Hz
5.1 Hz
184.5 Hz
6.6 Hz
10.9 Hz
ppm
124.35
129.8
139.2
151.9
150.25
166.4

61

62

What substituted pyridine with formula $C_5H_7N_3$ belongs to ^{13}C NMR Spectrum 62?

Recording Conditions

Solvent:	Hexadeuteriodimethylsulfoxide (septet at 39.7 ppm, not shown).
Concentration:	200 mg in 1 ml.
Temperature:	28° C.
Operating frequency:	22.63 MHz.
Decoupling:	*Top:* Proton off-resonance-decoupled (1,000 scans). *Bottom:* Proton noise-decoupled (500 scans).

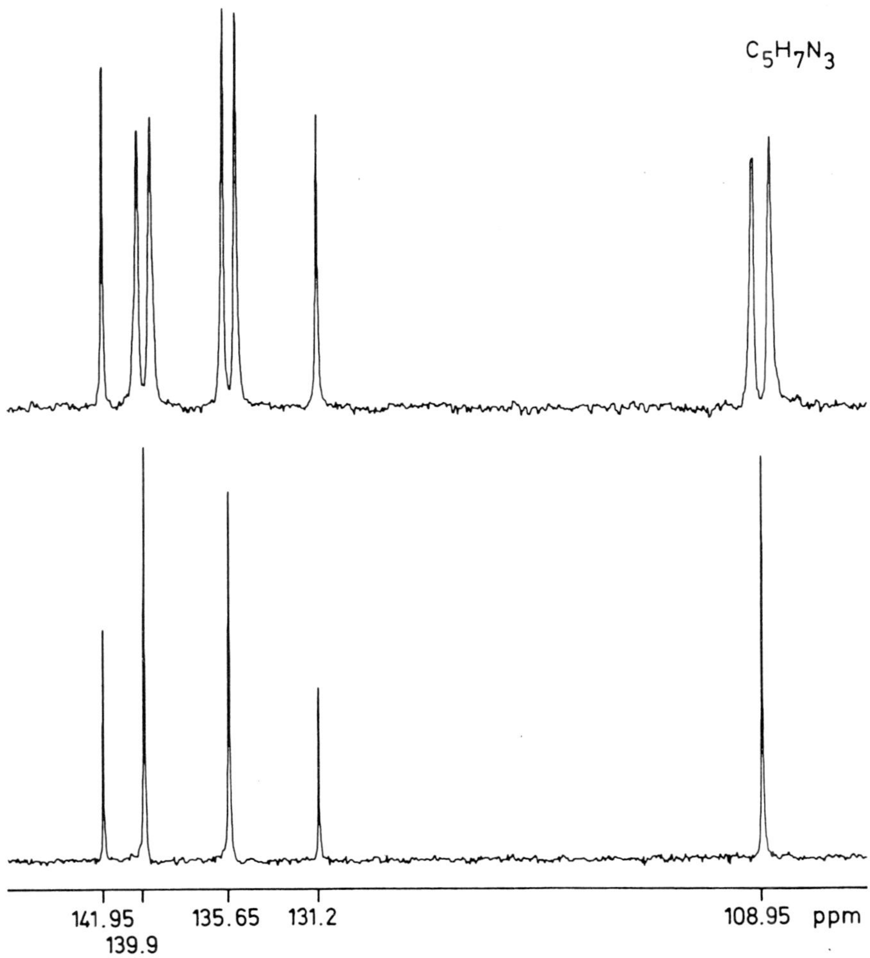

62

63

^{13}C NMR Spectra 63A and B were recorded for two pyrimidines with the same formula $C_4H_7N_5$. What pyrimidines are they? Use the chemical shifts and C-H couplings of unsubstituted pyrimidine, Fig. 8.2, p. 77.

Recording Conditions for Both Spectra

Solvent:	Hexadeuteriodimethylsulfoxide (septet at 39.7 ppm).
Concentration:	Saturated solutions.
Temperature:	28° C.
Operating frequency:	20 MHz.
Decoupling:	Coupled (20,000 scans each).

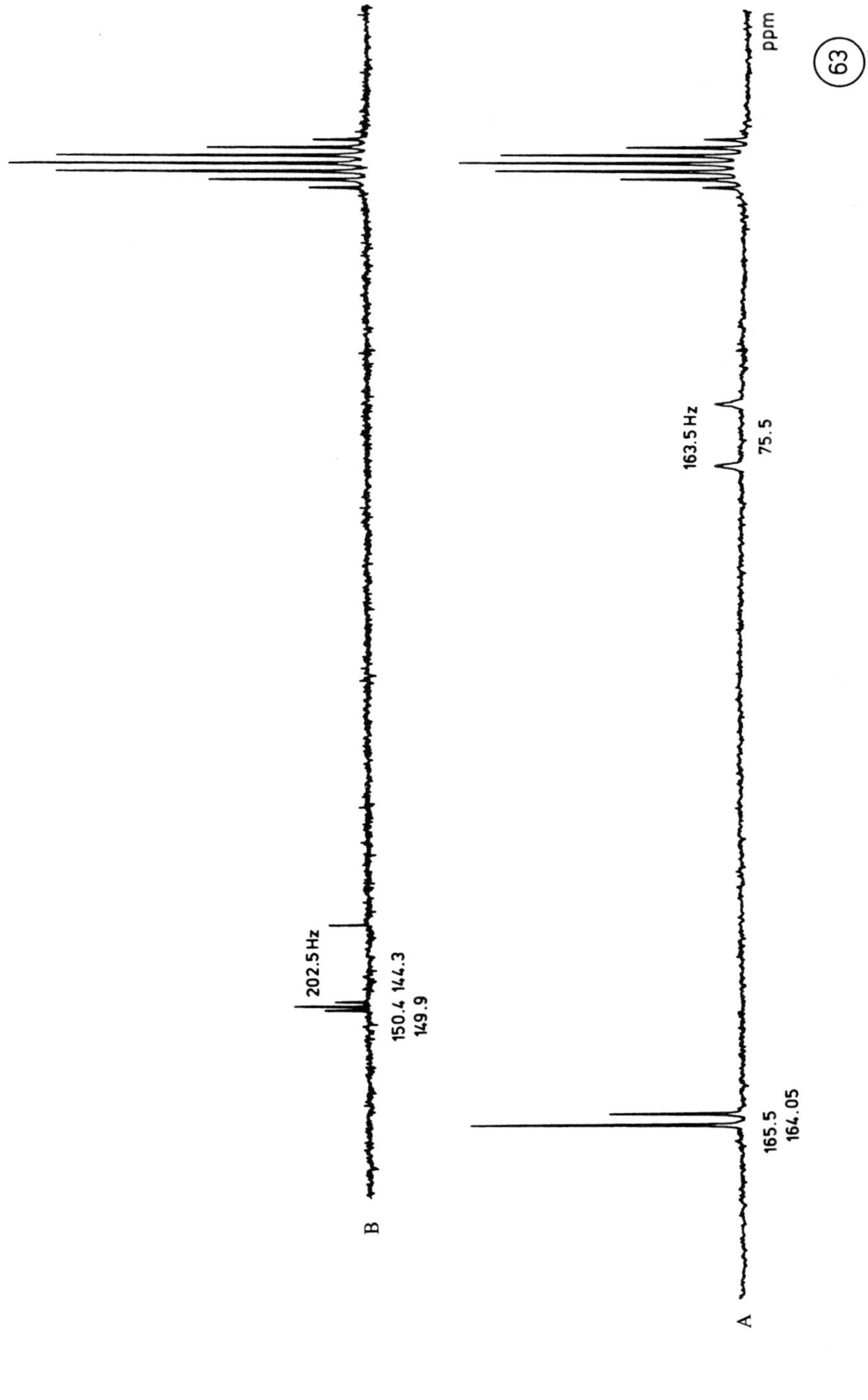

ppm
63
163.5 Hz
75.5
202.5 Hz
150.4 144.3
149.9
165.5
164.05
A
B

64

3-Amino-2-ethylacrolein (^{13}C chemical shifts: 190.55, 159.55, 117.8, 15.0, and 12.75 ppm in tetradeuteriomethanol) reacts with phenyl isothiocyanate affording a compound with formula $C_{12}H_{12}N_2OS$. Infer the structure of this product from ^{13}C NMR Spectrum 64.

Recording Conditions

Solvent:	Deuteriochloroform (triplet at 77.25 ppm).
Concentration:	Saturated solution.
Temperature:	30° C.
Operating frequency:	20 MHz.
Decoupling:	A. Decoupling and measurement frequencies pulsed alternately ("gated decoupling", 3,000 scans); an expanded portion of the spectrum limited by ▲ is shown above. B. Selective decoupling of the ethyl CH_2 protons with 0.05 W transmitter power (1,000 scans). C. Proton noise-decoupled only during the recording of the pulse interferograms ("inverse gated decoupling," 1,000 scans).
Remarks:	A waiting time of 40 s between exciting pulses was programmed to allow more complete relaxation of the nuclear spins. Slowly relaxing carbon nuclei give more intense signals under these conditions, and the signal intensities are more comparable. The selective decoupling of the ethyl CH_2 protons in Spectrum B is given as an assignment aid for the signals at 160.9 and 146.45 ppm.

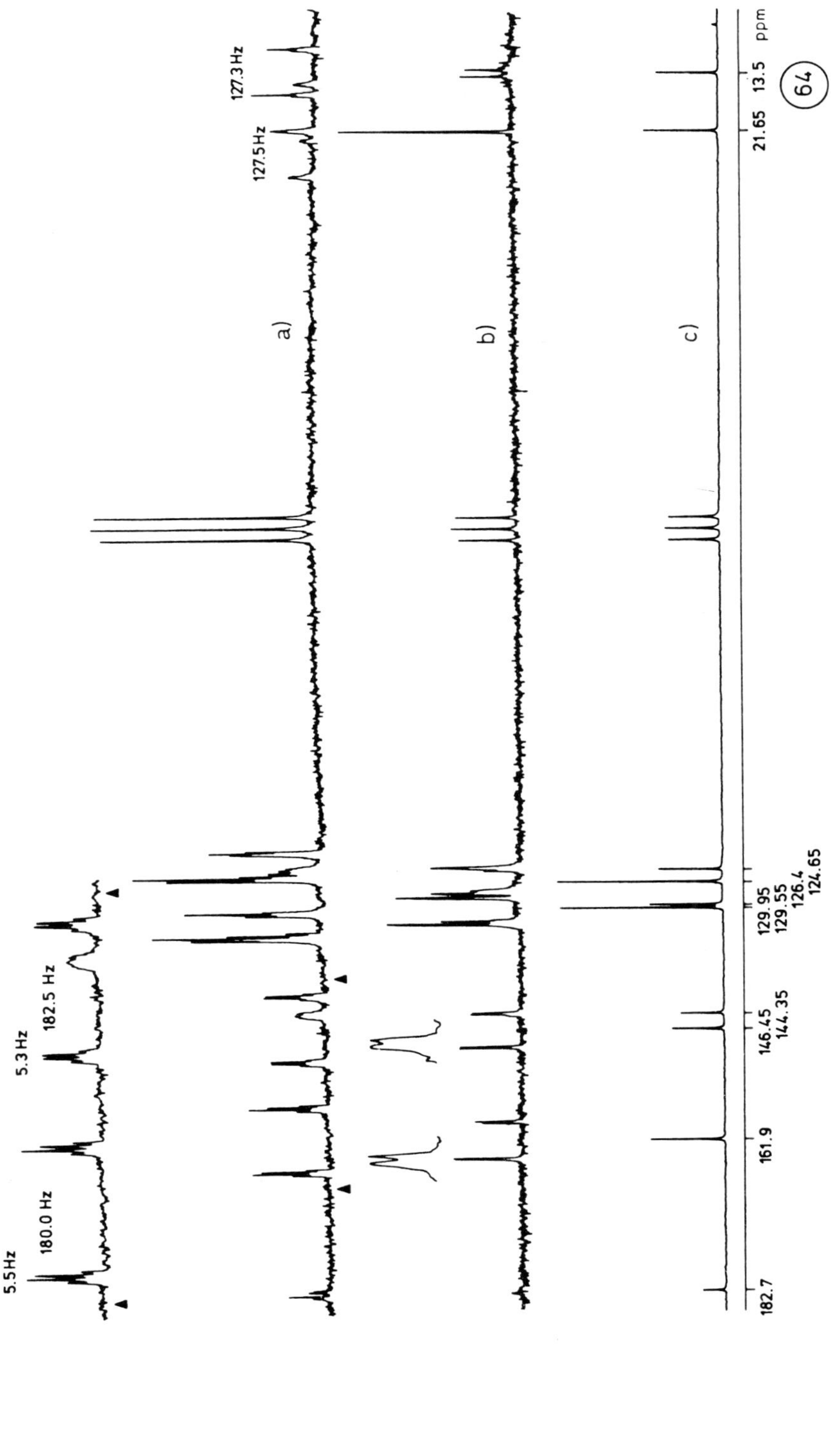

5.5 Hz
180.0 Hz
5.3 Hz
182.5 Hz
127.5 Hz
127.3 Hz
a)
b)
c)
182.7
161.9
146.45
144.35
129.95
129.55
126.4
124.65
21.65
13.5
ppm
64

65

6-Amino-4-oxo-2-thioxo-1,2,3,4-tetrahydropyrimidine reacts with 2-hydroxymethylenecycloheptanone, giving a 4-oxo-2-thioxo-1,2,3,4-tetrahydropyrido[2,3-d]pyrimidine with ^{13}C NMR Spectrum 65. How is the seven-membered ring condensed with the pyrido[2,3-d]-pyrimidine system? Use the carbonyl and thiocarbonyl resonances of 6-amino-4-oxo-2-thioxo-1,2,3,4-tetrahydropyrimidine (in hexadeuteriodimethylsulfoxide) and the C-H couplings and chemical shifts in 6-methyl-1,2,3,4-tetrahydropyrido[2,3-d]pyrimidine-2,4-dione for the signal assignment.

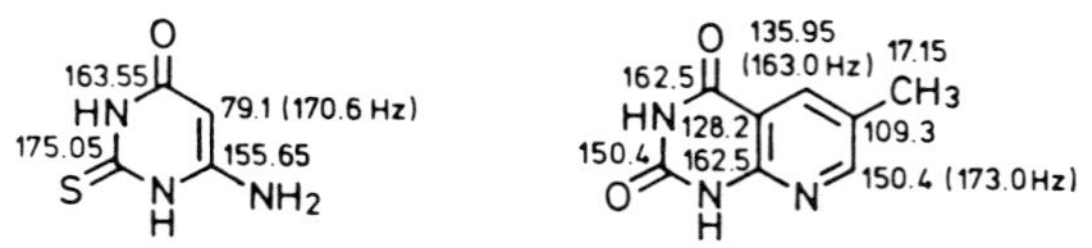

Recording Conditions

Solvent:	Hexadeuteriodimethylsulfoxide (septet at 39.7 ppm).
Concentration:	Nearly saturated solution.
Temperature:	30° C.
Operating frequency:	20 MHz.
Decoupling:	A. Decoupling and measuring frequencies pulsed alternately ("gated decoupling," 10,000 scans); the split signals are shown above, expanded by a factor of 2.
	B. Proton noise-decoupled (2,500 scans).

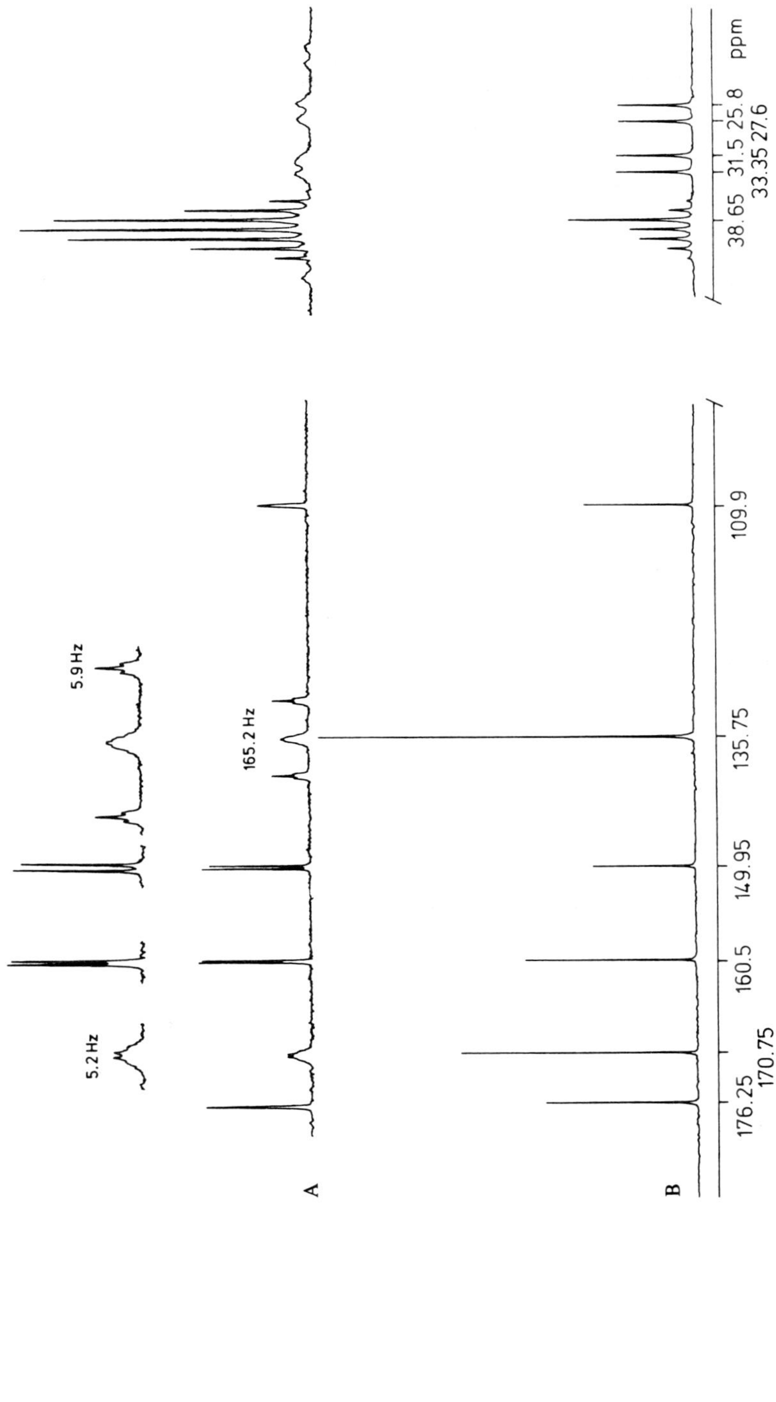
A
B
5.2 Hz
4.1 Hz
7.3 Hz
5.9 Hz
165.2 Hz
176.25
170.75
160.5
149.95
135.75
109.9
38.65
33.35
31.5
27.6
25.8
ppm

(65)

66

The reaction of potassium cyanide with carbon disulfide in dimethylformamide affords a product with the elementary composition $C_8N_4S_3$, after iodine oxidation, involving the intermediates sodium cyanodithioformate (A) and disodium cis-1,2-dicyanoethene-1,2-dithiolate (B):

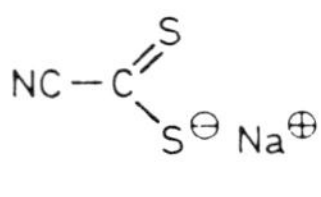

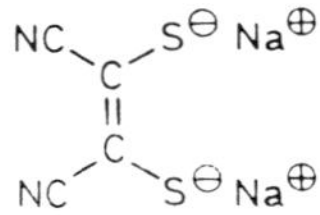

The end product is either a seven-membered heterocycle or one of two bicyclic sulfur-nitrogen ring systems.[1] Which of the three proposed structures fits ^{13}C NMR Spectrum 66 of this product?

Recording Conditions

Solvent:	1 ml ^{12}C-hexadeuteriodimethylsulfoxide (signal not visible).
Concentration:	60 mg in 1 ml (saturated solution).
Temperature:	25° C.
Operating frequency:	20 MHz.
Decoupling:	None (the compound has no protons).
Remarks:	Because of the expectedly long relaxation times, the interferograms were accumulated with a pulse width of 8.5 μs (90° pulse) and a waiting time of 20 s between pulses.

[1]See H. E. Simmons, R. D. Vest, D. C. Blomstrom, J. R. Roland, T. L. Cairns, *J. Am. Chem. Soc.* **84,** 4746 (1962).

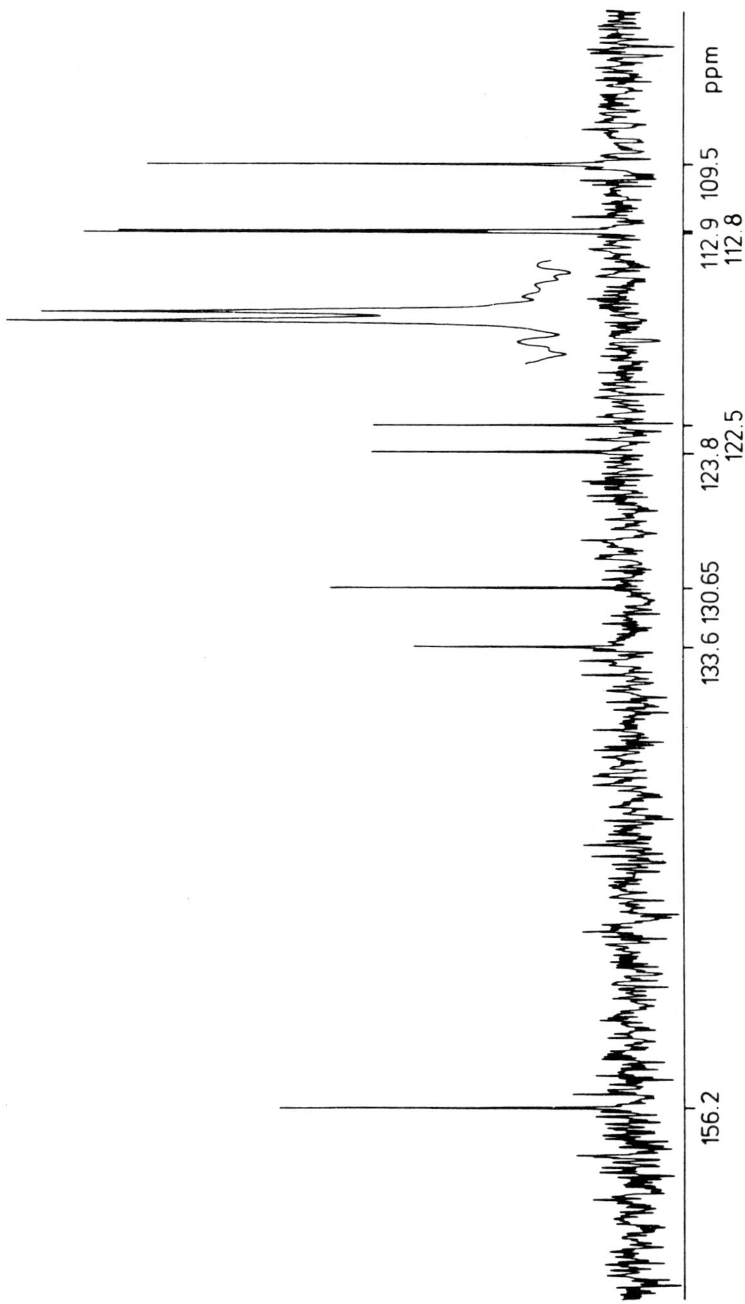

66

67

Spectrum 67 was recorded for a mixture of two ^{15}N-enriched amino acids (95% ^{15}N). One of the two amino acids is L-valine. Which signals correspond to this compound? What other amino acid is present in the mixture? Assign all the signals.

Recording Conditions

Solvent:	Distilled water.
Concentration:	100 mg in 1 ml.
External standard:	0.2 molar solution of sodium trimethylsilylpropane-sulfonate in deuterium oxide (0 ppm with respect to TMS); the signals of this standard are indicated in the spectrum by "St."
Temperature:	28° C.
Operating frequency:	22.63 MHz.
Decoupling:	Proton noise-decoupled (4,000 scans).

J

^{15}N-enriched amino acids
St. = standard signals

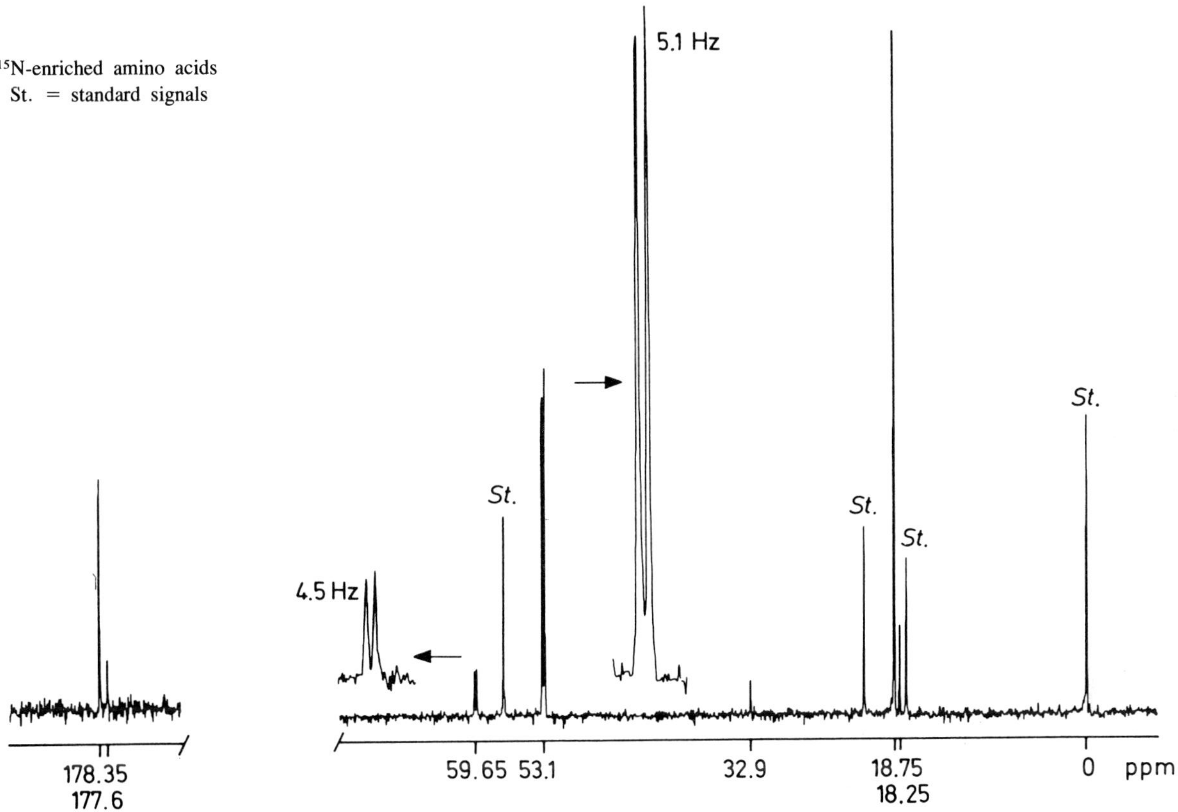

67

68

Which of the two ^{13}C NMR Spectra 68A and B corresponds to menthol, and which to isomenthol? Use the equatorial and axial methyl group increments found for methylcyclohexane.[1]

Recording Conditions for All Spectra

Solvent:	Deuteriochloroform (triplet at 77.25 ppm).
Concentration:	200 mg in 1 ml.
Temperature:	30° C.
Operating frequency:	20.0 MHz.
Decoupling:	A′-B′: Proton off-resonance-decoupled (500 scans).
	A-B: Proton noise-decoupled (100 scans).

[1]See Table 7.5 and D. K. Dalling, D. M. Grant, *J. Am. Chem. Soc.*, **89**, 6612 (1967) and **94,** 5318 (1972).

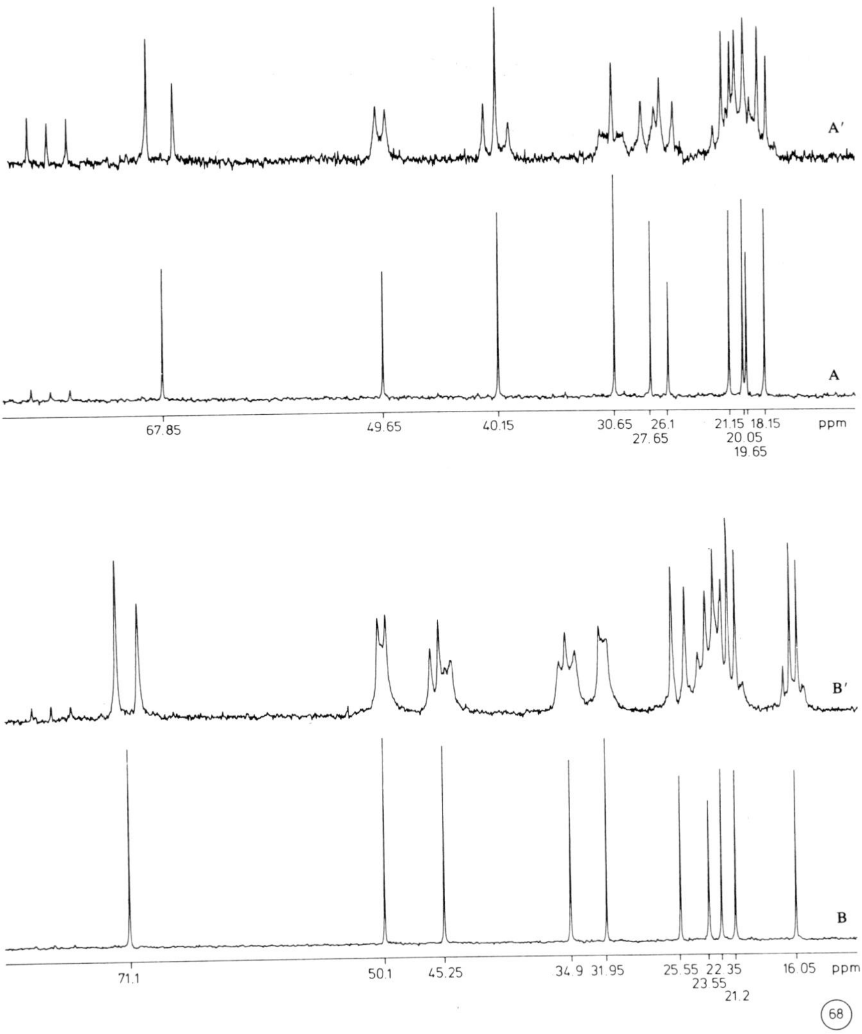
A′
A
67.85
49.65
40.15
30.65
26.1
27.65
21.15
18.15
20.05
19.65
ppm
B′
B
71.1
50.1
45.25
34.9
31.95
25.55
22.35
23.55
21.2
16.05
ppm

68

69

^{13}C NMR Spectrum 69 of D-xylose shows ten signals 48 hours after preparing the solution. What is the reason? Assign the spectrum quantitatively, assuming (as is possible under the conditions in which it was recorded) that the integrals of sufficiently separated signals (values above the resonances) may be compared.

Recording Conditions

Solvent:	Deuterium oxide.
Standard:	Methyl signal of sodium trimethylsilylpropanesulfonate (0 ppm, not shown).
Concentration:	1 molar.
Temperature:	30° C.
Operating frequency:	15.08 MHz.
Decoupling:	Proton noise-decoupled (1,000 scans).
Remarks:	The spectrum was recorded using the pulse sequence 90°, τ, 90°, τ, . . . for τ = 25 s, 48 hours after preparing the solution. Relaxation influences on the signal intensities are avoided by means of the waiting time between pulses. The signal assignment shown above the spectrum is based on D. E. Dorman, J. D. Roberts, J. Am. Chem. Soc., **92,** 1355 (1970).

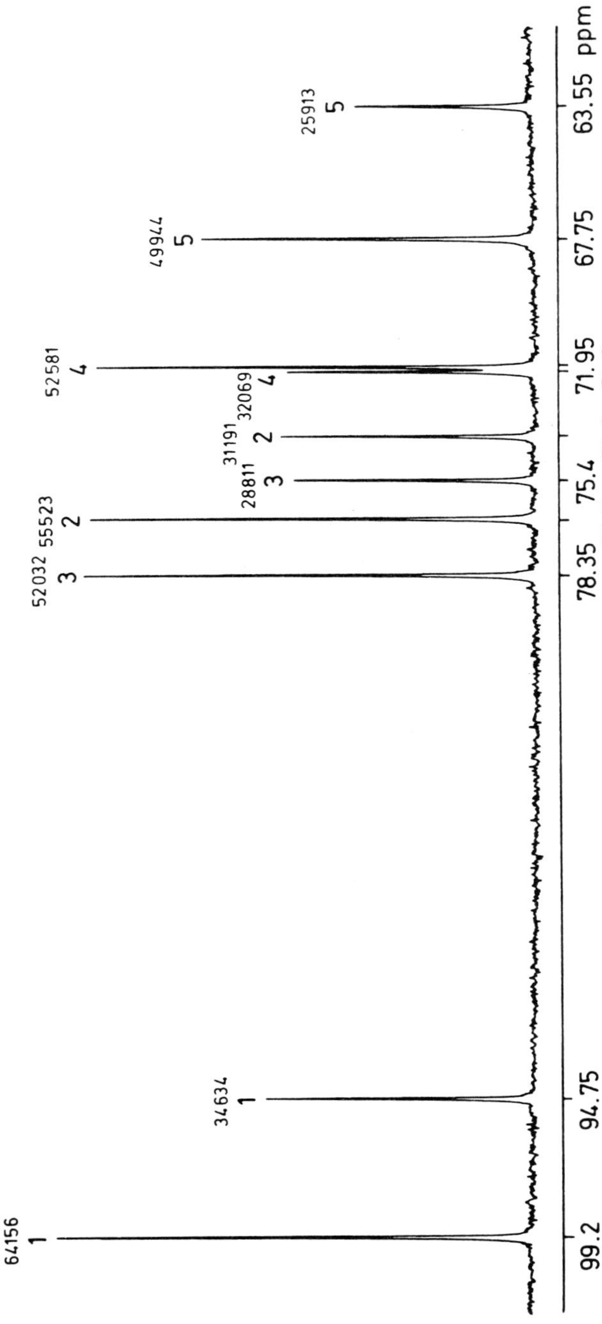

64156
1
34634
1
52032
3
55523
2
28811
3
31191
2
52581
4
32069
4
49944
5
25913
5
99.2
94.75
78.35
76.6
75.4
74.0
71.95
71.8
67.75
63.55 ppm

69

70

Assign the ^{13}C NMR spectra of adenosine (70A and B) and of adenosine-5′ monophosphate (70C). Use the ^{13}C chemical shifts of β-D-ribofuranose (Fig. 4.2, p. 33) and of adenine (below) for comparison.

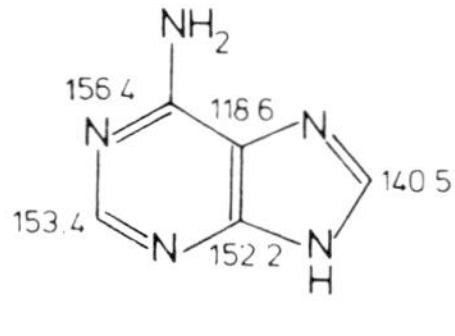

Recording Conditions

Solvent:	Hexadeuteriodimethylsulfoxide (septet at 39.7 ppm).
Concentration:	50 mg in 1 ml each time.
Temperature:	25° C.
Operating frequency:	20 MHz.
Decoupling:	A. Proton noise-decoupled (1,000 scans).
	B. Proton off-resonance decoupled (2,000 scans).
	C. Proton noise-decoupled.

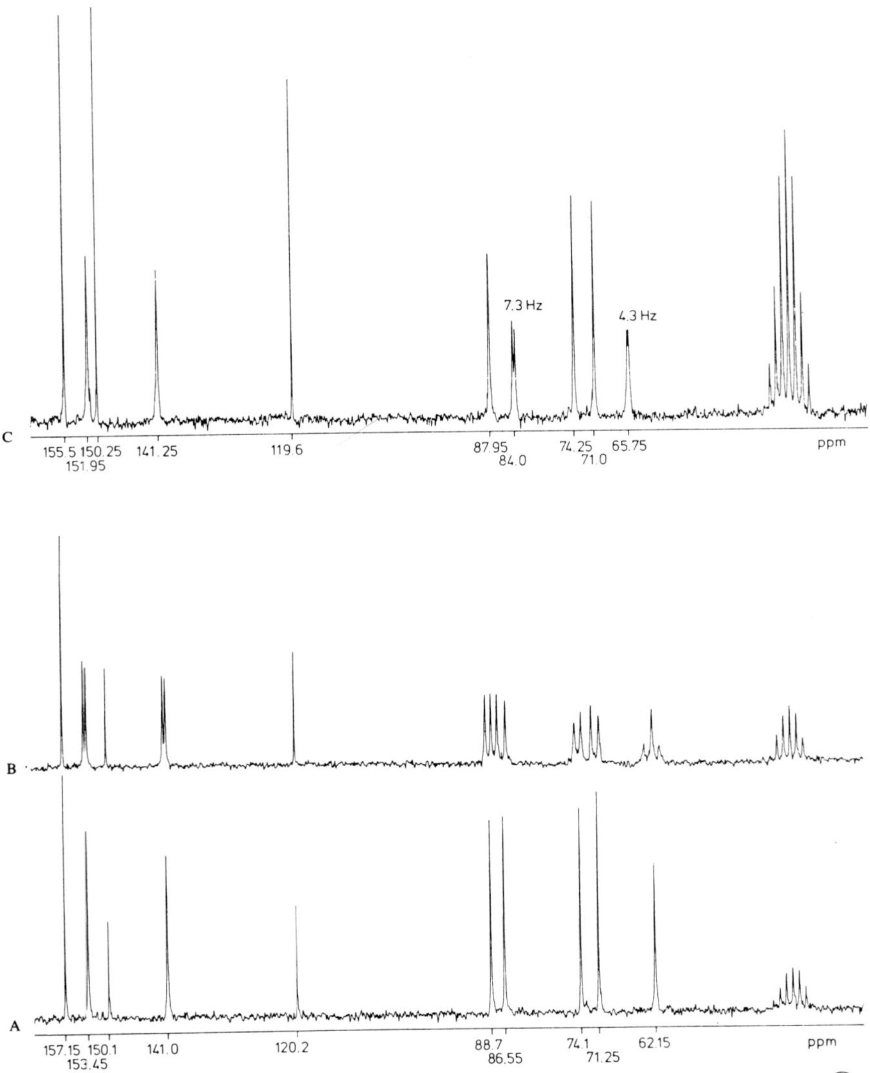
7.3 Hz
4.3 Hz
C
155.5 150.25 141.25 119.6 87.95 74.25 65.75 ppm
151.95 84.0 71.0
B
A
157.15 150.1 141.0 120.2 88.7 74.1 62.15 ppm
153.45 86.55 71.25

70

71

For what steroid derivative were ^{13}C NMR Spectra 71A-E recorded? Assign all ^{13}C chemical shifts and C-H coupling constants shown. The ^{13}C chemical shifts of some steroids may be used, in part, as assignment aids, as the compound in question belongs to the same class.[1]

Recording Conditions

Solvent:	Tetradeuteriomethanol (septet at 49.35 ppm).
Concentration:	Saturated solution.
Temperature:	25° C.
Operating frequency:	20 MHz.
Decoupling:	A. Proton noise-decoupled (2,000 scans). B. Part of Spectrum A, $\delta < 60$ ppm. C. Proton off-resonance decoupled partial spectrum for $\delta < 60$ ppm (4,000 scans). D. Part of Spectrum A, $\delta > 60$ ppm. E. Coupled partial spectrum for $\delta > 60$ ppm (10,000 scans).

[1]See T. A. Wittstruck, U. I. H. Williams, *J. Org. Chem.*, **38**, 1542 (1973).

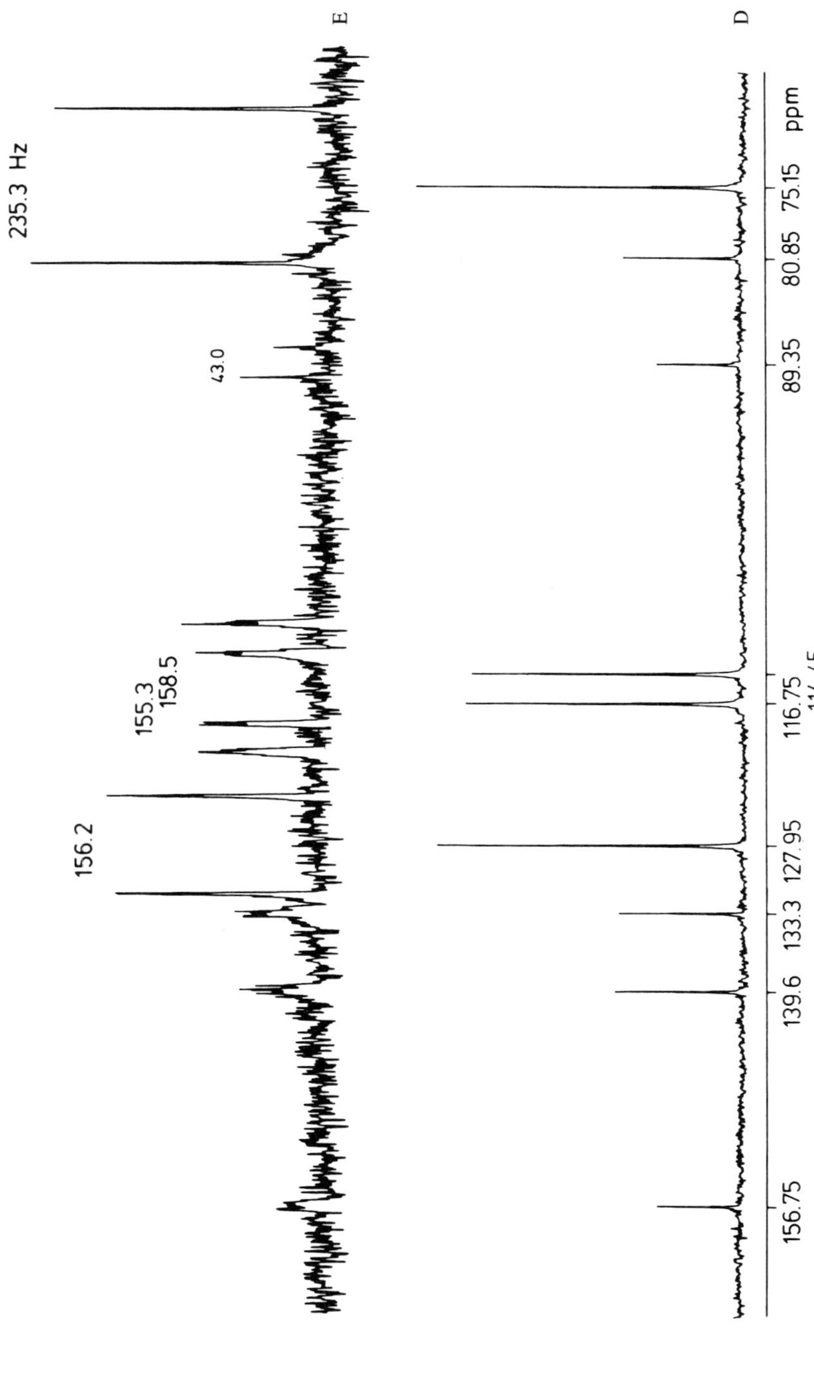
E
D
235.3 Hz
43.0
155.3
158.5
156.2
ppm
75.15
80.85
89.35
116.75
114.45
127.95
133.3
139.6
156.75

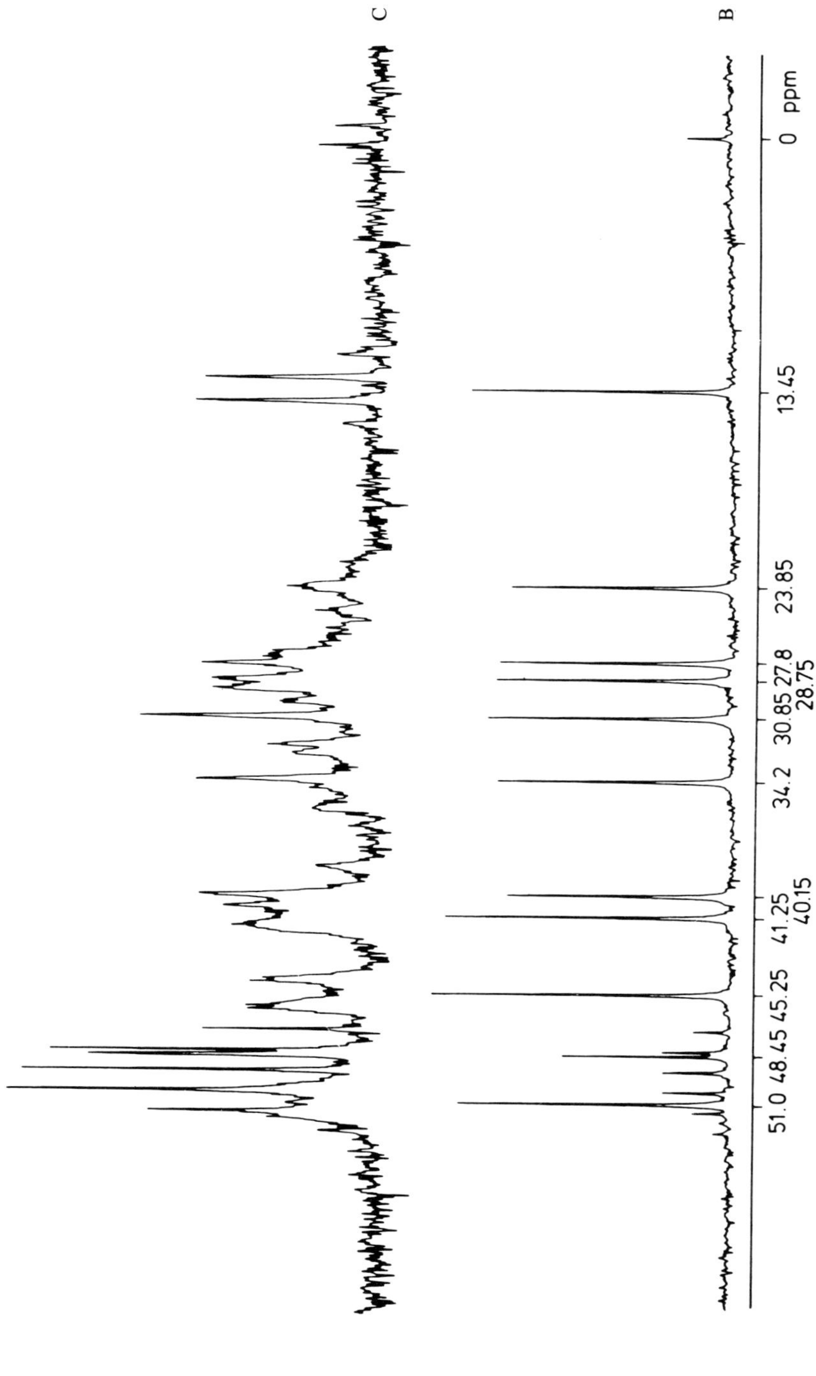
C
B
0 ppm
13.45
23.85
27.8
28.75
30.85
34.2
40.15
41.25
45.25
48.45
51.0

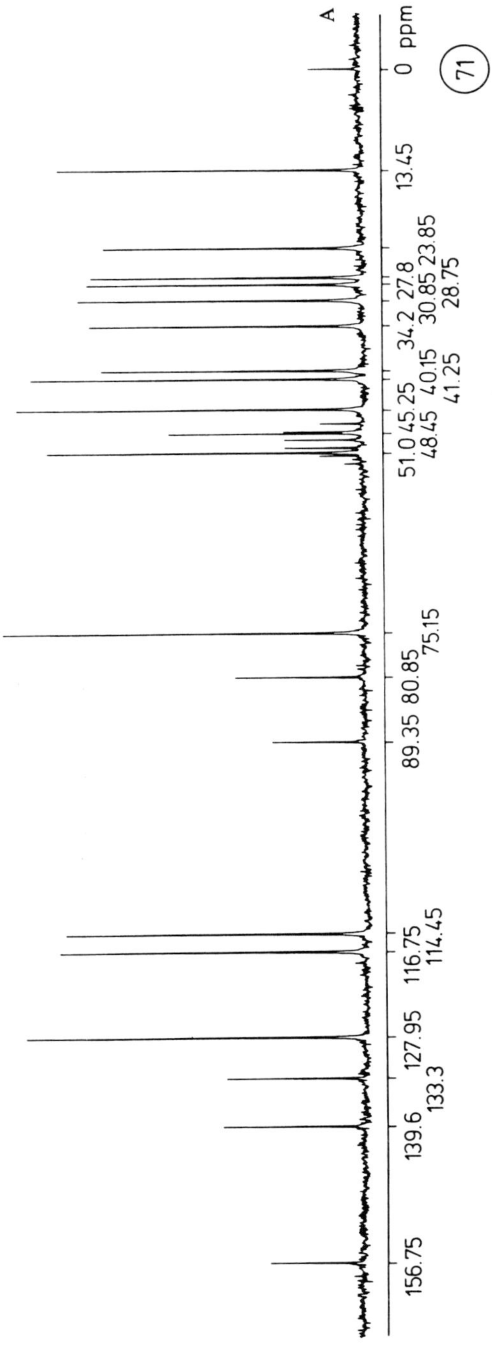
A
0 ppm
71
13.45
27.8
30.85 23.85
28.75
34.2
40.15
41.25
51.0 45.25
48.45
75.15
80.85
89.35
116.75
114.45
127.95
133.3
139.6
156.75

72

The proton resonances determined in ^{1}H NMR Spectrum 72A of the meadow-saffron alkaloid colchicine, $C_{22}H_{25}NO_6$, were noise-decoupled with high and low power in the ^{13}C NMR Spectra 72B and C. The aliphatic and aromatic protons were selectively decoupled in the series of spectra 72D-J and K-N, respectively.

Using Spectrum 72C, distinguish between carbon atoms with and without bonded hydrogen. By means of the chemical shifts, try to assign as many hydrogen-free carbon nuclei as possible. Then assign the signals of the carbon atoms bonded to hydrogen with the help of the decoupling experiments (D-N). Use the known assignment of the ^{1}H NMR spectrum of colchicine.[1]

Recording Conditions

Solvent:		Deuteriochloroform (triplet at 77.25 ppm).
Concentration:		200 mg in 1 ml (the same solution for ^{1}H and ^{13}C NMR spectra).
Temperature:		30° C.
Operating frequency:	A.	80 MHz (^{1}H).
	B-N.	20 MHz (^{13}C).
Decoupling:	A.	Coupled (1 scan).
	B.	Proton noise-decoupled, high power (2,000 scans).
	C.	Proton noise-decoupled, low power (4,000 scans).
	D-N.	Selective decoupling of the proton resonances shown in Spectrum A (F_2 = 5539 to 5997 Hz; 1,000 scans each experiment; transmitter power 0.05 W; no noise modulation).

[1]See N. S. Bhacca, D. P. Hollis, L. F. Johnson, E. A. Pier, *NMR Spectra Catalog*, Varian Associates, 1963, Spectrum 689.

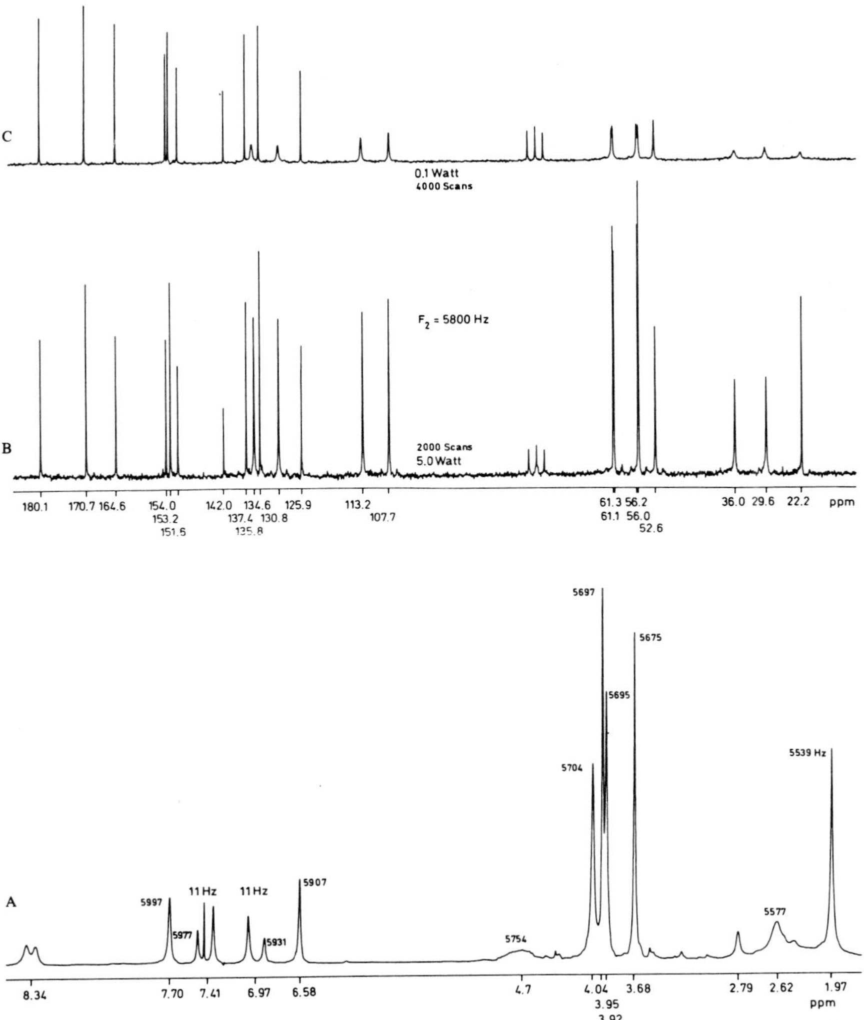
C
0.1 Watt
4000 Scans
F2 = 5800 Hz
B
2000 Scans
5.0 Watt
180.1 170.7 164.6 154.0 153.2 151.6 142.0 134.6 137.4 135.8 130.8 125.9 113.2 107.7 61.3 61.1 56.2 56.0 52.6 36.0 29.6 22.2 ppm
5697
5675
5695
5704
5539 Hz
A
5997
5977
11 Hz
11 Hz
5931
5907
5754
5577
8.34 7.70 7.41 6.97 6.58 4.7 4.04 3.95 3.92 3.68 2.79 2.62 1.97
ppm

72

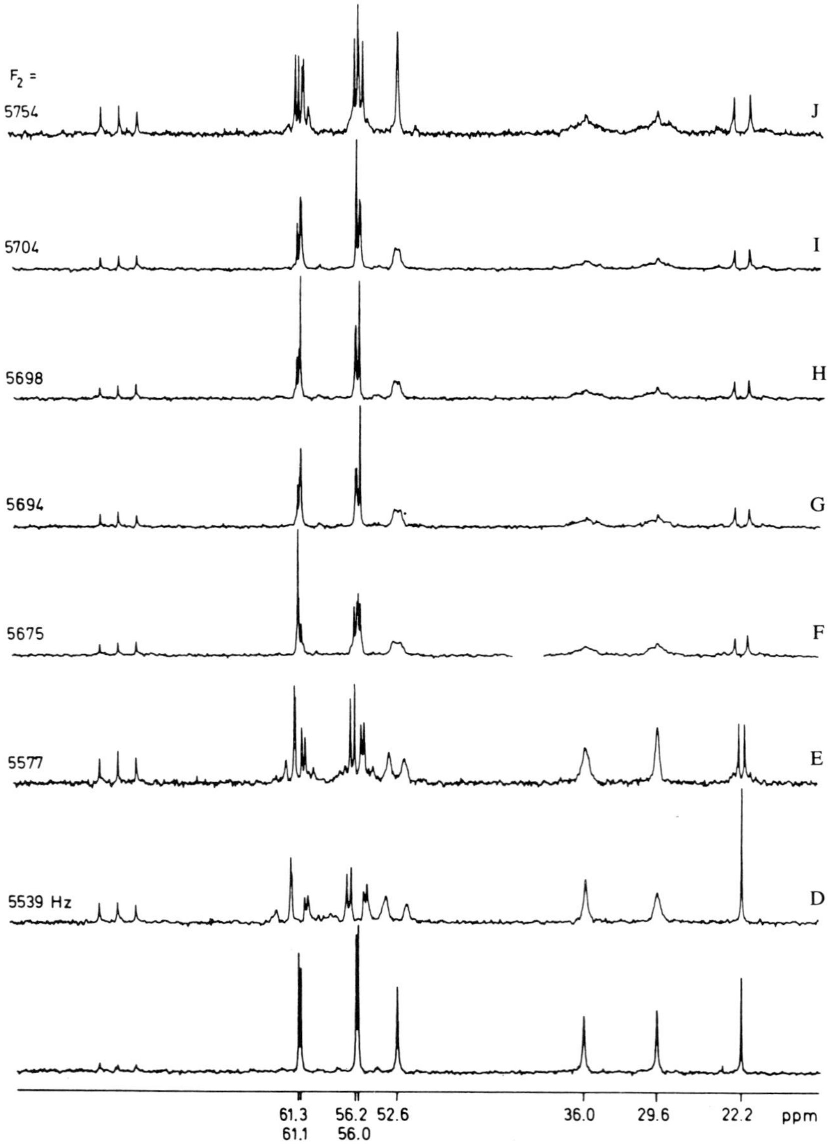
F2 =
5754
J
5704
I
5698
H
5694
G
5675
F
5577
E
5539 Hz
D
61.3
61.1
56.2
56.0
52.6
36.0
29.6
22.2
ppm

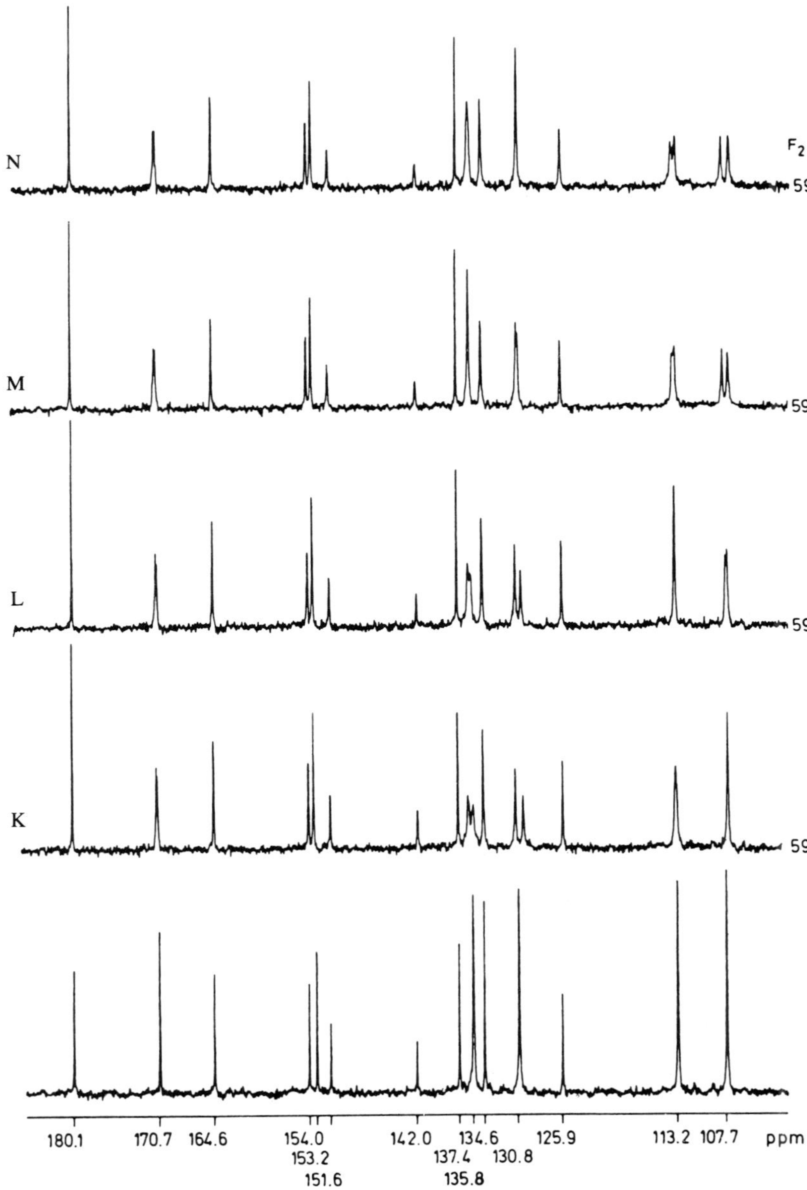
N
M
L
K
F2 =
5997
5977
5931
5907 Hz
180.1
170.7
164.6
154.0
153.2
151.6
142.0
137.4
135.8
134.6
130.8
125.9
113.2
107.7
ppm

73

The following assignments have been published for the ^{13}C chemical shifts of methyl lysergate[1]:

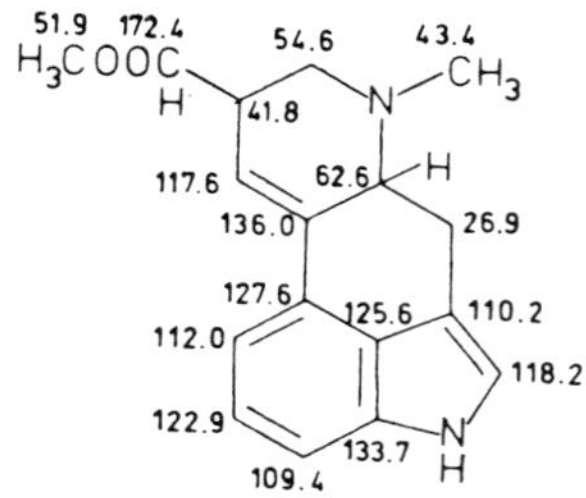

Analyse the ^{13}C NMR Spectrum 73 of lysergic acid diethylamide on the basis of this known assignment. Why do the N-ethyl groups give four signals?

Recording Conditions

Solvent:	Deuteriochloroform (triplet at 77.25 ppm).
Concentration:	15 mg in 1 ml.
Temperature:	30° C.
Operating frequency:	15.08 MHz.
Decoupling:	Proton noise-decoupled (45,000 scans).

[1]See N. J. Bach, H. E. Boaz, E. C. Kornfeld, C. J. Chang, H. G. Floss, E. W. Hagaman, E. Wenkert, *J. Org. Chem.*, **39**, 1272 (1974).

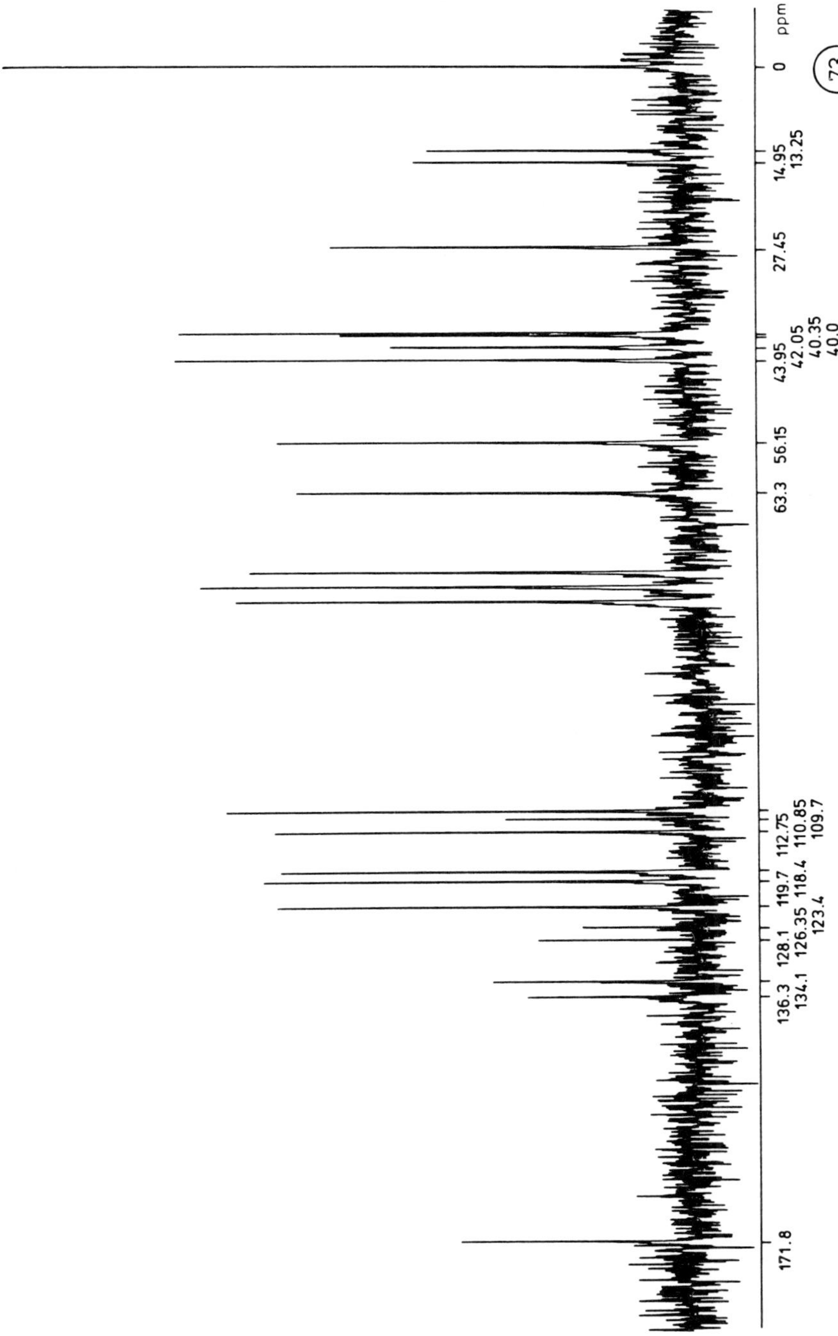
ppm
0
14.95
13.25
27.45
43.95
42.05
40.35
40.0
56.15
63.3
112.75
110.85
109.7
119.7
118.4
128.1
126.35
123.4
136.3
134.1
171.8

73

74

Upon acid and basic hydrolysis of the tropane alkaloid scopolamine,[1] scopoline is formed, with the elementary composition $C_8H_{13}O_2N$ and ^{13}C NMR Spectrum 74. What is the structure of scopoline? Does the N-methyl group have the syn- or anti-configuration?[2]

Recording Conditions

Solvent:	Deuteriochloroform (triplet at 77.25 ppm).
Concentration:	100 mg in 1 ml.
Temperature:	30° C.
Operating frequency:	15.08 MHz.
Decoupling:	*Top:* Proton off-resonance decoupled (4,000 scans). *Bottom:* Proton noise-decoupled (1,000 scans).

---

[1]See ^{13}C NMR spectrum and assignment in E. Wenkert, J. S. Bindra, C. J. Chang, D. W. Cochran, *Acc. Chem. Res.*, **7**, 46 (1974).

[2]See H. J. Schneider, L. Sturm, *Angew. Chem.*, **88**, 574 (1976).

$C_8H_{13}O_2N$

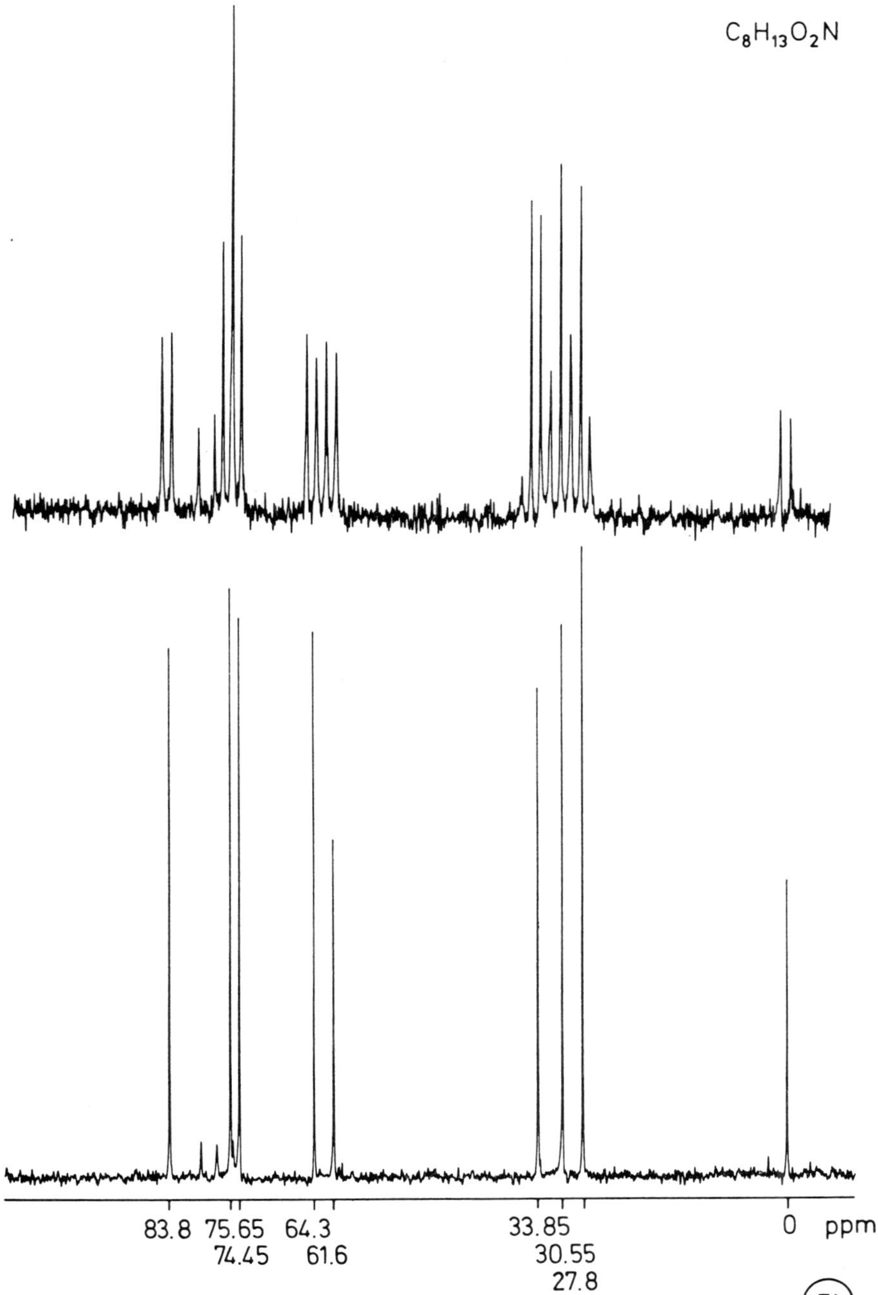

(74)

75

The hydrolysis of a glucoside tetraacetate isolated from birch bark, with the formula $C_{24}H_{36}O_{11}$ and ^{13}C NMR Spectrum 75A, yields an aglycone with the formula $C_{10}H_{18}O_2$ and ^{13}C NMR Spectrum 75B. Elucidate the structure of the aglycone using its formula and Spectrum B. Then, comparing chemical shifts of appropriate model compounds with Spectrum A, try to establish how the glucoside is joined together.

Recording Conditions

Solvent:	Deuteriochloroform (triplet at 77.25 ppm).
Concentration:	A. 80 mg in 1 ml.
	B. 20 mg in 1 ml.
Temperature:	30° C.
Operating frequency:	A. 15.08 MHz.
	B. 20.0 MHz.
Decoupling:	A. Proton noise-decoupled (12,000 scans).
	B. *Top:* Proton off-resonance decoupled (100,000 scans); *Bottom:* Proton noise-decoupled (20,000 scans).

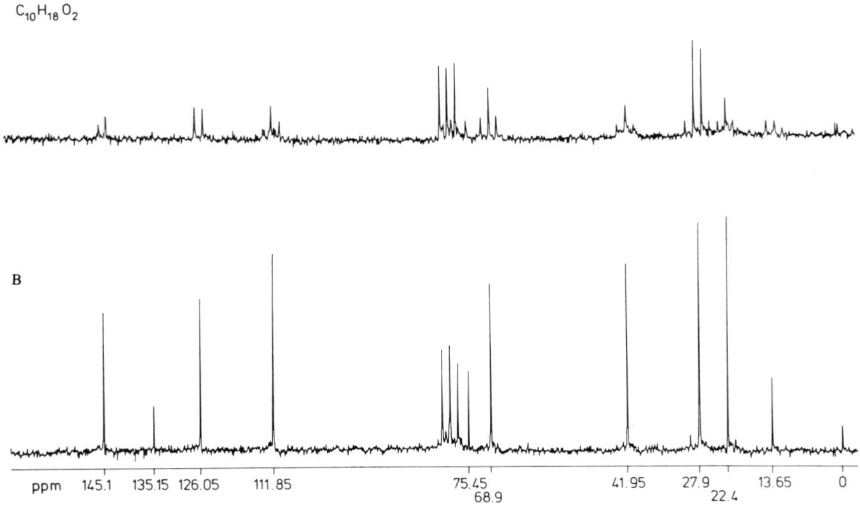
$C_{10}H_{18}O_2$
B
ppm 145.1 135.15 126.05 111.85 75.45 68.9 41.95 27.9 22.4 13.65 0

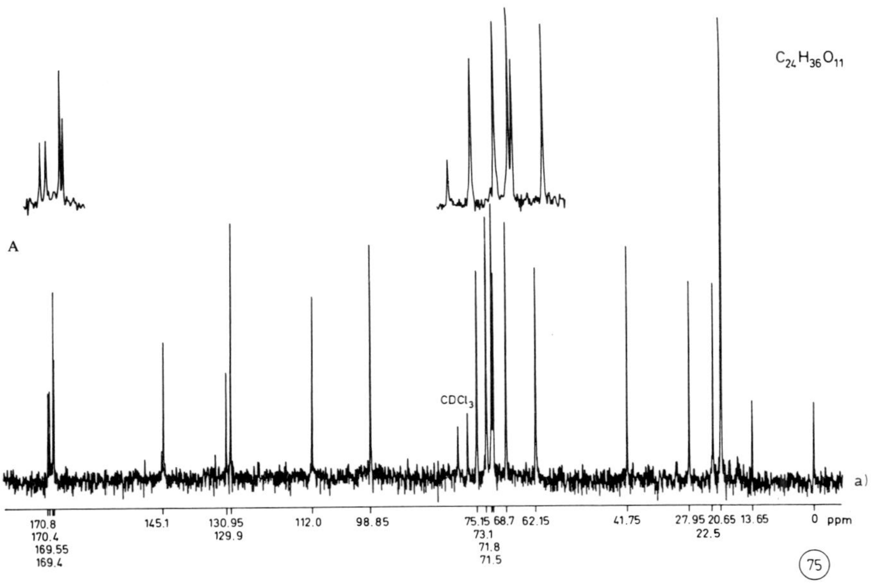
$C_{24}H_{36}O_{11}$
A
$CDCl_3$
a)
170.8 170.4 169.55 169.4 145.1 130.95 129.9 112.0 98.85 75.15 73.1 71.8 71.5 68.7 62.15 41.75 27.95 22.5 20.65 13.65 0 ppm

(75)

76

A natural product was isolated from Plantago arenaria seeds, with the elementary composition $C_{11}H_{17}N_3O$, according to high-resolution mass spectrometry (m/z 207.1367), and melting at 208-210° C. The proton noise- and off-resonance decoupled ^{13}C NMR spectra in deuteriochloroform afford the information about the molecule's carbon atoms that is listed in the table below. Using the table, identify partial structures and propose a structural formula in agreement with the given ^{13}C data.

Recording Conditions

See literature reference at the end of the solution to the problem, p. 387.

Ppm	Off-Resonance Multiplicity
179.9	singlet
158.6	singlet
142.6	doublet
116.2	triplet
65.2	singlet
63.2	doublet
42.4	doublet
40.8	triplet
30.0	triplet
20.7	quartet
11.6	quartet

Section B

Solutions to the Problems in the Interpretation of ^{13}C NMR Spectra

1

Interpretation of the Spectrum

ppm	Off-resonance Multiplicity	Intensity	Assignment
53.4	Triplet		$—CH_2—$
31.15	Singlet[a]	Weak	$—CR_3$
30.2	Quartet	Extremely strong	$—CH_3$
25.55	Quartet	Strong	$—CH_3$
24.85	Doublet[a]		$—CHR_2$

[a]Overlaps neighboring CH_3 quartet.

The doublet at 24.85 ppm and the strong quartet at 25.55 ppm correspond to an isopropyl group. Similarly, the combination of the singlet at 31.15 ppm and the extremely strong quartet at 30.2 ppm corresponds to a *t*-butyl group. If both moieties are joined through a methylene group (triplet at 53.4 ppm), 2,2,4-trimethylpentane (isooctane) results, in agreement with the following molecular formula:

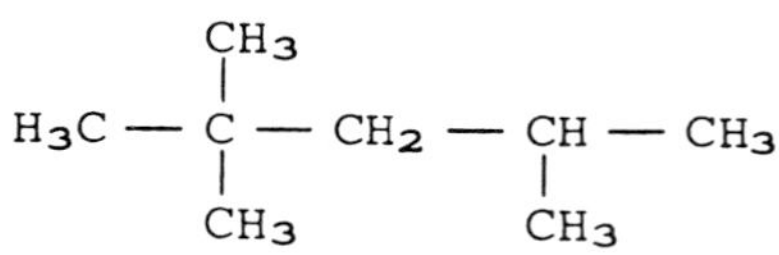

The large chemical shift of the methylene group (53.4 ppm) caused by steric interactions can be calculated according to Grant's and Paul's method using the methane chemical shift, two α increments, five β increments, and the steric corrections for a secondary carbon atom joined to quaternary and tertiary carbon atoms:

$$\begin{aligned} B &= -2.3 \text{ ppm} \\ 2A_\alpha &= 18.2 \text{ ppm} \\ 5A_\beta &= 47.0 \text{ ppm} \\ S_{2^\circ 4^\circ} &= -2.5 \text{ ppm} \\ S_{2^\circ 3^\circ} &= -7.5 \text{ ppm} \\ \hline & \ 52.9 \text{ ppm} \end{aligned}$$

2

Interpretation of the spectrum

ppm	Off-resonance Multiplicity	Assignment
36.45	Triplet	$—CH_2—$
24.35	Triplet	$—CH_2—$
21.65	Triplet	$—CH_2—$
13.6	Quartet	$—CH_3$

The triplets at 36.45, 24.35, and 21.65 ppm correspond to three methylene groups. The quartet at 13.6 ppm corresponds to a methyl group, so the spectrum agrees with the molecular formula of butyl mercaptan. For example, for the assignment of the CH_2 signals, the chemical shift of C-1 can be estimated as follows:

$$\overset{4}{CH_3}—\overset{3}{CH_2}—\overset{2}{CH_2}—\overset{1}{CH_2}—SH$$

^{13}C chemical shift of methane B	=	−2.3 ppm
Chemical shift increment owing to C-α	=	9.1 ppm
Chemical shift increment owing to C-β	=	9.4 ppm
Chemical shift increment owing to C-γ	=	−2.5 ppm
Additional increment owing to the α-sulfur atom $Z_{\alpha,—S—}$	=	10.6 ppm
Chemical shift δ_1 of C-1	=	24.3 ppm

The following assignment results from estimating the remaining methylene chemical shifts:

ppm	C-1	C-2	C-3	C-4
Estimated	24.3	36.7	21.7	13.3
Measured	24.35	36.45	21.65	13.6

3

The molecular formula $C_5H_{13}N$ fits the primary aliphatic amines with structures 1 to 8, based on the hydrocarbons pentane, 2-methylbutane, and 2,2-dimethylpropane. Amines 3, 5, 7, and 8 are out of the question because less than the five observed signals should be expected owing to molecular symmetry. If one calculates the ^{13}C chemical shifts of the α-carbon atoms in the remaining structures, using the α increment of the NH_2 group (28.3 ppm according to Table 7.2, p. 52), the best agreement is found for pentylamine.

No.	Structure	Basic Hydrocarbon with ^{13}C Chemical Shifts (ppm)[a]	Calculated ^{13}C Chemical Shift (ppm)
1	$CH_3-CH_2-CH_2-CH_2-CH_2-NH_2$	34.4 22.4 13.6 $CH_3-CH_2-CH_2-CH_2-CH_3$	*13.6+28.3=41.9*
2	$CH_3-CH_2-CH_2-CH-CH_3$ | NH_2	"	*22.4+28.3=50.7*
3	$CH_3-CH_2-CH-CH_2-CH_3$) | NH_2	"	
4	$CH_3-CH_2-CH-CH_2-NH_2$ | CH_3	11.4 31.1 29.7 21.9 $CH_3—CH_3—CH — CH_3$ | CH_3	*21.9+28.3=50.2*
5	NH_2 | $CH_3-CH_2-C-CH_3$) | CH_3	"	
6	★ $CH_3-CH-CH-CH_3$ | | H_2N CH_3	"	*31.1+28.3=59.4*
7	$H_2N-CH_2-CH_2-CH-CH_3$) | CH_3	"	
8	CH_3 | $CH_3-C-CH_2-NH_2$) | CH_3	CH_3 | 27.4 $H_3C-C-CH_3$ | 31.4 CH_3	

[a]Compare E. Breitmaier, W. Voelter, *^{13}C-NMR Spectroscopy: Methods and Applications,* "Monographs of Modern Chemistry," Vol. 5, H.F. Ebel, ed., Verlag Chemie GmbH, Weinheim, 2nd ed., 1978, p. 131.

The other measured ^{13}C chemical shifts also agree well with the values calculated for pentylamine:

CH_3 —	CH_2 —	CH_2 —	CH_2 —	CH_3		
13.6	22.4	34.4	22.4	13.6		ppm (gemessen)
0.0	*0.0*	*-5.1*	*11.3*	*28.3*		ppm *(Inkremente, vgl. Tab. 3, S.52)*
13.6	*22.4*	*28.3*	*33.7*	*41.9*		ppm *(berechnet)*
14.42	23.69	30.41	33.87	42.86		ppm (gemessen, vgl. Spektrum a)
CH_3 —	CH_2 —	CH_2 —	CH_2 —	CH_2 —	NH_2	

On going to the pentylammonium ion (Spectrum B), all the ^{13}C signals are shifted to higher fields. As is shown in Table 7.2, the $\overset{+}{N}H_3$ increment is largest for the β carbon atom. The following assignments and pH shifts result:

	CH_3	CH_2	CH_2	CH_2	CH_2	$\overset{+}{N}H_3$
$\delta_{R-\overset{+}{N}H_3}$	14.50	23.41	29.9	28.44	41.45	ppm
$\delta_{R-\overset{+}{N}H_3}-\delta_{R-NH_2}$	0.08	-0.28	-0.51	-5.43	-1.41	ppm

The cause of the pH shift is the electric field $\vec{E}$ of the $\overset{+}{N}H_3$ group and the corresponding field gradient $\partial E/\partial r$, which partly compensates the field effect proper at the α position. Therefore, the pH effect is larger at the β position (see Breitmaier and Voelter, *^{13}C NMR Spectroscopy*).

The following pseudocontact shifts Δ result from Spectrum C:

	CH_3	CH_2	CH_2	CH_2	CH_2	NH_2
	14.46	23.73	30.44	33.39	43.99	ppm
Δ	0.04	0.04	0.03	-0.48	1.13	ppm

Consequently, the pH effects are larger and more generally useful as assignment aids in the case of amines (in agreement with Table 7.2) than are the pseudocontact shifts.

4

Adding the increments and comparing the results with the measured chemical shifts to produce the best agreement leads to a preliminary assignment according to the following scheme:

Octane	CH_3	CH_2	CH_2	CH_2	CH_2	CH_2	CH_2	CH_3	
Measured values	14.0	23.0	32.4	29.7	29.7	32.4	23.0	14.0	ppm
OH increments					-6.2	10.1	49.0	10.1	ppm
Estimated values	14.0	23.0	32.4	29.7	23.5	42.5	72.0	24.1	ppm
Measured values	14.3	23.2	32.65	30.15	23.85	40.15	67.9	26.5	ppm
2-Octanol	CH_3	CH_2	CH_2	CH_2	CH_2	CH_2	CH(OH)	CH_3	
	C-8	7	6	5	4	3	2	1	

The zero transits τ_0 in Spectra Series 4 give the following T_1 values according to the equation $T_1 = \frac{\tau_0}{\ln 2}$:

C-	ppm	τ_0[s]	T_1[s]	CH_N	NT_1[s]
1	26.5	1.8	2.6	CH_3	7.8
2	67.9	2.4	3.5	CH	3.5
3	40.15	1.5	2.2	CH_2	4.4
4	23.85	1.5	2.2	CH_2	4.4
5	30.15	2.1	3.0	CH_2	6.0
6	32.65	2.7	3.9	CH_2	7.8
7	23.2	3.4	4.9	CH_2	9.8
8	14.3	3.8	5.5	CH_3	16.5

The T_1 values of the carbon atoms in the hydrophilic part of the molecule increase toward the hydrophobic end of the molecule. Because of hydrogen bonding, the part of the molecule around the OH group (C-1, C-2, C-3) is relatively immobile. Intramolecular mobility increases with increasing distance from C-2, and the T_1 values rise accordingly, especially for the terminal methyl groups. To make the influence of intramolecular mobility more noticeable in preponderantly dipolar relaxation processes, the NT_1 values ($T_1 \sim \frac{1}{N}$) are often compared, where N is the number of hydrogen atoms directly bonded to the carbon atom.

The increase of the T_1 and the NT_1 values from C-2 to C-8 confirms the assignment based on increment addition. For example, compare C-4 with δ = 23.85 ppm and NT_1 = 4.4 s, and C-7 with δ = 23.2 ppm and NT_1 = 9.8 s.

5

The spectrum shows four strong and five weaker signals. Because the sample is analytically pure, it consists of two structural isomers in a ratio of approximately 7:3, as follows from the signal intensities. On the basis of the off-resonance multiplets, which in part show higher-order structures, these compounds are 3- and 2-methylbutyl nitrites:

3-Methylbutyl nitrite

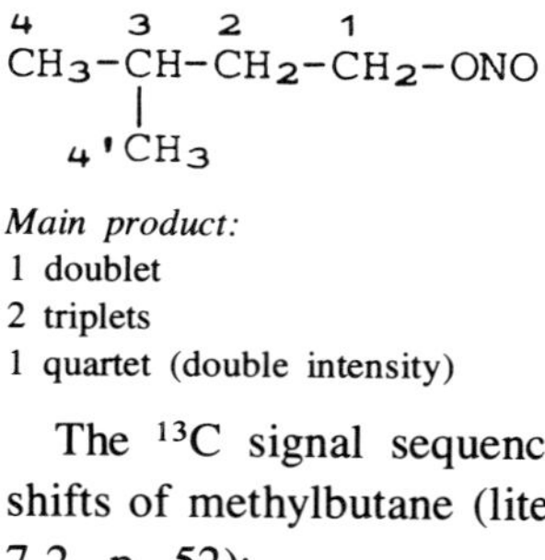

Main product:
1 doublet
2 triplets
1 quartet (double intensity)

2-Methylbutyl nitrite

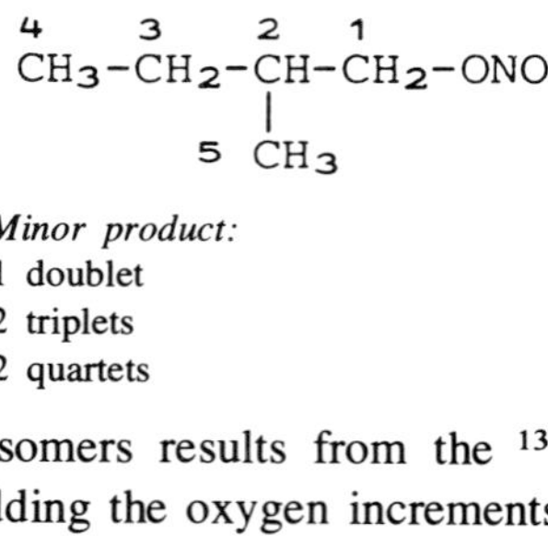

Minor product:
1 doublet
2 triplets
2 quartets

The ^{13}C signal sequence for both isomers results from the ^{13}C chemical shifts of methylbutane (literature) by adding the oxygen increments (see Table 7.2, p. 52):

2-Methylbutane

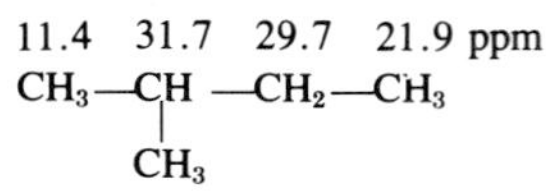

3-Methylbutyl nitrite Carbon	ppm	2-Methylbutyl nitrite Carbon	ppm
C-1 : *(11.4+49.0=60.4)*	67.4 T	C-1 : *(21.9+49.0=70.9)*	73.75 T
C-2 : *(31.7+10.1=41.8)*	38.55 T	C-2 : *(29.7+10.1=39.8)*	35.6 D
C-3 : *(29.7−6.2 =23.5)*	25.8 D	C-3 : *(31.7−6.2 =25.5)*	26.75 T
C-4,4′ : *(21.9+0.0 =21.9)*	22.7 Q	C-4 : *(11.4+0.0 =11.4)*	11.35 Q
		C-5 : *(21.9−6.2 =15.7)*	16.65 Q

6

Interpretation of the Spectrum

ppm	Off-resonance Multiplicity	Intensity	Assignment
164.6	Singlet	Weak	>C=O
114.7	Singlet	Weak	−C≡N
63.0	Triplet	Strong	$-CH_2-$
25.0	Triplet	Strong	$-CH_2-$
14.15	Quartet	Strong	$-CH_3$

The singlet at 164.6 ppm corresponds to a carbonyl carbon atom, such as is found in carboxylic acids or their derivatives. The methylene signal at 63.0 ppm and the methyl signal at 14.15 ppm correspond to an ethoxy group, so that the problem compound should be thought of as a carboxylic acid ethyl ester:

$$R—CO—O—CH_2—CH_3$$

Although the position of the second methylene signal at 25.0 ppm shows no strong substituent influence, the chemical shift of the quaternary carbon atom at 114.7 ppm corresponds to a nitrile group. In agreement with the molecular formula, ethyl cyanoacetate results.

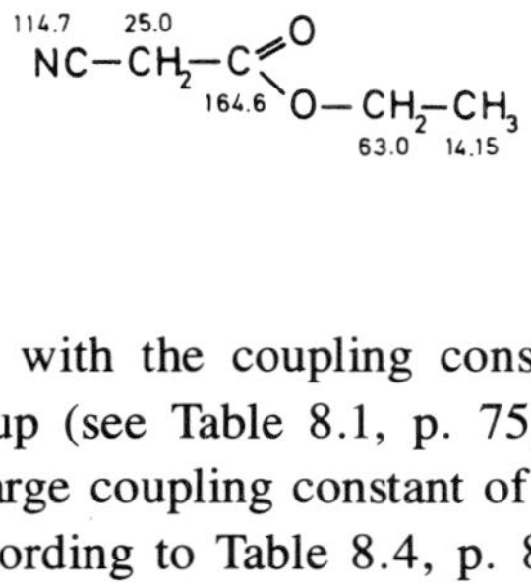

7

The quartet at 55.2 ppm with the coupling constant J_{CH} = 150 Hz corresponds to a methoxy group (see Table 8.1, p. 75). Another quartet at 116.5 ppm with the extremely large coupling constant of 264.6 Hz corresponds to a trifluoromethyl group, according to Table 8.4, p. 82. A quartet of long-range quartets at 159.05 ppm remains. The larger coupling constant (41.9 Hz), according to Table 8.4, corresponds to a C-F coupling across two bonds. The smaller quartet splittings (4.4 Hz) can be attributed to three methoxy protons. Therefore, the carbon atom with δ = 159.05 ppm is flanked by CF_3 and OCH_3; according to its ^{13}C chemical shift, it is a carbonyl carbon, which also agrees with the given molecular formula ($C_3H_3F_3O_2 - CF_3 - CH_3O = CO$, a double-bond equivalent of C_3H_8). Therefore, the compound is methyl trifluoroacetate.

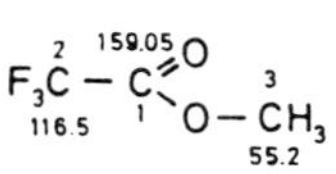

Couplings:

C-1: J_{CCF} = 41.9 Hz
J_{COCH} = 4.4 Hz
C-2: J_{CF} = 264.6 Hz
C-3: J_{CH} = 150.0 Hz

8

In comparison with δ-hydroxylevulinic acid, $C_5H_8O_4$, the problem compound $C_6H_6O_5$ has two hydrogen atoms too few and one carbon atom too many. Therefore, during hydrolysis a dehydration has taken place ($-H_2O$) but no decarboxylation has occurred ($+CO_2$) because

$$C_6H_6O_5 = C_5H_8O_4 + CO_2 - H_2O.$$

According to this, saponification affords first hydroxyacetylsuccinic acid, which then loses water directly or after enolisation. Two lactones, (1) and (2), with formula $C_6H_6O_5$ should thus be considered:

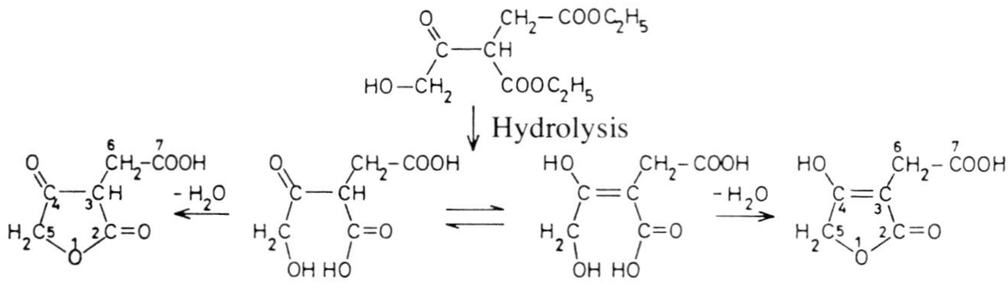

1 2

Interpretation of the ^{13}C-NMR Spectrum

ppm	Multiplicity	Coupling Constant J_{CH} [Hz]	Assignment [Lactone (2)]
174.95	Long-range triplet	Small	C-2 and C-7
171.75	Long-range triplet	Large	C-4
94.3	Long-range triplet	Small	C-3
67.1	Triplet	149	C-5
27.15	Triplet	127	C-6

The long-range triplet at 94.3 ppm corresponds to an sp^2 carbon atom not bonded to hydrogen, fitting C-3 of lactone (2). In the spectrum of lactone (1), C-3 would be expected to give a doublet in the sp^3 carbon resonance range.

9

According to the molecular formula C_4H_9NO, the carbonyl signal at 169.95 ppm does not correspond to a carboxyl group (two oxygens) but to a carboxamide function ($\rangle N{-}C\langle^{O}$). The spectrum also contains three methyl quartets, two of which must be N-methyl groups, judging from their chemical shifts (37.7 and 34.6 ppm) [$-N(CH_3)_2$, see dimethylformamide in Table 5.1,

p. 38]. The remaining quartet corresponds to the methyl portion of an acetyl group (see acetic acid, Table 5.1). These moieties can be pieced together to form dimethylacetamide.

Because of the partial π character of the C-N bond, one N-methyl group is *cis* and the other is *trans* with respect to the carbonyl oxygen atom. The *trans* methyl group is less shielded (37.7 ppm).[1]

10

Interpretation of the Spectrum Considering the Molecular Formula

ppm	Multiplicity	C-H Coupling Constants	Assignment
113.35	Doublet of long-range multiplets	J_{C-H} = 172 Hz	–C(H)(O–)(O–) or $\geq$C–H
53.0	Quartet of long-range doublets (strong)	J_{C-H} = 139 Hz	2 –O–CH_3
37.45	Quartet of long-range quartets (strong)	J_{C-H} = 135 Hz	–N(CH_3)(CH_3)
			$C_5H_{13}NO_2$

The functional groups found are parts of N,N-dimethylformamide dimethylacetal:

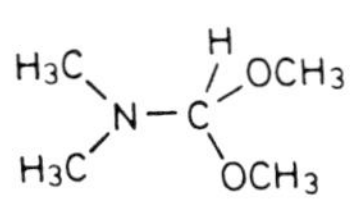

The extremely large C-H coupling of 172 Hz for an sp^3 carbon atom is a result of the two directly bonded oxygen atoms. Similarly, the chemical shift of 113.35 ppm is typical of acetal carbon atoms. Unlike N,N-dimethylfor-

[1]See F. A. L. Anet, A. J. R. Bourn, *J. Am. Chem. Soc.* **87,** 5350 (1965); and G. C. Levy, G. L. Nelson, *J. Am. Chem. Soc.* **94,** 4897 (1972).

mamide and N,N-dimethylacetamide (see Spectrum 9), the $N(CH_3)_2$ group can rotate freely after acetalization of the formyl group. The N-methyl resonances therefore coincide.

11

The ^{13}C-NMR spectrum shows only eight resonances for the molecule's 14 carbon atoms. Two pairs of signals (63.35 and 62.95, 15.75 and 15.65 ppm) attract attention because of their small chemical-shift differences with the same off-resonance multiplicities (triplets and quartets, respectively). The consequent molecular symmetry and the relative signal intensities, off-resonance multiplicities, and ^{13}C chemical shifts make the following groups recognizable:

ppm	Relative Intensity	Off-resonance Multiplicity	Number and Kind of Group	
104.95	Strong	Doublet	(Acetal)[a]	2 $-CH(O-)_2$
63.35	Strong	Triplet	(Ethoxy group)[a]	2 $-OCH_2-$
62.95	Strong	Triplet		2 $-OCH_2-$
50.15	Weak	Doublet	(with substituent effect)	$>CH-$
24.95	Weak	Doublet	(Isopropyl)[b]	$>CH-$
22.0	Strong	Quartet		2 $-CH_3$
15.75	Strong	Quartet	(ethoxy group)[a]	2 $-CH_3$
15.65	Strong	Quartet		2 $-CH_3$
CH sum based on the off-resonance multiplicities and intensities			$C_{14}H_{30}$	

[a]See Figure 3.4A, p. 22
[b]See starting compound.

Piecing together the building blocks found in this way, and considering the reaction route, 2-isopropylmalonic dialdehyde tetraethyl acetal arises as the solution:

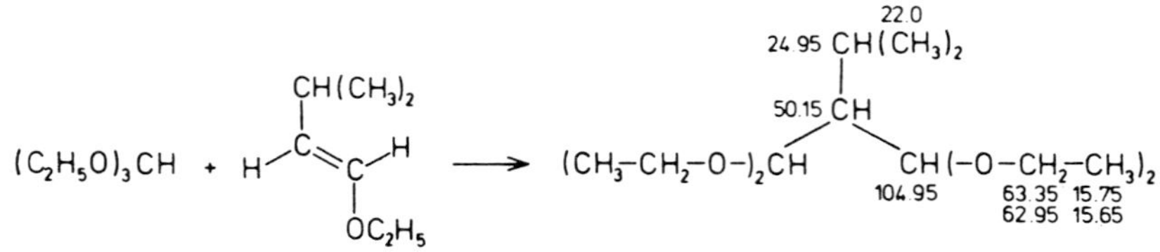

The appearance of pairs of signals for the methylene and methyl carbon atoms (63.35/62.95 and 15.75/15.65 ppm) is a consequence of the *diastereotopism* of the two pairs of ethoxy carbon atoms in each acetal group, as the Newman projection of one of the three possible stable conformations shows:

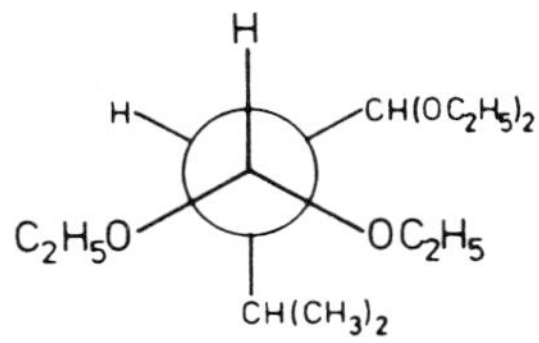

12

According to the ^{13}C NMR spectrum, all four carbon atoms in the molecule are chemically different. Considering the two oxygen atoms present, the following assignment results:

ppm	Off-resonance Multiplicity	Assignment
205.5	Singlet	>C=O
77.65	Triplet	$-O-CH_2-$
58.8	Quartet	$-O-CH_3$
25.85	Quartet	$-CH_3$

Molecular formula according to the ^{13}C NMR spectrum: $C_4H_8O_2$.

Joining the building blocks thus found, methoxyacetone is obtained:

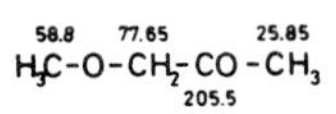

The methyl resonances of methoxyacetone may be calculated from the methyl chemical shift in acetone (30.7 ppm according to Table 5.1, p. 38) and the α- or γ-increments for ether oxygen according to Table 7.10, p. 61:

	$-O-CH_2-$		$-O-CH_3$
	30.7		30.7
Z_α =	49.0	Z_γ =	−6.2
	79.7 ppm		24.5 ppm

13

Interpretation of the Spectrum

ppm	Multiplicity	C-H Coupling Constants (in Hz)	Assignment	Origin
109.7	Singlet with poorly resolved long-range coupling	—	Quaternary C (ketal)	Methyl vinyl ketone
76.2	Triplet of long-range triplets	145.0	$-CH_2-(CH_2)-NO_2$	Nitromethane
65.15	Triplet (very strong)	147.0	$-O-CH_2-CH_2-O-$	Ethylene glycol
35.9	Triplet with poorly resolved long-range coupling	127.0	$-CH_2-$	Methyl vinyl ketone
24.0	Quartet	126.0	$-CH_3$	Methyl vinyl ketone
22.7	Triplet of poorly resolved long-range triplets	126.5	$-CH_2-(CH_2)-\overset{\vert}{\underset{\vert}{C}}-$	Methyl vinyl ketone

Interpretation of the spectrum, considering the starting compounds (origin of the carbon atoms), leads to the following assignment and structure for A′:

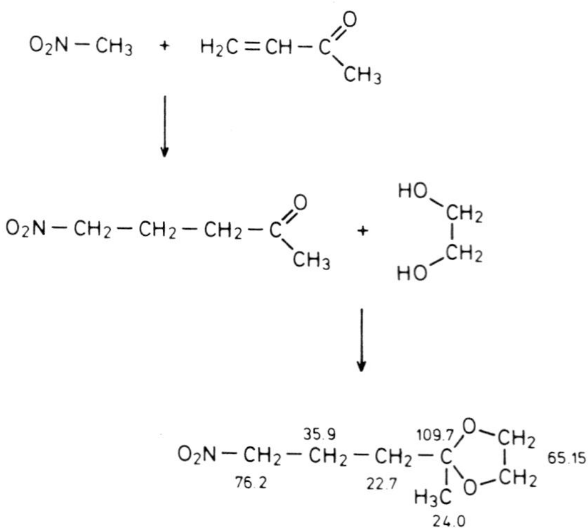

2-methyl-2-(3-nitropropyl)-1,3-dioxolan

14

Ethyl acetoacetate exists as an equilibrium mixture of its keto and enol forms. The keto form predominates and gives the stronger signals. Considering the intensities, the following assignment results:

ppm	Off-resonance Multiplicity	Intensity	Assignment	Keto (%)	Enol (%)
202.3	Singlet	56,021	K C-3	92.2	—
177.55	Singlet	4,723	E C-3	—	7.8
174.3	Singlet	4,126	E C-1	—	9.5
168.8	Singlet	39,467	K C-1	90.5	—
90.5	Doublet	2,987	E C-2	—	5.0
61.65	Triplet	44,513	K C-5	94.3	—
60.75		2,671	E C-5	—	5.7
50.5	Triplet	56,429	K C-2	95.0	—
30.1	Quartet	57,795	K C-4	94.8	—
21.1	Quartet (weak)	3,190	E C-4	—	5.2
14.5		5,724	E C-6	—	9.7
14.35	Quartet	53,209	K C-6	90.3	—

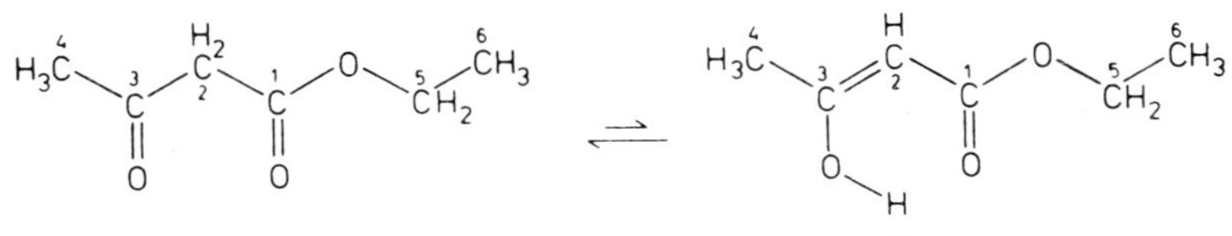

K E

Ethyl acetoacetate is therefore enolized to an extent of 5 to 10% in acetone. On one hand, the reason for this considerable dispersion is the overlapping of some signals (E C-6 and K C-6, K C-4, and the central signal of the hexadeuterioacetone septet). On the other hand, it is assumed that influences of spin-lattice relaxation on the intensities of the C-1 and C-3 signals may be neglected, notwithstanding the fact that the long waiting times between the pulses (120 s) are less than sufficient.

For methyl acetoacetate, ^{13}C NMR spectroscopy has revealed the existence of a *trans*-enol form coexisting with the keto and the *cis*-enol forms (see R. Matusch, *Angew. Chem.* **87,** 283, 1975).

15

The ^{13}C NMR spectrum shows twice as many signals (8) as expected (4). In addition to the carbodiimide function (131.2 ppm, broadened), a weak signal can be observed at 117.4 ppm with a chemical shift that fits a nitrile group (see Fig. 6.2, p. 43). All the other carbodiimide carbon atoms [N—CH_2—, —CH_3, —$Si(CH_3)_3$] give pairs of signals with an intensity ratio of about 45:55. This result suggests the presence of two isomers that can interconvert through a silatropic rearrangement (see I. Ruppert, *Angew. Chem.* **89,** 336, 1977).

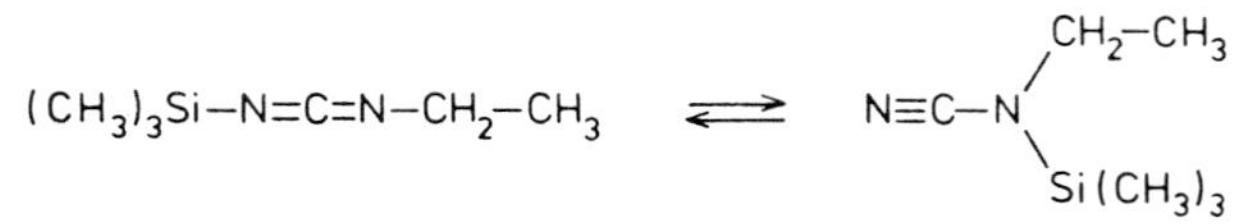

Carbodiimide Cyanamide

The methyl carbon atoms of the trimethylsilyl and ethyl groups in the cyanamide are shielded by the γ effect of the other substituent, whereas the trimethylsilyl group affects the N-methylene resonance through a β effect (weaker shielding, see Table 7.2, p. 52). The following signal assignments can be reached, as well as the conclusion that the carbodiimide form predominates slightly (about 55%) in the equilibrium mixture.

Assignment of the Spectrum

Form	ppm	Multiplicity	Coupling Constants (Hz)	Coupling Partner	Carbon Atom
Carbodiimide	131.2	Cannot be determined reliably due to signal broadening			⟍N=C=N⟍
	40.4	Triplet of long-range quartets	141.5 4.5	2 directly bonded H —CH_3	⟍N—CH_2—
	17.4	Quartet of long-range triplets	127.5 3.0	3 directly bonded H —CH_2—	—CH_3
	0.8	Quartet of long-range septets	120.5 1.7	3 directly bonded H $(CH_3)_2$ on Si	—$Si(CH_3)_3$
Cyanamide	117.4	Long-range triplet	6.0	—CH_2—	—C≡N
	42.4	Triplet of long-range quartets	142.0 4.5	2 directly bonded H —CH_3	⟍N—CH_2— ⟋
	15.2	Quartet of long-range triplets	128.0 3.0	3 directly bonded H —CH_2—	—CH_3
	−1.45	Quartet of long-range septets	120.5 1.7	3 directly bonded H $(CH_3)_2$ on Si	—$Si(CH_3)_3$

16

Natural rubber is *cis*-, and gutta-percha is *trans*-polyisoprene. Since $\delta_{trans} > \delta_{cis}$ for the α carbon atoms next to the double bond in an alkene (see Section 7.4, p. 54), the gutta-percha methyl group should be more strongly shielded, whereas the C-1 methylene should be less shielded than in natural rubber. Therefore, gutta-percha has Spectrum B. The C-4 methylene resonance is barely affected by the configurational change because it interacts with a *cis*-alkyl group in both isomers.

The C-3 signal is stronger than the C-2 resonance owing to the directly bonded proton (nuclear Overhauser effect), affording a basis for the assignment of the alkene signals.

Gutta-percha

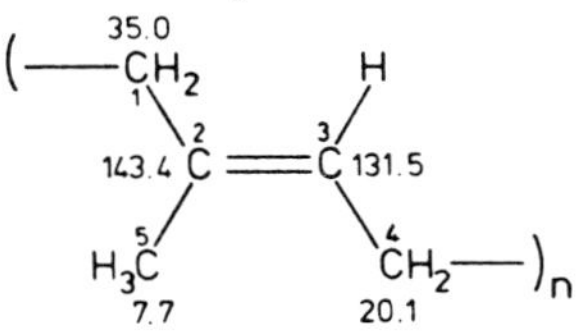

Natural rubber

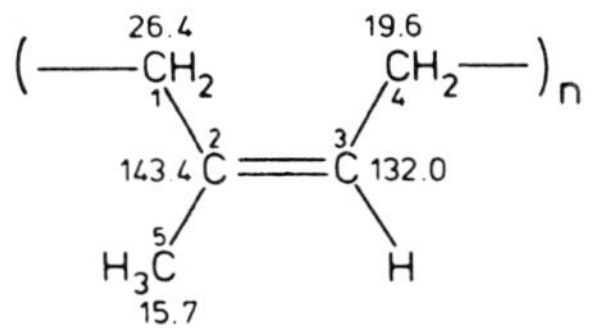

17

The following assignment results from the known structure of the salt,

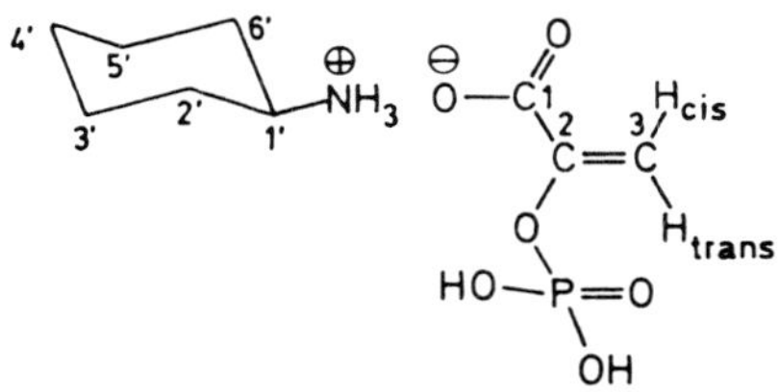

using the substituent increments for the ammonium group (Table 7.4, p. 54), the chemical shift of cyclohexane (27.6 ppm) and a comparison of the signal intensities:

ppm	Carbon atom
168.25	C-1 (carbonyl)
146.4	C-2 (enol)
111.0	C-3 (=CH_2, stronger than the other signals in the 100-170 ppm range owing to the nuclear Overhauser effect)
52.65	C−1′
32.65	C−2′,6′
26.65	C−4′
26.2	C−3′,5′

Because the partial ^{13}C spectrum shows direct proton couplings with magnitudes that fit sp^2 carbon atoms better than sp^3 carbons (they are larger than 160 Hz), the phosphoenolpyruvic acid carbon bonded to hydrogen atoms (C-3) is earmarked. The following coupling scheme results from the ^{31}P spectrum with the phosphorus-carbon coupling constant of 4.5 Hz and the ^{13}C partial spectrum:

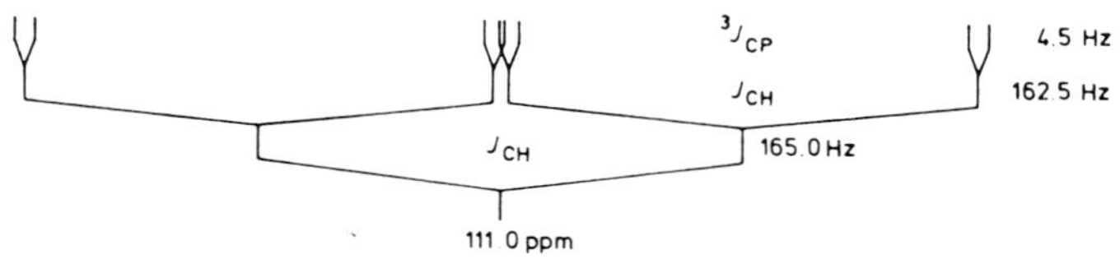

Which of the two C-H couplings correponds to the *cis*- and which to the *trans*-proton, could be decided by substitution (e.g., with deterium) of one of these hydrogen atoms.

18

The number of signals (expected, 5; found, 10) suggests that the analytically pure sample is a mixture of isomers. Because the sample contains a double-bond equivalent (compare C_5H_{10} with C_5H_{12}), these could be *cis-trans* isomers.

The pairs of signals at 68.0 and 15.6, and at 64.9 and 15.1 ppm (off-resonance triplets and quartets) belong to one ethoxy group each (see Fig. 3.1, p. 16). If one subtracts the corresponding partial formula C_2H_5O from the given molecular formula $C_5H_{10}O$, C_3H_5 is left. On the basis of the off-resonance multiplicities (two pairs of doublets in the olefinic region, one pair of quartets in the methyl region), this residue is a propenyl group. (Why is the allyl group, which also fits the formula C_3H_5, not considered?) The problem compound is thus a *cis-trans* mixture of ethyl propenyl ether.

The following assignment results from the known relations $\delta_{trans} > \delta_{cis}$ for carbon atoms in the α position with respect to the double bond (Table 7.8, p. 59) and from the intensity relationship of the remaining pairs of signals.

ppm	Off-resonance Multiplicity	Carbon/isomer	Percentage
148.4	Doublet	C-1/trans	25.7
146.95	Doublet	C-1/cis	74.3
101.3	Doublet	C-2/cis	75.2
98.45	Doublet	C-2/trans	24.8
68.0	Triplet	C-4/cis	73.9
64.9	Triplet	C-4/trans	26.1
15.6	Quartet	C-5/cis	72.8
15.1	Quartet	C-5/trans	27.2
12.75	Quartet	C-3/trans	28.5
9.35	Quartet	C-3/cis	71.5

The fact that the *cis* form predominates here follows not only from the integration of the ^{1}H NMR spectrum—in which the assignment can be assured because of the relationship $^3J_{HH(trans)} > {}^3J_{HH(cis)}$—but also from the coupled ^{13}C NMR spectrum: the stronger quartet at 9.35 ppm exhibits the larger coupling constant $^3J_{CH}$, corresponding to the *trans* configuration of the methyl carbon atom and the vinyl proton in the *cis* isomer (see Table 8.2A, p. 78).

The weak shielding of C-1 is a consequence of the inductive effect of the oxygen atom:

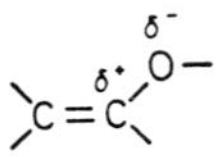

The increased shielding of C-2 can be explained by resonance:

19

Spectrum Assignment

ppm	Off-resonance Multiplicity	Intensity	Assignment
193.04	Doublet	Extremely strong	—CH=O
169.25	Singlet	Weak	>C=O
143.25	Doublet	Strong	=CH—N
119.05	Singlet	Medium	=CH—
22.95	Quartet	Strong	—CO—CH_3
7.85	Quartet	Strong	—CH_3

An aldehyde function in conjugation with a C-C double bond (doublet at 193.05 ppm see Section 7.8, p. 65) and an acetyl group (singlet at 169.25 ppm and quartet at 22.95 ppm, compare with dimethyl acetamide, Exercise 9, p. 272) can be recognized in the spectrum. Because the molecule contains only two oxygen atoms (molecular formula), which have already been located as components of the acetyl and formyl groups, and because methyl ketones give signals at $\delta > 190$ ppm, an N-acetyl function is present. Furthermore, the spectrum shows no N-methyl resonance but instead a shielded C-methyl carbon atom (quartet at 7.85 ppm), suggesting an -NH-acetyl group. The spectrum also shows the presence of two olefinic carbon atoms (=CH— doublet at 143.25 ppm and =C— singlet at 119.05 ppm). The moieties found in this way add up to the given molecular formula.:

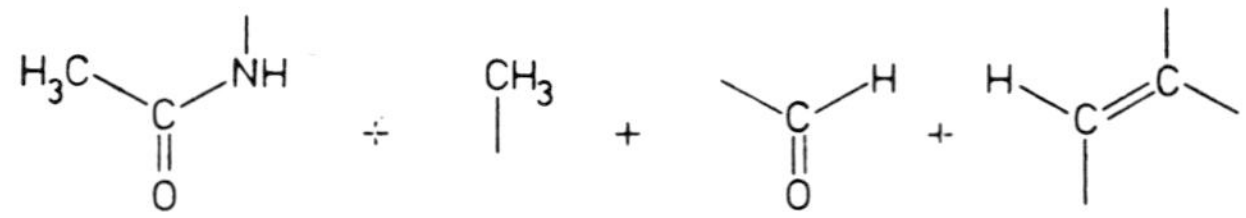

Joining them together, 3-acetylamino-2-methylacrolein results, with the following signal assignment:

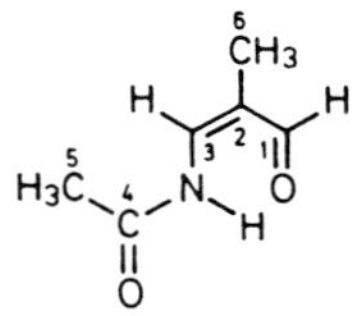

C-1	193.05	ppm
C-2	119.05	ppm
C-3	143.25	ppm
C-4	169.25	ppm
C-5	22.95	ppm
C-6	7.85	ppm

The weak shielding of C-3 can be explained by conjugation with the carbonyl group (vinylogy) and by the inductive effect of the enamine nitrogen atom. The configuration of the substituents cannot be determined with this spectrum.

20

The ^{13}C NMR spectrum shows four instead of two signals. Ketene is therefore present at its dimer (2+2 cycloaddition product), in agreement with the ^{13}C chemical shifts and C-H coupling constants:

Diketene

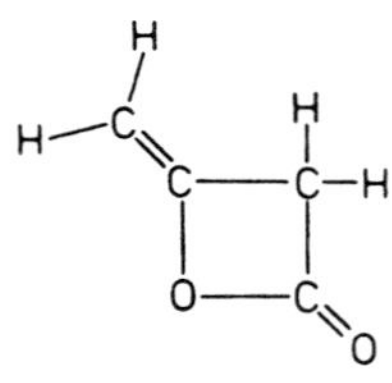

		Assignment		
ppm	Carbon Atom	Multiplicity	Coupling Constant(s) (Hz)	Coupling Partner
167.2	$-O-C(=O)-$	Long-range triplet	$^{2}J_{CH}=7.1$	$-CH_2-$
150.05	$=C<$	Long-range double doublet of long range triplets	$^{2}J_{CH}=7.0$ and 5.5	$H_2C=$
			$^{2}J_{CH}=5.0$	$-CH_2-$
86.9	$H_2C=$	Double doublet (triplet) of long-range doublets	$J_{CH}=163.0$	Directly bonded protons
			$^{3}J_{CH}=1.8$ and 1.5	$=C-CH_2-$
42.95	$-CH_2-$	Double doublet (triplet)	$J_{CH}=142.5$	Directly bonded proton
			$^{3}J_{CH}=7.3$ (*trans*) and 2.3 (*cis*)	H *cis*, H *trans* $H_2C=C-CH_2-$

The large methylene C-H coupling constant (142.5 Hz) is typical of the C-H bond in strained rings (see oxetane, Table 8.1, Section 8.1), and the exomethylene coupling constant of 163.0 Hz lies within the range of C-H couplings in sp^2 C-H carbon atoms.

21

Spectrum a does not show the three signals expected for cyclopentadiene. Four strong doublets are visible in the olefinic chemical shift region, and two triplets and four doublets are visible in the aliphatic region. On the basis of their chemical shifts and multiplicities, these ten signals correspond to dicyclopentadiene, produced by 4+2 cycloaddition of two cyclopentadiene molecules.

Using the C-H coupling constants (Spectrum C), it is possible to decide between the two methylene carbons. The value for the C-7 methylene group in norbornadiene is 135.5 Hz. (see H. Friebolin et al., *NMR-Spektroskopie, Eine Einführung mit Übungen,* Verlag Chemie GmbH, Weinheim, 1974, Taschentext Bd. 15). This value approaches the C-H coupling constant of the triplet at 50.8 ppm (136 Hz), which therefore corresponds to C-7 of the norbornene moiety. The *endo* linkage of the norbornene and cyclopentene rings follows from the chemical shift of this triplet (50.8 ppm) because values of about 45 ppm are typical for an *exo* linkage (γ effect; see H. Günther, *NMR-Spektroskopie, Eine Einführung,* Georg Thieme Verlag, Stuttgart, 1973; and Section 7.3, p. 54).

On the basis of the given spectra, a more complete assignment is not possible.

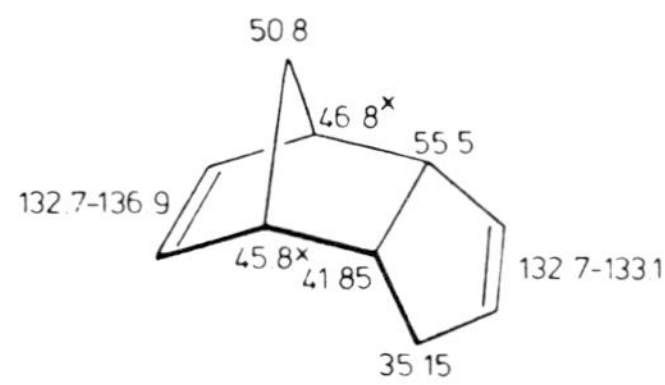

[a]Assignment interchangeable.

22

The quartet at 23.95 ppm (J_{CH} = 127.7 Hz) and the long-range multiplet at 169.95 ppm correspond to the methyl and carbonyl carbon atoms of an acetyl group (see N,N-dimethylacetamide, Exercise 9. The singlet at 51.45 ppm suggests the presence of a substituent at C-1 (= 3, 5, or 7) of the adamantane skeleton, which on the basis of its chemical shift should be an acetylamino group, in agreement with the molecular formula ($C_{12}H_{19}NO$ — $C_{10}H_{15}$ (adamantyl) — C_2H_3O = NH). $C_{12}H_{19}NO$ is thus 1-acetylaminoadamantane.

The eightfold splitting of the carbonyl signal (169.95 ppm) is a consequence of the long-range coupling with the methyl protons ($^2J_{CH}$ = 6.0 Hz, see Table 8.2A, p. 78) and the amino proton ($^2J_{CH}$ = 3.2 Hz). A complete assignment can be made by considering the multiplicities in the coupled Spectrum B.

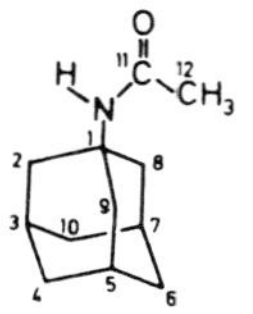

ppm	Multiplicity	Coupling Constant J_{CH} [Hz]	Assignment
169.94	Long-range quartet of long-range doublets	$^2J_{C\text{-}11\text{-}H\text{-}12} = 6.0$ $^2J_{C\text{-}11\text{-}N\text{-}H} = 3.2$	C-11
51.45	Singlet	—	C-1
41.2	Triplet	$J_{C\text{-}2,8,9\text{-}H\text{-}2,8,9} = 133.3$	C-2,8,9
36.05	Triplet	$J_{C\text{-}4,6,10\text{-}H\text{-}4,6,10} = 127.9$	C-4,6,10
29.05	Doublet	$J_{C\text{-}3,5,7\text{-}H\text{-}3,5,7} = 125.9$	C-3,5,7
23.95	Quartet	$J_{C\text{-}12\text{-}H\text{-}12} = 127.7$	C-12

The acetylamino increments can be calculated from the ^{13}C chemical shifts of 1-acetylaminoadamantane by subtracting the values for the corresponding carbon atoms in unsubstituted adamantane:

$$Z_\alpha = 51.45 - 28.35 = 23.1$$
$$Z_\beta = 41.2 - 37.78 = 3.4$$
$$Z_\gamma = 29.05 - 28.35 = 0.7$$
$$Z_\delta = 36.05 - 37.89 = -1.75$$

The increments for the other substituents can be obtained in a similar fashion:

	Cl	Br	J	OH	CH_3
Z_α	40.25	37.65	22.15	39.9	1.5
Z_β	10.0	11.4	14.6	7.4	6.85
Z_γ	3.35	4.0	4.65	2.4	3.1
Z_δ	-2.2	-2.4	-2.2	-1.6	-0.8

It is noteworthy that, in contrast with Table 7.2, p. 52, the δ carbon is shielded by the substituent instead of the γ carbon. This shielding can be explained as a consequence of the Van der Waals interaction between the substituent at the α carbon and the hydrogen atoms at the δ position. Because of the larger separation, the corresponding interaction with the γ hydrogen is weaker.

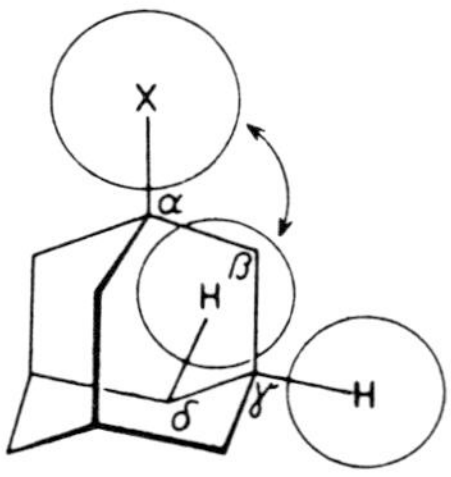

The weak α effect of 1.5 ppm in 1-methyladamantane is a result of steric crowding of the alkyl groups. This is also found, for instance, in *t*-butyl groups in which the quaternary carbon atoms resonate near 30 ppm (see Exercise 1, p. 262).

23

The temperature dependence of the ^{13}C NMR spectrum is a consequence of the inversion of the cyclohexane ring. At 40° C this process is too rapid with respect to the NMR time scale defined by the chemical shift differences. For this reason, only four *averaged* signals rather than eight can be observed for the nonequivalent carbon atoms of *cis*-1,2-dimethylcyclohexane:

I 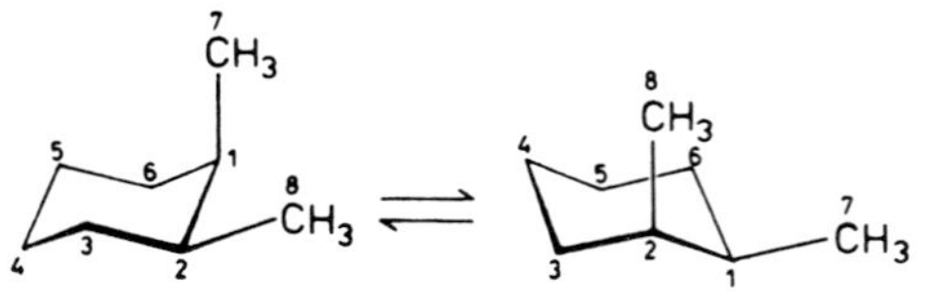 II

On the contrary, this inversion is "frozen" at −100° C, so that all eight expected signals appear in the spectrum. Between −60° and −40° C, the ring inversion rate constant k_C attains values approaching the chemical shift differences (e.g., C-2/C-1, C-3/C-6, or C-4/C-5), and the signals broaden and coalesce. The coalescence temperatures cannot be determined exactly because the signals are too close to each other. It can be seen, however, that the pair of signals at 35.5 and 33.45 ppm coalesces at a lower temperature (about −60° C) than the pair at 20.3 and 11.75 ppm (about −40° C).

The signal assignment at −100° C can be found using Table 7.5, p. 55:

C-1		C-2		C-3	
27.6		27.6		27.6	
+ 1.4	($Z_{\alpha a}$)	+ 6.0	($Z_{\alpha e}$)	+ 9.0	($Z_{\beta e}$)
+ 9.0	($Z_{\beta e}$)	+ 5.4	($Z_{\beta a}$)	− 6.4	($Z_{\gamma a}$)
− 3.4	($Z_{\alpha a \beta e}$)	− 2.9	($Z_{\beta a \alpha e}$)	− 0.8	($Z_{\beta \gamma a}$)
34.6		36.1		29.4	

C-4		C-5		C-6	
27.6		27.6		27.6	
+ 0.0	($Z_{\gamma e}$)	− 6.4	($Z_{\gamma a}$)	+ 5.4	($Z_{\beta a}$)
− 0.1	($Z_{\delta a}$)	− 0.2	($Z_{\delta e}$)	+ 0.0	($Z_{\gamma e}$)
−	($Z_{\gamma e \delta a}$)	−	($Z_{\gamma a \delta e}$)	+ 1.6	($Z_{\beta a \gamma e}$)
27.5		21.0		34.6	

Using the calculated values and noting the identical intensities of the signal pairs at 35.5 and 33.45 ppm and at 20.3 and 11.75 ppm (methyl resonances), the following assignment is reached:

Conformation I Carbon	− 100° C Measured/calculated		40° C Measured
C-1	33.45	34.6	35.05
C-2	35.5	36.1	
C-3	28.7	29.4	32.05
C-4	27.3	27.5	24.2
C-5	21.0	21.0	
C-6	34.05	34.6	
C-7	11.75	—	16.25
C-8	20.3	—	

Calculation of the inversion barrier $\Delta G^{\ddagger}$:
For a chemical shift difference $\Delta\nu$ between two coalescing signals, the ring inversion-rate constant at the coalescence temperature T_C is

$$k_C \approx \frac{\pi}{\sqrt{2}} \Delta\nu.$$

Knowing T_C and k_C, Eyring's equation can be used to calculate the free energy of activation $\Delta G^{\ddagger}_{T_C}$ at the coalescence temperature (see Reference on page 138):

$$\Delta G^{\ddagger}_{T_C} = 4.57\ T_C\ (10.32 + \lg \frac{T_C}{k_C}) \cdot 10^{-3} \frac{\text{kcal}}{\text{mole}}$$

Signal pair at 35.5 and 33.45 ppm:

$$\Delta\nu \approx 35.5 - 33.45 \text{ ppm} \approx 41 \text{ Hz bei } 20 \text{ MHz}$$

$$T_C \approx 213\ ^{\circ}\text{K}\ (-60\ ^{\circ}\text{C})$$

$$k_C \approx \frac{\pi}{\sqrt{2}}\ 41 \approx 91\ \text{s}^{-1}$$

$$\Delta G^{\ddagger}_{213} \approx 4.57 \cdot 213(10.32 + \lg \frac{213}{91}) \cdot 10^{-3} \frac{\text{kcal}}{\text{mol}}$$

$$\approx 4.57 \cdot 213 \cdot 10.69 \cdot 10^{-3} \approx 10.4 \frac{\text{kcal}}{\text{mol}}$$

or $$\Delta G^{\ddagger}_{213} \approx 10.4 \cdot 4.18 \approx 43.5 \frac{\text{kJ}}{\text{mol}}$$

Signal pair at 20.3 and 11.7 ppm (methyl groups):

$$\Delta\nu \approx 20.3 - 11.7 \text{ ppm} \approx 172 \text{ Hz bei } 20 \text{ MHz}$$

$$T_C \approx 233\ ^{\circ}\text{K}\ (-40\ ^{\circ}\text{C})$$

$$k_C \approx \frac{\pi}{\sqrt{2}} \cdot 172 \approx 382\ \text{s}^{-1}$$

$$\Delta G^{\ddagger} \approx 4.57 \cdot 233\ (10.32 - \lg \frac{233}{382}) \cdot 10^{-3} \frac{\text{kcal}}{\text{mol}}$$

$$\approx 10.75 \frac{\text{kcal}}{\text{mol}}$$

or $$\Delta G^{\ddagger}_{233} \approx 45 \frac{\text{kJ}}{\text{mol}}$$

Therefore, the inversion barrier lies between 43.5 and 54 kJ/mole.

24

Chemical shifts of 65 to 70 ppm are typical of the hydroxylated carbon atoms of secondary (rarely primary) alcohols. The number of signals (expected, 6; found, 12) indicates that the 4-*t*-butylcyclohexanol obtained by reduction must be a mixture of *cis-trans* isomers:

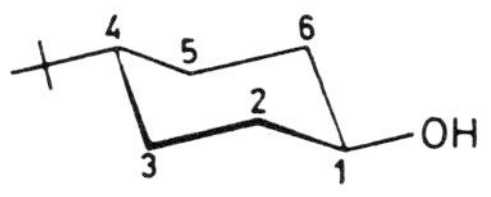

trans

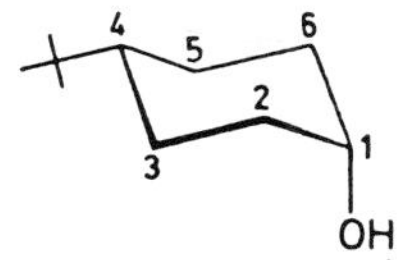

cis

Since $\delta_{eq(OH)} > \delta_{ax(OH)}$, the *trans* isomer with the stronger CHOH signal at 70.7 ppm predominates over the *cis* compound with the weaker resonance at 65.1 ppm. This intensity ratio (*trans* > *cis*), molecular symmetry, and the substituent effects of —OH and —$C(CH_3)_3$ allow the following assignment to be made:

ppm	Isomer	Carbon	Evaluation of Relative Intensities
70.7	*trans*	1	67% [220 · 100/(200+108)]
65.1	*cis*	1	33%
48.85	*cis*	4	36%
48.05	*trans*	4	64%
36.6	*trans*	2,6	65%
34.05	*cis*	2,6	35%
32.9	*cis*	*t*-butyl-C	Weak
32.55	*trans*	*t*-butyl-C	Weak
27.9	*trans*	*t*-butyl-CH_3	Overlapping
27.8	*cis*	*t*-butyl-CH_3	Overlapping
26.85	*trans*	3,5	66.5%
21.5	*cis*	3,5	33.5%

From the intensities of all the ring carbon resonances, it follows that the mixture consists of about 65% *trans* and 35% *cis*-4-*t*-butylcyclohexanol.

25

Unlike C_6H_{14}, C_6H_8O contains three double-bond equivalents. In the ^{13}C NMR spectrum, a carbonyl group (198.65 ppm) and an olefinic C=C double bond (150.6 and 129.85 ppm) are recognizable. The remaining double-bond equivalent, together with the three methylene resonances (38.2, 25.75, and 22.95 ppm), suggest a cyclohexenone structure.

The carbonyl carbon (198.65 ppm) is shielded with respect to its counterpart in cyclohexanone (208.5 ppm, see references below). Furtheremore, one olefinic carbon resonates at an extremely low field (150.6 ppm). In consequence, the carbonyl group and the C=C double bond are conjugated (resonance effect of the C=O group on the C=C double bond; see Section 7.8, p. 65). Therefore, the compound is 2-cyclohexenone.

The methylene signal assignment can be made most easily by comparing the chemical shifts with the values for cyclohexanone and cyclohexene (see F. J. Weigert, J. D. Roberts, *J. Am. Chem. Soc.* **92**, 1347, 1970; and D. E. Dorman, M. Jautelat, J. D. Roberts, *J. Org. Chem.* **36**, 2757, 1971:

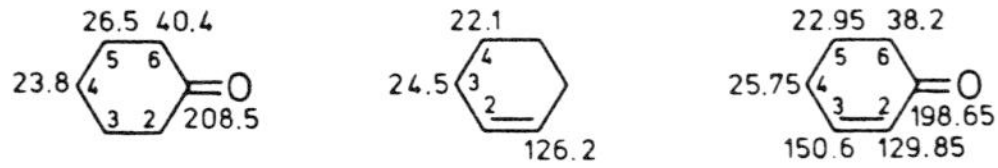

26

Cyclooctatetraene dimerizes easily. Three structures—I, II, and III—were proposed for the dimers, each of which is expected to have a different number of olefinic and saturated carbon resonances:

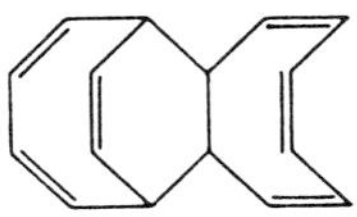
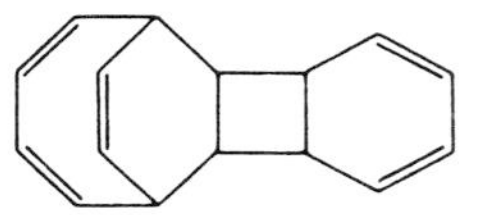
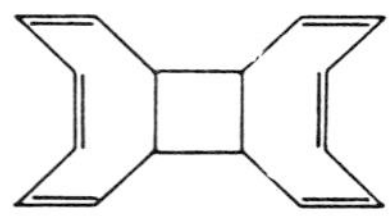

I	II	III
6 x δ > 120 ppm	5 x δ > 120 ppm	3 x δ > 120 ppm
2 x δ < 50 ppm	3 x δ < 50 ppm	1 x δ < 50 ppm

Comparing these spectral predictions, it ensues that the four additional signals mentioned, with the intensity ratio of about 1:1:1:1, have chemical shifts that support Structure III. The following assignment assumed for the carbon resonances rests, among other things, on selective proton decoupling experiments (see suggested reading in Exercise 26).

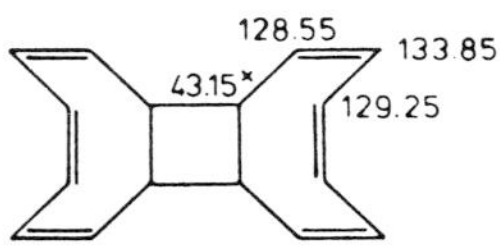

$^{x}J_{CH}$ = 138 Hz

The cycloaddition of cyclooctatetraene to dienophiles, like its dimerization, can proceed from both valence isomers. Thus, using tetracyanoethylene as a dienophile, the adducts IV and V can be formed, for which again, different ^{13}C NMR spectra are expected:

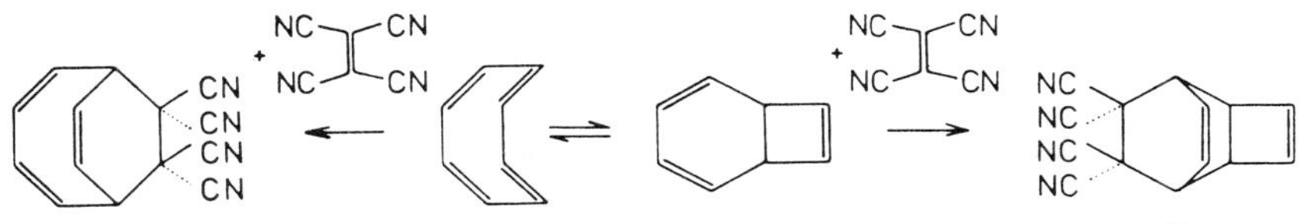

IV V

Valence isomers of cyclooctatetraene

3 x δ > 120 ppm	2 x δ > 120 ppm
2 x δ < 50 ppm	3 x δ < 50 ppm
(without CN resonances)	(without CN resonances)

The cycloaddition of tetracyanoethylene to cyclooctatetraene itself is unlikely because the tub conformation of this molecule does not present the dienophile with the requisite flat 1,3-diene system. This is confirmed in ^{13}C NMR Spectrum C: in addition to the two nitrile resonances, it shows two signals for olefinic carbons ($\delta > 120$ ppm) and three more for saturated carbons ($\delta < 50$ ppm), suggesting Structure V.

For the assignment of the ^{13}C chemical shifts in Spectrum C, one can first locate the carbons not bonded to hydrogen, recognizable by their weaker signals (no nuclear Overhauser effect) at 113.65, 112.25, and 42.9 ppm. According to Fig. 6.2, p. 43, the signals at 113.65 and 112.25 ppm correspond to the nitrile carbon atoms. The *endo*/*exo* assignment is difficult because γ effect A is probably larger than γ effect B in this case:

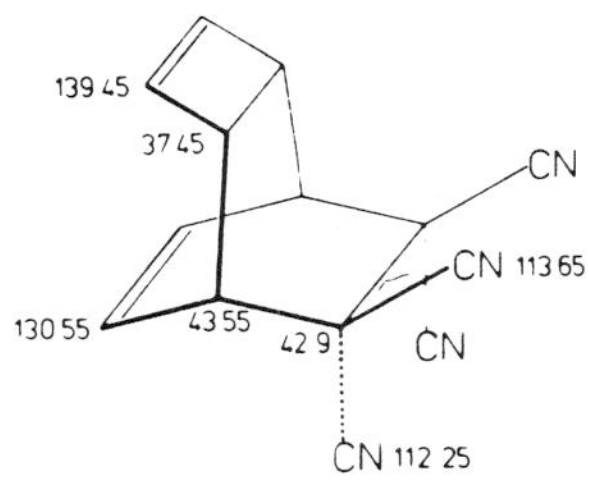

V

Decoupling experiments D through F make a distinction between the carbon atoms bonded to hydrogen possible: The cyclobutene ring protons appear as singlets at 3.21 and 5.95 ppm (τ values 6.79 and 4.05, see Reference). If one decouples selectively at 3.21 ppm in Spectrum D and at 5.95 ppm in Spectrum E, the ^{13}C resonances at 37.45 and 139.45 ppm appear as sharp singlets that therefore belong to the cyclobutene ring. Finally, selective decoupling of the olefinic protons of the bicycle [2.2.2] octene system in V at 6.38 ppm affords a singlet at 130.55 ppm in Spectrum F, whereas all other C-H carbon atoms appear as off-resonance doublets.

27

Comparison of the signal intensities gives the ratio 1:2:1. Thus the three kinds of carbon atoms found in the spectrum have a 4:8:4 distribution in the molecule $C_{16}H_{16}$.

According to its molecular formula, the compound contains nine double-bond equivalents (as compared with a saturated hydrocarbon with the formula $C_{16}H_{34}$), which can be fitted by two benzene rings ($2 \times 4 = 8$ double-bond equivalents) and an additional ring (one double-bond equivalent).

Because only two signals in the ratio 1:2 or 4:8 appear in the aromatic region, both benzene rings are *p*-substituted with the same substituent. This yields the partial formula $C_{12}H_8$; therefore, a C_4H_8 residue is left after subtracting from $C_{16}H_{16}$. This corresponds to four methylene groups with δ = 35.95 ppm, connecting the *para* positions of both benzene rings:

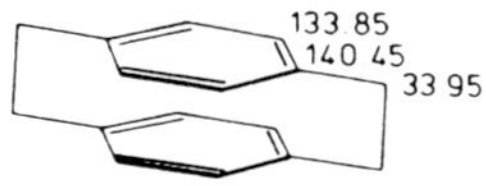

p-cyclophane

The chemical shift of the aromatic C-H carbons is somewhat higher than in alkylbenzenes (133.85 ppm instead of 129 ± 1 ppm, according to Table 7.10, p. 61), possibly due to distortion of the coplanarity of the benzene ring atoms. The large chemical shift of the methylene carbons can be explained as a consequence of the distorted ring current field of both benzene rings.

28

Both spectra show four signals each in the olefinic and aromatic carbon resonance region. Considering the somewhat different line widths (see, for instance, the signals at 147.03 and 126.37 ppm), the intensity ratios 1:2:2:1 (B) and 1:4:1 (A) are found, as is known to be the case for monosubstituted benzenes (e.g., Fig. 3.5B, p. 25). When the signals at 41 and 43 ppm are included, the ^{13}C NMR spectra are seen to correspond to two samples of polystyrene.

In spectrum A, the resonances at 43 ppm and at 146.43 ppm show a conspicuous structure (splitting), which is not observable for the corresponding signals in Spectrum B (sharp signals). This suggests that these carbon atoms are nonequivalent. The different shielding can only be a consequence of different stereoregularity (*tacticity*) in both samples, which is depicted as follows and symbolized using the accepted classification (see references in Exercise 28):

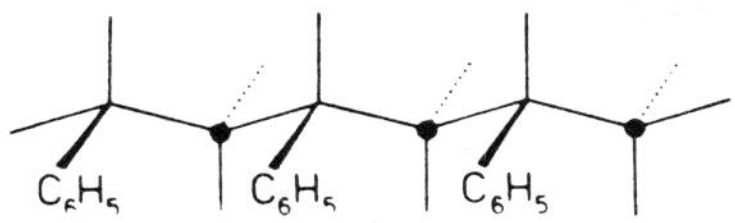

meso-meso- or
mm-triad

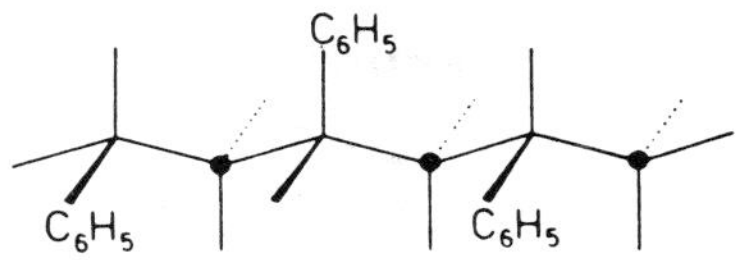

racem-racem- or
rr-triad

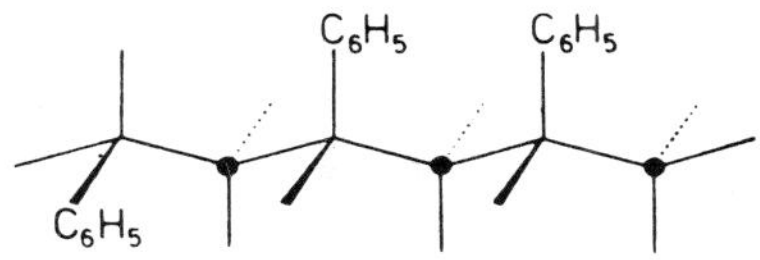

racem-meso- or
rm-triad

With these models, one can see that the CH_2 carbons (•) are shielded to different extents in the three possible triads. The same is true for the quaternary phenyl carbons because they are also in the α position with respect to the asymmetric methine carbon. Therefore, in the ^{13}C NMR spectrum of the *isotactic* material with the *mmmmmm* . . . . sequence, uniformly shielded methylene and C-1 phenyl carbon signals should appear. On the other hand, a *triad splitting* of these carbon resonances is expected in the *atactic* polymer with all possible statistically important sequences. It follows that Spectrum B was recorded for *isotactic* polystyrene, and *Spectrum* A was recorded for *atactic* polystyrene (triad splitting at 43 ppm, also suggested at 146.4 ppm). The following assignment is possible:

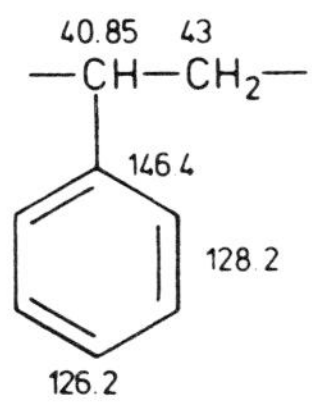

Atactic

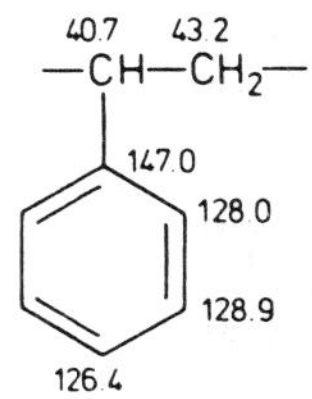

Isotactic

(See Table 7.10, ethyl increments)

29

The quartets at 21.25 and 20.95 ppm with the approximate intensity ratio 1:2 correspond to three methyl groups, two of which are equivalent. The doublet at 85.55 and the long-range doublet at 81.75 ppm, on the basis of their chemical shifts and extremely large coupling constants, belong to an ethynyl group. If one subtracts the moieties found in this way ($3 \times CH_3 + 1 \times C_2H = C_5H_{10}$) from the given molecular formula $C_{11}H_{12}$, C_6H_2 remains. This part of the molecule must be bonded to all four substituents; replacing these formally by hydrogens results in C_6H_6. The compound is therefore a tetrasubstituted benzene. Since only four rather than six carbon resonances lie in the aromatic region, the substituents are distributed symmetrically: *i.e.*, in the 1,2,4,6- or 1,3,4,5- positions:

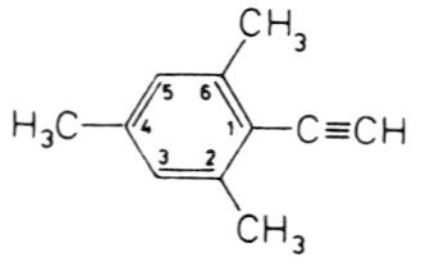

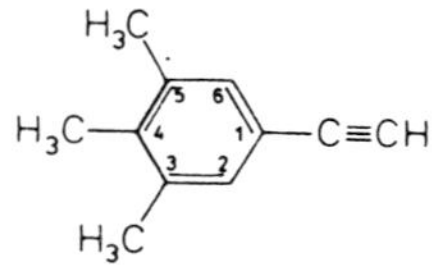

In the case of 3,4,5-trimethylphenylacetylene, the quaternary acetylene carbon should appear as a double triplet because of long-range coupling with the two *ortho* protons. Since this signal is only a plain doublet, the *ortho* positions are substituted. Therefore, the unknown compound is mesitylacetylene. The ^{13}C chemical shifts calculated for mesitylacetylene from the chemical shifts of the mesitylene rings carbons (C-C: 137.6 ppm, C-H: 127.4 ppm) and the ethynyl group increments (Table 7.10, p. 61) are in good agreement with the measured values and confirm the assignment:

Result of increment estimate

C-1 : 121.3 ppm
C-2,6: 141.4 ppm
C-3,5: 127.8 ppm
C-4 : 137.4 ppm

Assignment

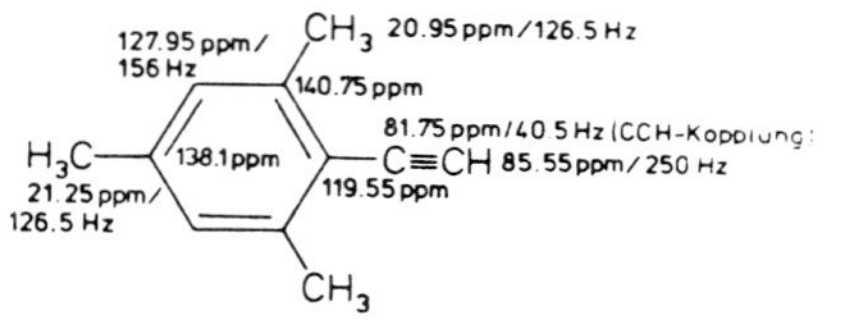

In this example, the dependence between the s character of the hybrid bond orbitals of the coupled carbon atom and its C-H coupling constants can be recognized.

30

Assignment of SpectrumA[1]

ppm	Intensity	Assignment
129.5	1.00	C-4,5
126.5	0.25	C-1,2
115.5	0.87	C-3,6
19.0	0.27	C-7

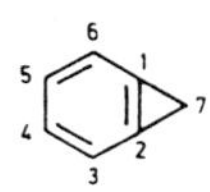

[1]See J. Prestien, H. Günther, *Angew. Chem.* **86,** 278 (1974).

In Spectrum (B) of the product obtained from 7-^{12}C-7,7-dichloronorcarene, the C-7 signal is obscured by noise. Therefore, dehydrohalogenation proceeds without rearrangement by Route A.

31

According to the molecular formula, the compound contains six double-bond equivalents (compare $C_{15}H_{20}$ and $C_{15}H_{32}$). Four of these belong to a monosubstituted benzene ring, as indicated by the four signals between 128.4 and 138.7 ppm (doublet, doublet, doublet, singlet; intensity ratio 1:2:2:weak). The two remaining double-bond equivalents correspond to two carboxyl groups (171.6 ppm), as shown by the intensity ratio of about 2:1:1 for the singlets at 171.6, 138.7, and 63.8 ppm.

In the region of the saturated carbon resonances, in addition to the already-mentioned quaternary carbon atom at 63.8 ppm, two O-ethyl groups and a C-ethyl group can be distinguished (triplet-quartet pairs at 61.75/14.15 and 29.8/9.5 ppm with the intensity ratio 2:1).

The following recognized building blocks add up to the known molecular formula $C_{15}H_{20}O_4$ and can be assembled to give the diethyl ester of phenylethylmalonic acid:

2 carbonyls	C_2O_2
1 phenyl	C_6H_5
1 quaternary carbon	C
2 O-ethyls	$C_4H_{10}O_2$
1 C-ethyl	C_2H_5

In the next assignment, the *ortho* and *meta* carbon resonances in the phenyl ring are interchangeable.

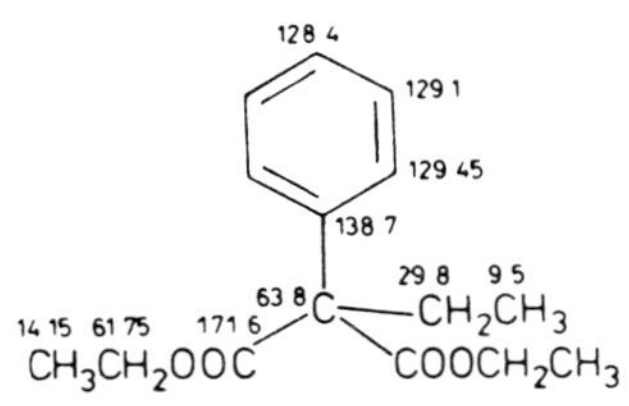

32

According to the differential equation for spin-lattice relaxation,[1]

$$\frac{dM_t}{dt} = \frac{M_o - M_t}{T_1} \quad \text{or} \quad d\ln(M_o - M_t) = \frac{dt}{T_1},$$

---

[1][See, for example, E. Breitmaier, K.-H. Spohn, S. Berger, *Angew. Chem.* **87**, 152 (1975), and Section 9.1, p. 00].

the spin-lattice relaxation time T_1 can be found by plotting the integral differences $(M_o - M_\tau)$ determined for each M_τ semilogarithmically against the respective interval τ. The integrals for τ = 450, 160, 80, or 40 s can be taken as the equilibrium value M_o, because these values differ only within the limits of experimental error. As an example, the *ortho* carbons at 123.3 ppm give for τ = 3 s, $M_o - M_\tau$ = 33799.4 − 8208.3 = 25591.1. Proceeding in a similar fashion for the other intervals, and doing the same for the signals at 129.5 and 134.8 ppm, three straight regression lines result (as shown in the figure on p. 297.) T_1 can be found most simply by parallel displacement of the straight lines to e.10^x (lg → ln conversion) and by intersection of the parallel lines obtained in this way with the decade 10^x.

Using the known assignment (see Fig. 8.3, p. 81), the result is as follows:

C-2,6	(123.3 ppm):	T_1 = 14.3 s
C-3,5	(129.5 ppm):	T_1 = 13.1 s
C-4	(134.8 ppm):	T_1 = 8.5 s

Therefore, the *ortho* and *meta* carbons relax much more slowly than the *para* carbon. The reason is the preferred rotation of monosubstituted benzenes around the axis running through the substituent and the *para* carbon atom. In this way, the orientation of the *ortho* and *meta* C-H bonds relative to the magnetic field vector varies too rapidly for a rapid dipole-dipole relaxation. On the contrary, the *para* C-H bond maintains its orientation when the molecule spins that way and is influenced only by the molecule's less preferred rotations. These rotations have lower frequencies and are more favorable to a rapid spin-lattice relaxation.

The C-1 signals are extremely broad owing to the ^{13}C-^{14}N coupling (quadrupole coupling). This circumstance, and the low intensity related to the absence of a nuclear Overhauser effect, lead to large fluctuations of the integrals. For this reason, one must do without a T_1 determination for C-1.

33

Using signal multiplicities, coupling constants, and intensities (namely, signals at 139.15 and 138.15 ppm), the following assignment can be tabulated:

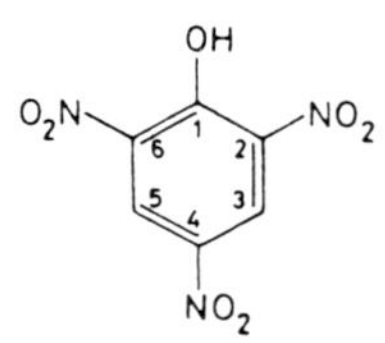

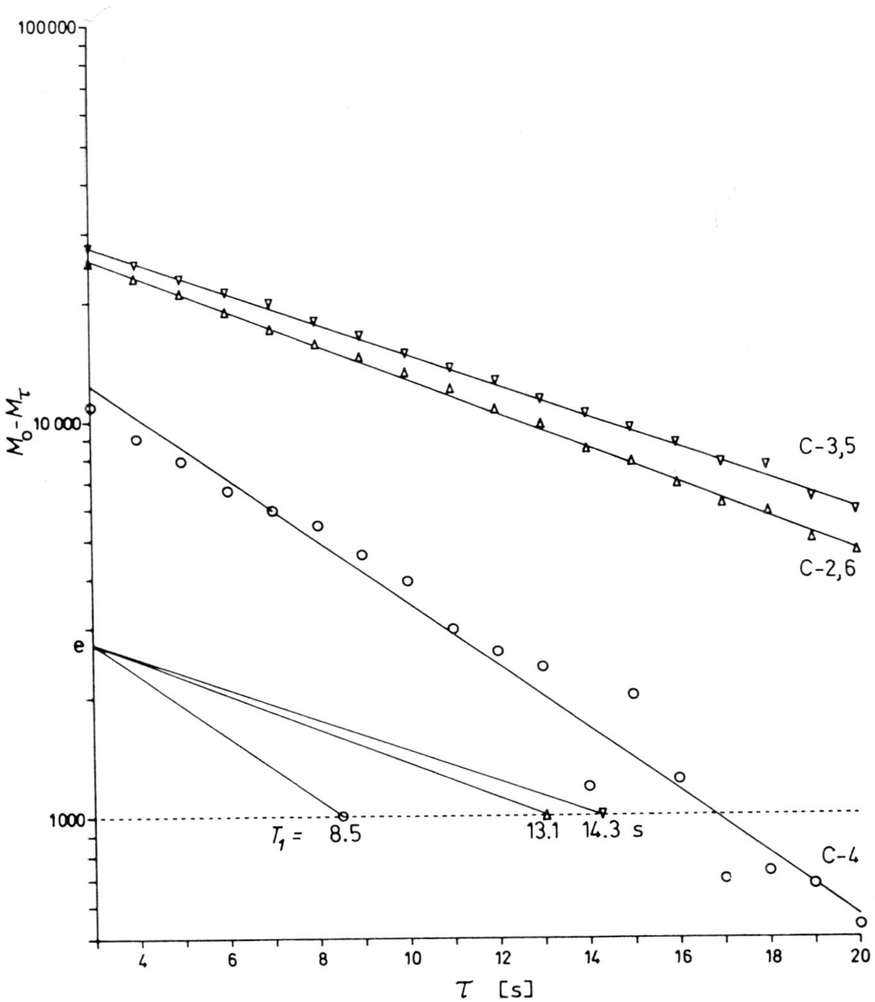

ppm	Multiplicity	Coupling Constant(s)	Assignment
153.2	Long-range triplet	$J_{C\text{-}1\text{-}H\text{-}3,5} = 5.2$ Hz	C-1
139.15	Unresolved long-range multiplet	—	C-2,6
138.15	Unresolved long-range multiplet	—	C-4
126.2	Doublet of long-range doublets	$J_{C\text{-}3\text{-}H\text{-}3} = J_{C\text{-}5\text{-}H\text{-}5} = 175.4$ Hz	C-3,5
		$J_{C\text{-}3\text{-}H\text{-}5} = J_{C\text{-}5\text{-}H\text{-}3} = 5.5$ Hz	

34

Interpretation of the Spectrum

ppm	Multiplicity	C-H Coupling Constant(s)	Assignment	Origin
201.45	Long-range sextet	$J_{C\text{-}C\text{-}H}$ = 6.2 Hz	$\diagdown$C=O	Acetone
152.05	Unresolved long-range multiplet (strong)	—	2 $\geqslant$C—NO_2	Trinitro-chlorobenzene
147.5	Long-range triplet	$J_{C\text{-}C\text{-}H}$ = 2.0 Hz	$\geqslant$C—NO_2	Trinitro-chlorobenzene
132.0	Long-range triplet	$J_{C\text{-}C\text{-}H}$ = 5.1 Hz	$\geqslant$C—CH_2—	Trinitro-chlorobenzene
124.0	Doublet of long-range doublets (extremely strong)	$J_{C\text{-}H}$ = 177.9 Hz[a] $J_{C\text{-}C\text{-}C\text{-}H}$ = 5.1 Hz[a]	2 $\geqslant$C—H	Trinitro-chlorobenzene
43.2	Triplet	$J_{C\text{-}H}$ = 132.3 Hz	—CH_2—	Acetone
29.7	Quartet	$J_{C\text{-}H}$ = 128.7 Hz	—CH_3 $C_9H_7N_3O_7$	Acetone

[a]In good agreement with the corresponding values for picric acid (Spectrum 33).

Multiplicities, coupling constants, and a look at the intensities of the signals at 152.05, 147.5, and 124.0 pm, allow an assignment that is consistent with the given molecular formula and fits the structure of 2,4,6-trinitrophenylacetone, which is one of the products one would expect from the reaction of 2,4,6-trinitrochlorobenzene and acetone.

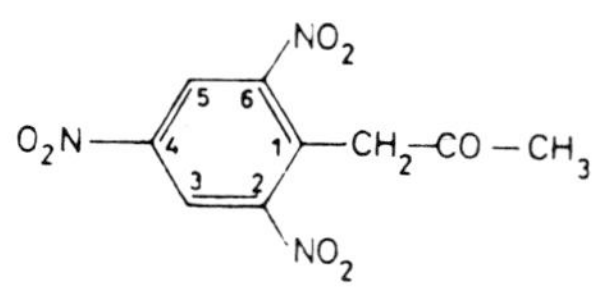

Although C-3,5 of the 2,4,6-trinitrophenyl ring in picric acid (Spectrum 33, 126.2 ppm) and in 2,4,6-trinitrophenylacetone (124.0 ppm) differ only slightly, the C-2,6 and C-4 resonances appear 10 ppm and more upfield in picric acid. Strong shielding is a consequence of the OH group's mesomeric electron-releasing ability. The inductive effect of both substituents can be recognized in the C-1 chemical shifts (picric acid: 153.2 ppm, OH effect; 2,4,6-trinitrophenylacetone: 132.0 ppm, alkyl effect).

35

Interpretation of the Spectrum

ppm	Multiplicity	C-H Coupling Constants	Assignment	Origin
205.55	Long-range sextet	$J_{C\text{-}C\text{-}H}$ = 5.9 Hz (CO with CH_2 and CH_3)	>CO	Acetone
133.35	Poorly resolved long-range multilet (strong)	—	2 >C—NO_2	Trinitrophenyl group
127.6	Doublet of long-range doublets (very strong)	$J_{C\text{-}H}$ = 166.2 Hz $J_{C\text{-}C\text{-}C\text{-}H}$ = 4.4 Hz (C-3,5 with H-5,3 and H-1)	2 >C—H	Trinitrophenyl group
121.5	Poorly resolved long-range multiplet	—	>C—NO_2	Trinitrophenyl group
47.0	Triplet (overlapping with DMSO-D_6 septet	$J_{C\text{-}H}$ = 130.1 Hz	—CH_2—	Acetone
34.5	Doublet of long-range triplets	$J_{C\text{-}H}$ = 145.6 Hz $J_{C\text{-}C\text{-}C\text{-}H}$ = 4.5 Hz (C-1 with H-3,5)	>C—H	Trinitrophenyl group
29.75	Quartet	$J_{C\text{-}H}$ = 127.2 Hz	—CH_3	Acetone

In the ^{13}C-NMR spectrum only three carbon signals (with intensity ratio ≈ 2 : 2 : 1) appear in the aromatic chemical shift region of the trinitrophenyl group. However, an additional signal appears in the aliphatic region which corresponds to an sp^3 carbon atom on the basis of its chemical shift (34.5 ppm), multiplicity, and C-H coupling constant (145.6 Hz), supporting the following structure for the Meisenheimer salt:

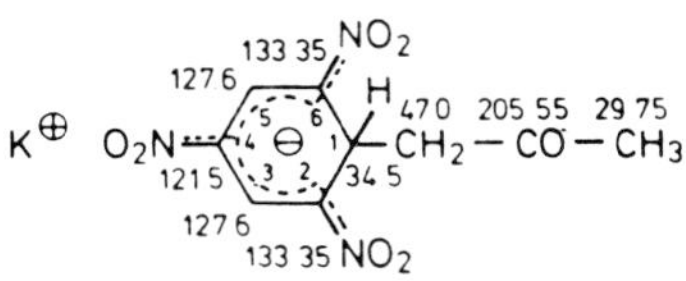

36

By adding the OH and $CH(CH_3)_2$ substituent increments given in Table 7.10, p. 61 to the ^{13}C chemical shift of C-3 in toluene (128.9 ppm), the following estimate can be made for the phenolic carbon atom ("C-3") of thymol:

$$
\begin{aligned}
\delta_3 = \delta_{3(\text{Toluol})} &= 128.9 \text{ ppm}\\
+Z_1(\text{OH}) &= 26.9 \text{ ppm}\\
+Z_o(\text{CH(CH}_3)_2 &= -\ 2.0 \text{ ppm}\\
\hline
\delta_3 = &\quad 153.8 \text{ ppm}
\end{aligned}
$$

All the carbon chemical shifts of thymol and carvacrol can be estimated in a similar way. The results are summarized in the following table, opposite the values obtained from Spectra A and B. Spectrum A therefore corresponds to thymol, and Spectrum B, to carvacrol. The discrepancies between the measured and the estimated values are considerable for the phenolic carbon atoms, probably as a consequence of hydrogen bonding.

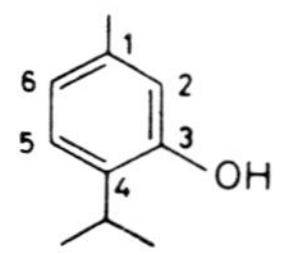

Thymol

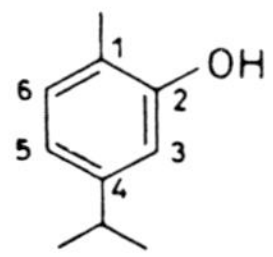

Carvacrol

Atom/estimated Value	Experimental Values A	Experimental Values B	Estimated Value/ atom
C-3/153.8	152.4	153.45	156.5 /C-2
C-1/137.45	136.5	148.4	147.45/C-4
C-4/133.55	132.0	131.1	131.2 /C-6
C-5/128.5	126.3	121.6	123.25/C-1
C-6/122.0	122.05	119.1	119.3 /C-5
C-2/117.0	116.5	113.5	114.5 /C-3

37

Interpretation of the Spectrum

ppm	Off-resonance Multiplicity	Intensity	Assignment
142.5	Singlet	Medium	-N-C=
131.15	Doublet	Very strong	=CH-
127.4	Doublet	Strong	=CH-
127.15	Singlet	Medium	=C-
122.25	Singlet	Medium	=C-
115.15	Doublet	Very strong	=CH-
20.4	Quartet	Medium strong	$-CH_3$
17.05	Quartet	Strong	$-CH_3$

The six signals in the 115.15–142.5 ppm region suggest the presence of a benzene ring. The two quartets at 17.05 and 20.4 ppm correspond to two methyl groups occupying different ring positions. Considering the molecular formula $C_8H_{11}N$, after subtracting the methyl groups and three ring protons, an amino group is left as the third ring substituent. Therefore, the singlet at 142.5 ppm corresponds to the quaternary C-1, which is most strongly deshielded by the amino group. The symmetrically substituted 2,6- and 3,5-dimethylanilines can be ruled out on the basis of the number of signals (two methyl and six aromatic carbon resonances). Using the ^{13}C chemical shifts of toluene and the increments for NH_2 and CH_3, the ring carbon chemical shifts can be estimated for the four unsymmetrically substituted dimethylanilines, as shown in the following table. The best agreement between measured and estimated values is found for 2,4-dimethylaniline. To assign the C-4 and C-5 resonances, the signal intensities in the proton-decoupled spectrum (nuclear Overhauser effect at C-5, with 127.4 ppm) must be considered.

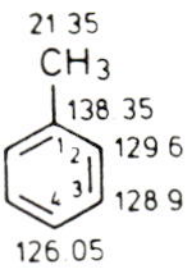

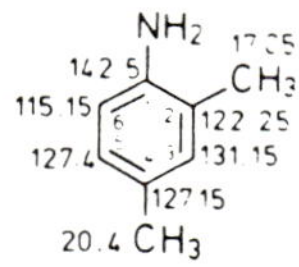

-Dimethyl-aniline	δ_1	δ_2	δ_3 (ppm)	δ_4	δ_5	δ_6
2,3-	148.8	125.95	140.25	120.1	125.8	113.65
2,4-	145.7	125.95	131.5	128.7	127.95	116.5
2,5-	148.8	122.85	130.9	120.0	136.65	117.1
3,4-	148.1	117.2	140.25	129.4	130.8	113.65
	142.5	*122.25*	*131.15*	*127.15*	*127.4*	*115.15*

$$\delta_1 = 129.6 + Z_{1NH_2} + Z_{pCH_3} = 129.6 + 19.2 - 3.1 = 145.7$$
$$\delta_2 = 138.35 + Z_{oNH_2} + Z_{mCH_3} = 138.35 - 12.4 + 0.0 = 125.95$$
$$\delta_3 = 129.6 + Z_{mNH_2} + Z_{oCH_3} = 129.6 + 1.3 + 0.6 = 131.5$$
$$\delta_4 = 128.9 + Z_{pNH_2} + Z_{1CH_3} = 128.9 - 9.5 + 9.3 = 128.7$$
$$\delta_5 = 126.05 + Z_{mNH_2} + Z_{oCH_3} = 126.05 + 1.3 + 0.6 = 127.95$$
$$\delta_6 = 128.9 + Z_{oNH_2} + Z_{mCH_3} = 128.9 - 12.4 + 0.0 = 116.5$$

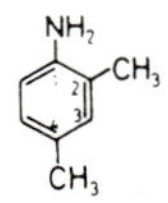

$$\delta_1 = 129.6 + Z_{1NH_2} + Z_{mCH_3} = 129.6 + 19.2 + 0.0 = 148.8$$
$$\delta_2 = 138.35 + Z_{oNH_2} + Z_{oCH_3} = 138.35 - 12.4 + 0.6 = 125.95$$
$$\delta_3 = 138.35 + Z_{mNH_2} + Z_{oCH_3} = 138.35 + 1.3 + 0.6 = 140.25$$
$$\delta_4 = 129.6 + Z_{mCH_3} + Z_{pNH_2} = 129.6 + 0.0 - 9.5 = 120.1$$
$$\delta_5 = 128.9 + Z_{pCH_3} + Z_{mNH_2} = 128.9 + 3.1 + 0.0 = 125.8$$
$$\delta_6 = 126.05 + Z_{oNH_2} + Z_{mCH_3} = 126.05 - 12.4 + 0.0 = 113.65$$

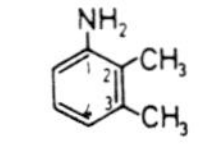

$$\delta_1 = 129.6 + Z_{1NH_2} + Z_{mCH_3} = 129.6 + 19.4 + 0.0 = 148.8$$
$$\delta_2 = 138.35 + Z_{oNH_2} + Z_{pCH_3} = 138.35 - 12.4 - 3.1 = 122.85$$
$$\delta_3 = 129.6 + Z_{mNH_2} + Z_{mCH_3} = 129.6 + 1.3 + 0.0 = 130.9$$
$$\delta_4 = 128.9 + Z_{pNH_2} + Z_{oCH_3} = 128.9 - 9.5 + 0.6 = 120.0$$
$$\delta_5 = 126.05 + Z_{1CH_3} + Z_{mNH_2} = 126.05 + 9.3 + 1.3 = 136.65$$
$$\delta_6 = 128.9 + Z_{oNH_2} + Z_{oCH_3} = 128.9 - 12.4 + 0.6 = 117.1$$

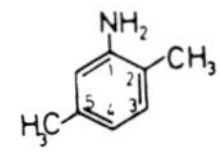

$$\delta_1 = 128.9 + Z_{1NH_2} + Z_{pCH_3} = 128.9 + 19.2 - 3.1 = 148.1$$
$$\delta_2 = 129.6 + Z_{oNH_2} + Z_{mCH_3} = 129.6 - 12.4 + 0.0 = 117.2$$
$$\delta_3 = 138.35 + Z_{mNH_2} + Z_{oCH_3} = 138.35 + 1.3 + 0.6 = 140.25$$
$$\delta_4 = 129.6 + Z_{pNH_2} + Z_{1CH_3} = 129.6 - 9.5 + 9.3 = 129.4$$
$$\delta_5 = 128.9 + Z_{mNH_2} + Z_{oCH_3} = 128.9 + 1.3 + 0.6 = 130.8$$
$$\delta_6 = 126.05 + Z_{oNH_2} + Z_{mCH_3} = 126.05 - 12.4 + 0.0 = 113.65$$

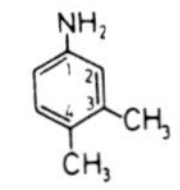

38

The alkylation took place on the sulfur, as the more intense downfield off-resonance triplet lies at 33.35 ppm, corresponding to an —S—CH_2— group. Values of about 50 ppm are typical for N—CH_2— groups (see Fig. 6.2, p. 43, and the chemical shift increments in Table 7.2, p. 52).

Because only two signals with the intensity ratio 2:1 appear in the aliphatic resonance region, two *o*-aminothiophenol molecules were *S*-alkylated. Therefore, the reaction product is 1,3-bis[2-aminophenylthio]propane. The assignment of the ^{13}C chemical shifts can be made on the basis of the off-resonance multiplicities (S = singlet, D = doublet, T = triplet) and the effects of the larger amino group substituents on the benzene carbon chemical shifts (deshielding of C-1, shielding at the *ortho* and *para* positions, see Table 7.2).

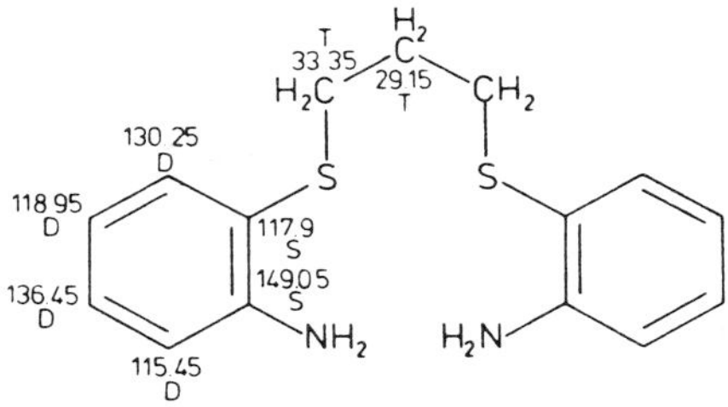

39

N-Phenylthiourea can exist as any one of the following tautomeric forms, a, b, and c, or as an equilibrium mixture of these:

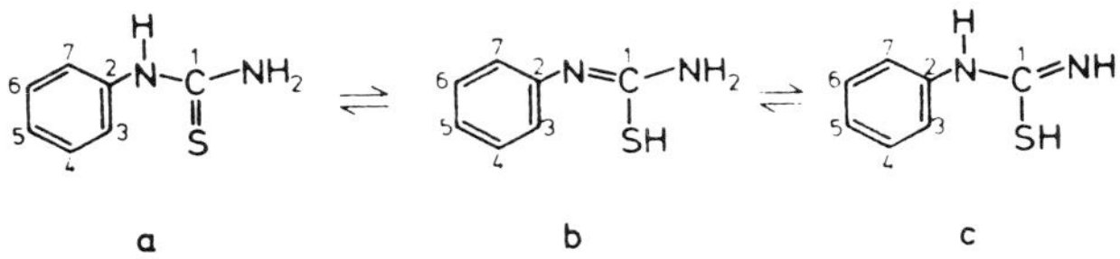

1	: 182.2 ppm		1	: 157.15 ppm	
2	: 139.8 ppm		2	: 141.2 ppm	
3,7	: 123.95 ppm		3,7	: 118.7 ppm	
4,6	: 129.55 ppm		4,6	: 129.4 ppm	
5	: 125.35 ppm		5	: 122.05 ppm	

In addition to five strong signals, the noise-decoupled sectrum shows another five low-intensity signals with similar chemical shifts. The strong singlet at 182.2 ppm corresponds to the thiocarbonyl carbon C-1 in tautomer a, whereas the singlet at 157.15 ppm corresponds to the much more strongly shielded azomethine carbon C-1 of the isothiourea forms b and c. Accordingly, the four strong *N*-phenyl ring signals in the 139.8–123.95 ppm region should be assigned to the more abundant tautomer a, and the four weak signals in the 141.2–122.05 ppm region belong to b or c. Assignment of the aromatic carbon signals can be made using the multiplicities and coupling constants in the coupled spectrum and the known C-H coupling constants in monosubstituted benzenes (Table 8.2B, p. 78).

Thus the spectrum shows that the *N*-phenylthiourea exists in two tautomeric forms in hexadeuteriodimethylsulfoxide. Tautomer a (ca. 90%) is more stable than b or c (ca. 10%), and whether a or b is actually present could be decided by labeling one of the nitrogen atoms with ^{15}N.

Assignment of Spectrum 39

ppm	Multiplicity	Coupling Constant $J_{C\text{-}H}$ [Hz]	Assignment a	Assignment b,c
182.2	Singlet	—	C-1	—
157.15	Singlet	—	—	C-1
141.2	Long-range triplet	$J_{C\text{-}2\text{-}H\text{-}4,6} \approx 7$	—	C-2
139.8	Long-range triplet	$J_{C\text{-}2\text{-}H\text{-}4,6} = 6.9$	C-2	—
129.55	Doublet of long-range doublets	$J_{C\text{-}4\text{-}H\text{-}4} = 161.5$ $J_{C\text{-}4\text{-}H\text{-}6} = 7.0$	C-4,6	—
129.4	Not clearly visible because of overlap		—	C-4,6
125.35	Doublet of long-range triplets	$J_{C\text{-}5\text{-}H\text{-}5} = 161.5$ $J_{C\text{-}5\text{-}H\text{-}3,7} = 6.7$	C-5	—
123.95	Doublet of long-range multiplets	$J_{C\text{-}3\text{-}H\text{-}3} = 161.0$	C-3,7	—
122.05	—	—	—	C-5
118.7	Broad doublet	$J_{C\text{-}3\text{-}H\text{-}3} \approx 161$	—	C-3,7

40

Amines add to isocyanates to produce ureas. Accordingly, 3-amino-2-methylacrolein and phenyl isocyanate should react as follows:

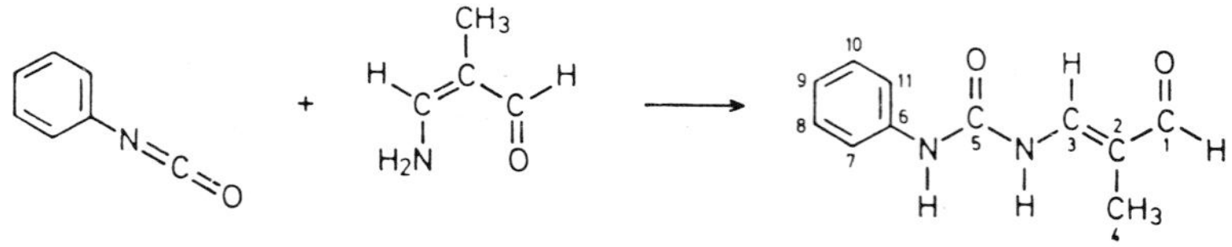

The measured ^{13}C chemical shifts correspond to the 2-formylpropenylphenylurea obtained in this way. It is possible to recognize the 3-amino-2-methylacrolein C-1 (192.1 ppm), C-2 (116.9 ppm), and C-4 (7.95 ppm) signals, as shown in the following table. The urea carbonyl carbon (C-5) appears as a singlet in the off-resonance proton-decoupled spectrum. The enamine carbon (C-3) resonance lies near by, but as an off-resonance doublet. The C-3 and C-5 resonances can thus be assigned unequivocally. The remaining signals agree well with an *N*-acylated aniline, as can be seen by comparing them with the benzene ring chemical shifts after adding the increments for an *N*-acetylamino group (Table 7.10, p. 61).

Assignment of the Spectrum

ppm	Origin	Off-resonance Multiplicity	Assignment
192.10	3-amino-2-methylacrolein	Doublet	C-1 (—CH=O)
151.30	Phenyl isocyanate	Singlet	C-5 (>C=O)
145.50	3-amino-2-methylacrolein	Doublet	C-3
139.75	Phenyl isocyanate	Singlet	C-6
129.65	Phenyl isocyanate	Doublet (very strong)	C-8,10
123.60	Phenyl isocyanate	Doublet	C-9
119.30	Phenyl isocyanate	Doublet (very strong)	C-7,11
116.90	3-amino-2-methylacrolein	Singlet	C-2
7.95	3-amino-2-methylacrolein	Quartet	C-4 ($—CH_3$)

41

The ^{13}C NMR spectrum shows the following functional groups, which also agree with the molecular formula:

1. $—O—C_2H_5$ (Off-resonance triplet for $O—CH_2—$ at 63.25 ppm and off-resonance quartet for $—CH_3$ at 14.7 ppm; see, for example, diethyl ether and ethanol, Fig. 2.2, p. 15, and Fig. 4.1, p. 32)

2. $—N(H)—CO—CH_3$ (Off-resonance quartet for $—CH_3$ at 23.8 ppm and singlet at 168.2 ppm; see, for example, the $—CO—CH_3$ resonances of dimethylacetamide in Exercise 9, p. 273)

In the case of the acetyl group, it cannot be bonded to carbon, as one would expect $\delta > 190$ ppm (see the examples in Section 7.8, p. 65). An *O*-acetyl group can be excluded because the molecular formula only admits two oxygen atoms, and these atoms are already accounted for as parts of the ethoxy and carbonyl groups. On the basis of the elementary composition, only an NH-acetyl group is possible.

By subtracting the partial formulas found ($—NH—CO—CH_3$ + $—O—C_2H_5$ = $C_4H_9NO_2$) from the given molecular formula, $C_{10}H_{13}NO_2$, C_6H_4 results, in agreement with a disubstituted benzene. Because only four (and not six) signals appear in the aromatic resonance region, with a relative intensity ratio of 1:1:2:2 (the stronger signals are off-resonance doublets), the benzene ring is symmetrically (1,4) disubstituted. Therefore, the compound is *p*-ethoxyacetanilide (phenacetin).

To assign the ring carbon signals, one can take the chemical shifts of anisole (Section 6.2, p. 41), add the acetylamino group increments—thus estimating the chemical shifts of *p*-methoxyacetanilide as a reference substance—and carry over the signal sequence to the given spectrum:

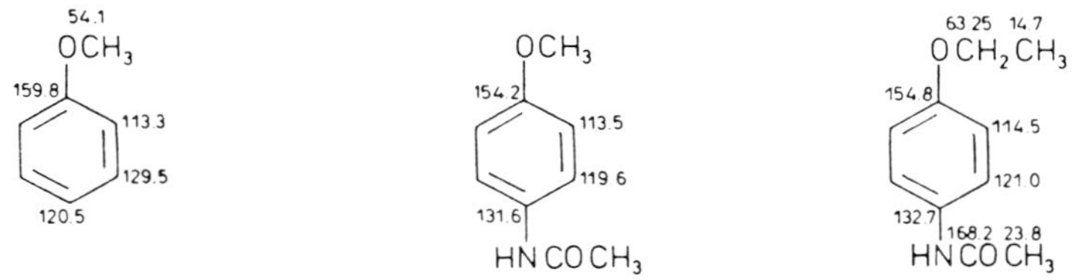

Anisole (measured values) | *p*-methoxyacetanilide (estimated values) | *p*-ethoxyacetanilide (measured values)

$\delta_1 = 159.8 + Z_{p\,NHCOCH_3} = 154.2$ ppm $\delta_{3,5} = 129.5 + Z_{o\,NHCOCH_3} = 119.6$ ppm

$\delta_{2,6} = 113.3 + Z_{m\,NHCOCH_3} = 113.5$ ppm $\delta_4 = 120.5 + Z_{1\,NHCOCH_3} = 131.6$ ppm

42

^{1}H-NMR spectrum (A):

Because only one intense singlet appears at 1.34 ppm, in addition to signals in the aromatic region and those corresponding to the acetylamino group (NH, 11.4; acetyl CH_3, 2.28 ppm), the alkyl residue introduced is a *t*-butyl group.

Two signals in the aromatic region (8.62 and 7.66 ppm) constitute an *ortho*-coupled AB system with $J_{AB} = 8.7$ Hz. Another proton (M) appears at 8.16 ppm and exhibits a *meta* coupling, which can also be observed in the B portion of the AB system at 7.66 ppm ($J_{BM} = 2.5$ Hz). Therefore, the compound is an *o,p*-disubstituted acetanilide.

The position of the substituents results from the fact that the *ortho* effect of the nitro group on the proton chemical shift is 0.95 ppm, whereas that of the *t*-butyl group is only 0.01 ppm (compare H. Günther, *NMR-Spektroskopie,* Georg-Thieme-Verlag, Stuttgart, 1973, p. 104). Adding these increments to the proton chemical shift of benzene (7.27 ppm), one obtains 8.23 ppm. This result agrees best with the measured value of 8.16 ppm for the *meta*-coupled aryl proton. Therefore, the reaction product, $C_{12}H_{16}N_2O_3$, is 2-nitro-4-*t*-butylacetanilide:

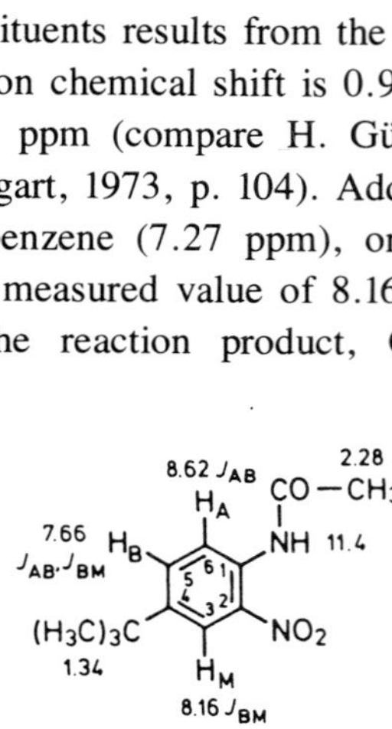

^{13}C-NMR spectra (B–E):
In ^{13}C NMR Spectrum B, the assignment of the acetyl and *t*-butyl resonances is straightforward (acetyl CO, 169.2; acetyl CH_3, 25.55 ppm). The assignments of the benzene carbon resonances can be reached using the substituent increments for $NHCOCH_3$, NO_2, and *t*-C_4H_9, according to Table 7.10, p. 61:

Estimated and measured values of the ^{13}C chemical shifts of 2-nitro-4-*t*-butylacetanilide

	Estimated	Measured
C-1	131.2	133.15
C-2	137.8	137.10
C-4	145.7	147.75

Owing to the small chemical shift differences, the assignment of the signals at 122.7 and 122.9 ppm remains open. This problem can be solved using the selectively proton-decoupled ^{13}C-NMR spectra C–E:

Spectrum	Perturbed ^{1}H-Signal	Decoupled ^{13}C-Signal	Assignment
C	7.66 ppm/6001 Hz, H-5	134.05 ppm	C-5
D	8.16 ppm/6041 Hz, H-3	122.7 ppm	C-3
E	8.62 ppm/6076 Hz, H-6	122.9 ppm	C-6

Altogether, the following assignment results:

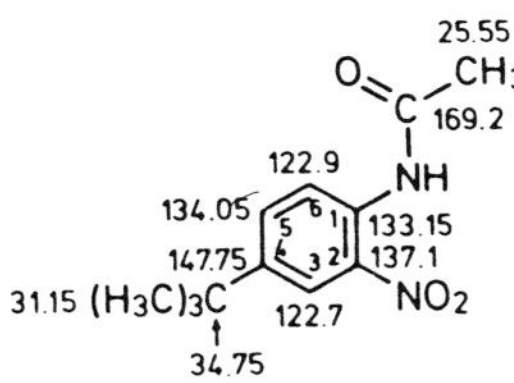

43

The compound contains a conjugated aldehyde group (doublet at 192.45 ppm, J_{CH} = 173 Hz, see Section 7.8, p. 65) and a methoxy group (quartet at 56.45 ppm, J_{CH} = 145.1 Hz; compare anisole, Section 6.2, p. 41). In addition, the spectrum shows six signals between 110 and 153 ppm, three of which are singlets and three are doublets with J_{CH} = 160–163 Hz (Spectrum B). Because, on the basis of its molecular formula, the compound contains five double bond equivalents and one of them belongs to the aldehyde group, the six signals indicate a trisubstituted benzene (four double-bond equivalents). The third substituent must be a hydroxyl group since $C_8H_8O_3$ − (CH_3O + CHO + C_6H_3) = OH. Thus the aromatic comound is a hydroxymethoxy-benzaldehyde. The *meta* position of the methoxy group and the *para* position of the hydroxy group, corresponding to the structure of vanillin, result for C-1, C-3, and C-5 from the chemical shift increments given in Table 7.10, p. 61. For example:

$$\delta_1 = 128.5 + Z_{1_{CHO}} + Z_{m_{OCH_3}} + Z_{p_{OH}} = 128.5 + 7.5 + 0.9 - 7.6$$
$$= 129.3 \text{ ppm}$$

Carbon atoms C-2, C-3, and C-4 show larger discrepancies between the estimated chemical shifts (in italics) and measured ^{13}C chemical shifts:

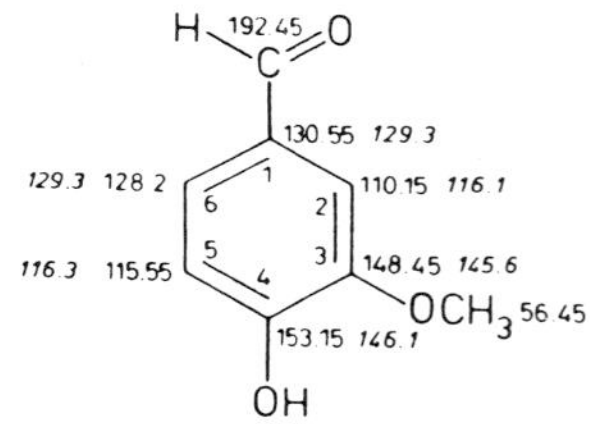

The assignment can be made, however, analyzing the coupled spectra B and C. Thus C-2 shows a long-range coupling with H-6 (7.3 Hz) and another with the aldehyde proton (3.2 Hz), whereas C-5 does not exhibit any long-range splitting. Similarly, two *meta* couplings (9.6 and 7.3 Hz) and an *ortho* coupling (2.7 Hz) can be observed for C-4, whereas C-3 shows only a *meta* coupling, obscured by further fine splitting because of the *ortho* proton and the OCH_3 protons. The complete analysis of spectra A through C can be tabulated as follows:

Assignment of Spectrum 43

ppm	Carbon Atom	Coupling Constants (Hz)
192.45	—CH=O	J_{CH} = 173.0 $^3J_{CH}$ = 4.8
153.15	C-4	$^3J_{C\text{-}4\text{-}H\text{-}2}$ = 9.6 $^3J_{C\text{-}4\text{-}H\text{-}6}$ = 7.3 $^2J_{C\text{-}4\text{-}H\text{-}5}$ = 2.7
148.45	C-3	

Assignment of Spectrum 43

ppm	Carbon Atom	Coupling Constants (Hz)
130.55	C-1	$^{2}J_{C\text{-}1\text{-}H\text{-}CHO}$ = 23.8 $^{3}J_{C\text{-}1\text{-}H\text{-}5}$ = 7.8 $^{2}J_{C\text{-}1\text{-}H\text{-}2}$ = 1.3
128.2	C-6	J_{CH} = 162.1 $^{3}J_{C\text{-}6\text{-}H\text{-}2}$ = 7.3 $^{3}J_{C\text{-}6\text{-}H\text{-}CHO}$ = 1.8
115.55	C-5	J_{CH} = 163.0
110.15	C-2	J_{CH} = 160.7 $^{3}J_{C\text{-}2\text{-}H\text{-}6}$ = 7.3 $^{3}J_{C\text{-}2\text{-}H\text{-}CHO}$ = 3.2
56.45	—OCH_3	J_{CH} = 145.1

44

Interpretation of the Spectra

ppm		Assignment	Relative Intensity According to Spectrum B
194.35	>C=O	Conjugated	1
165.5	⋛C—O—	(Phenol ether, phenol)	1
132.75	⋛C—H	(large nuclear Overhauser enhancement in Spectrum A)	2
126.7	⋛C—		1
114.65	⋛C—H	(Large nuclear Overhauser enhancement in Spectrum B)	2
55.6	—OCH_3	(phenol ether)	1
Number of nonequivalent carbon atoms			8

The molecule $C_{16}H_{14}O_4$ contains ten double bond equivalents, as compared to $C_{16}H_{34}$. It consists of two symmetrical halves, $C_8H_7O_2$, showing only six (and not eight) resonances in the ^{13}C NMR spectrum rather than 16. Considering the ^{13}C chemical shifts, a conjugated carbonyl carbon (194.35 ppm; one double-bond equivalent) and a methoxyl group (55.6 ppm) are recognizable. Subtracting the corresponding partial formula $C_2H_3O_2$ from the formula of one-half the molecule, $C_8H_7O_2$, C_6H_4 remains, for a disubstituted benzene (four double-bond equivalents). Because the substituents —CO— and —OCH_3 are different, whereas only four signals with the relative intensity ratio 1:2:2:1 appear in the ^{13}C NMR spectrum, the half-molecule $C_8H_7O_2$ is a *para*-disubstituted benzene and is therefore a *p*-methoxybenzoyl group. Thus the compound $C_{16}H_{14}O_4$ is 4,4′-dimethoxybenzil.

Considering the influence of the methoxy group upon the chemical shift of benzene (see Table 7.10, p. 61, shielding at the *ortho* and *para* positions, hardly any effect at the *meta* positions), one reaches the following assignment, which is also consistent with the signal intensities:

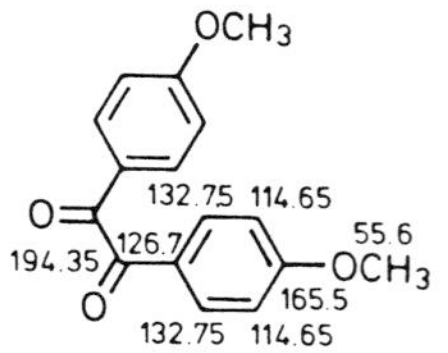

45

Ten signals are expected for 3-dimethylamino-3-phenylpropanoic acid ethyl ester (one carboxyl carbon, four nonequivalent aromatic carbons, and five nonequivalent aliphatic carbons). If the increased intensity of the signal at 128.85 ppm is due to an additional aromatic carbon, the spectrum shows nine signals too many, belonging to an impurity. Two of them are located in the aliphatic region, and six, in the olefinic-aromatic region. It follows that the impurity contains an extra C=C double bond (two signals) in addition to the phenyl ring (four signals).

The two additional aliphatic signals at 60.6 and 14.4 ppm appear close to the ethoxy resonances of 3-dimethylamino-3-phenylpropanoic acid ethyl ester (60.45 and 14.15 ppm) and therefore belong to an ethoxy group. The extra ester carbonyl group signal at 167.55 ppm (compare with 172.45 ppm) suggests an α, β-unsaturated ester (conjugation effect; see Section 7.8, p. 65). The impurity is thus ethyl cinnamate, produced in this case by thermal β-elimination of dimethylamine from 3-dimethylamino-3-phenylpropanoic acid ethyl ester.

Based on the signal intensities and a comparison of the chemical shifts of both esters, the following assignment is possible:

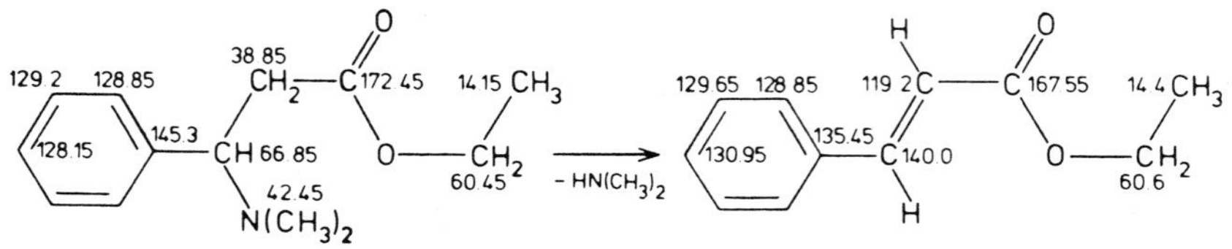

The assignment of the olefin resonances is supported by the fact that the carbon atom in the α position with respect to the carbonyl group is shielded (119.2 ppm), whereas the one in the β position is deshielded (140.0 ppm), as described for α, β-unsaturated carbonyl compounds in Section 6.2, p. 43.

The *trans* configuration of ethyl cinnamate is known from the value of the coupling constant for the neighboring protons (J_{HH} = 16 Hz).

46

The ^{13}C chemical shifts of a phenylazo group are recognizable in the spectrum:

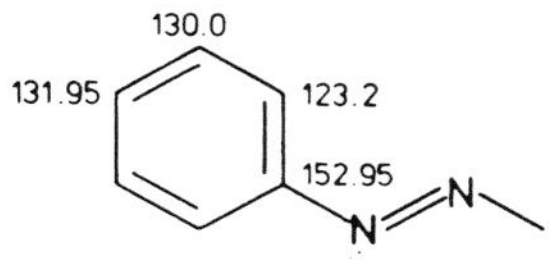

In the aromatic region another four signals appear: 151.45 ppm (singlet), 145.75 ppm (singlet), 130.05 ppm (doublet), and 123.35 ppm (doublet). The intensities of the resonances at 130.05 and 123.35 ppm are comparable to those of the phenylazo group resonances. Therefore, these signals belong to two pairs of equivalent carbon atoms, as in the case of the *ortho* and *meta* carbon atoms in the phenylazo group. The second phenyl ring of the azobenzene derivative is thus *para*-substituted. An *ortho* or *meta* linkage of the substituent would result in six additional aromatic resonances (four of them off-resonance doublets) instead of four owing to the loss of molecular symmetry.

The substituent contains two methylene groups joined to each other (off-resonance triplets at 35.1 and 30.45 ppm with additional higher-order couplings, see Section 3.4, p. 18) and a carboxyl function (singlet at 174.65 ppm). The elementary composition rules out a carboxamide or carboxylic acid halide function, which could also be postulated on the basis of its ^{13}C chemical shift. The compound is thus 3-(4-phenylazophenyl)-propanoic acid.

Comparison of the chemical shifts allows the following assignment to be made. Azobenzene and ethylbenzene are adequate as reference substances (see J. B. Stothers, *Carbon-13 NMR Spectroscopy,* Academic Press, New York, 1972, p. 98), using the carboxyl group increments in Table 7.2, p. 52. The assignments are interchangeable in the cases where the chemical shift differences are extremely small.

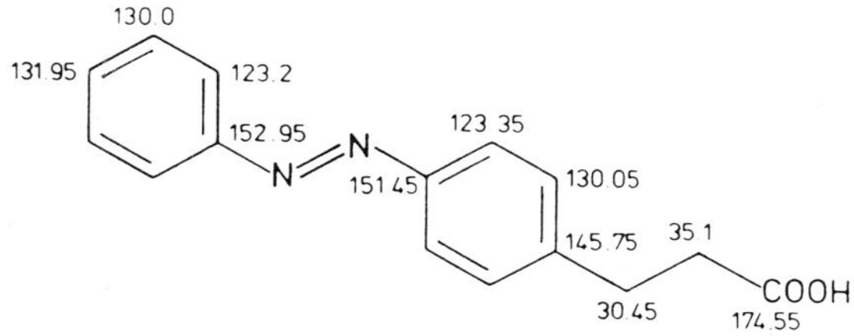

47

The primary product, $C_{20}H_{16}O_2$, is a secondary alcohol, as the doublet at δ = 59.0 ppm with J_{CH} = 149.3 Hz shows. Only seven separate signals are found for the molecule's 20 carbon atoms, and there is no resonance in the carbonyl region. Consequently, a phenylacetylene molecule adds to each of the two aldehyde groups of *trans*-2-butenedial:

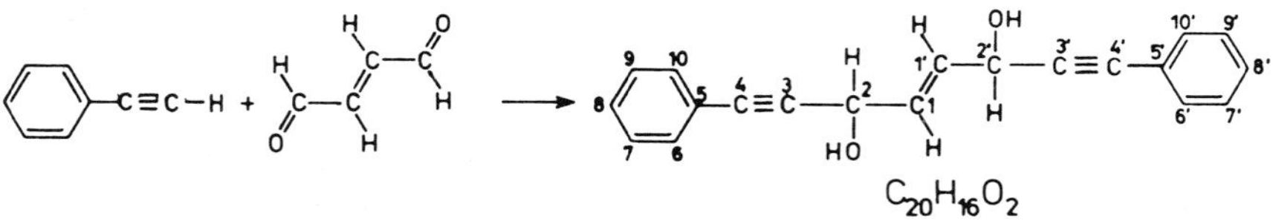

The signal at 59.0 ppm is lacking in the spectrum of the dehydrogenation product. Instead, a carbonyl resonance appears at 177.2 ppm, and one of the sp^2 CH signals is shifted downfield (from 131.85 to 142.25 ppm). The same happens with an sp carbon signal (from 88.3 to 94.0 ppm), in which no C-H long-range coupling (9.5 Hz in Spectrum A) is recognizable. Therefore, the dialcohol was dehydrogenated to the diketone:

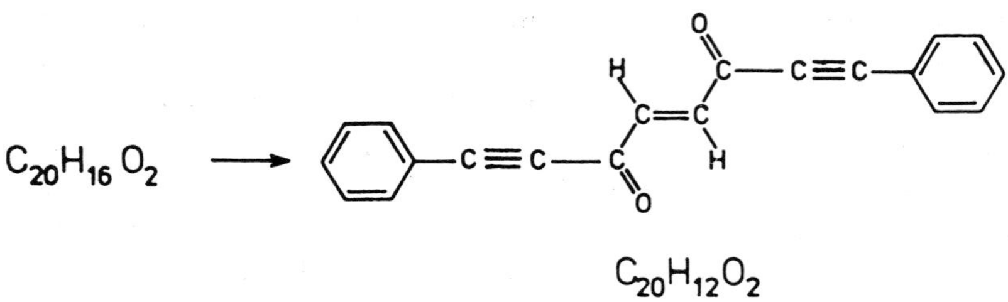

Using the chemical shifts and coupling constants and considering the clearly increased intensity of the signal at 131.85 ppm in Spectrum A, the assignment can be tabulated:

Assignment of Spectra A and B

Compound	Carbon	ppm	Multiplicity	Coupling Constant
A. $C_{20}H_{16}O_2$	1	131.85	Doublet (see C-6,10)	ca. 14.0 Hz (see C-6,10)
	2	59.0	Doublet of long-range doublets	149.3 Hz with H-2 6.6 Hz with H-1 and OH or H-1′
	3	88.3	Long-range doublet	9.5 Hz with H-2
	4	85.75	Singlet	— —
	5	122.3	Long-range triplet	6.9 Hz with H-7,9 (Overlap with C-1)
	6 (+10)	131.85	Doublet of long-range doublets of doublets	163.2 Hz with H-6 5.2 Hz with H-8,10 (Overlap with C-1)
	7 (+ 9)	128.35	Doublet of long-range doublets	ca. 14.0 Hz with H-7 ca. 6.6 Hz with H-9 (Overlap)
	8	128.7	Doublet of long-range triplets	ca. 164.0 Hz with H-8 ca. 6.6 Hz with H-6,10 (Overlap)
B. $C_{20}H_{12}O_2$	1	142.25	Doublet of long-range doublets	166.9 Hz with H-1 3.7 Hz with H-1′
	2	177.2	Long-range doublet of doublets	5.1 Hz with H-1 and H-1′
	3	94.0	Singlet	— —
	4	86.35	Singlet	— —
	5	119.35	Long-range triplet	6.0 Hz with H-7,9
	6 (+10)	133.35	Doublet of long-range doublet of doublets	164.0 Hz with H-6 5.9 Hz with H-8,10
	7 (+ 9)	128.85	Doublet of long-range doublets	163.2 Hz with H-7 7.4 Hz with H-9
	8	131.4	Doublet of long-range doublets	162.5 Hz with H-8 7.3 Hz with H-6,10

48

On the basis of the C—H couplings, the following tabulated assignment is reached. The finding that $^3J_{CH} > {}^2J_{CH}$ in benzene derivatives (see Table 8.2, p. 78) is useful as an assignment criterion. C-10 has two protons in the *o*-position, in contrast to C-9, and thus will profit from a larger nuclear Overhauser enhancement. Therefore, the signal at 136.35 ppm can be related to C-10 and the very weak signal at 132.65 ppm can be related to C-9.

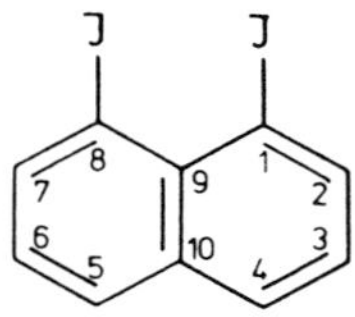

ppm	Multiplicity	Coupling Constants [Hz]	Carbon	Coupling Protons
144.7	Doublet of long-range doublets of doublets	166.7 8.7 2.7	C-2(7)	H-2(7) H-4(5) H-3(6)
136.35	Long-range multiplet	Not completely interpretable because of overlap	C-9	(H-3,6 H-4,5)
132.65	Not recognizable, because the signal is too weak		C-10	(H-2,7)
131.65	Doublet of long-range doublets of doublets	163.0 7.3 5.5	C-4(5)	H-4(5) H-2(7) H-5(4)
127.5	Doublet	164.3	C-3(6)	H-3(6)
96.55	Long-range doublet	11.0	C-1(8)	H-3(6)

The substituent influences of both iodine atoms on the chemical shifts of naphthalene (Table 7.13, p. 66) deviate considerably from the iodine increments for C-2(7) and C-4(5) (Table 7.10, p. 61).

	$\delta_{1,8\text{-diiodonaphthalene}}$ $-\delta_{\text{naphthalene}}$	$\delta_{\text{iodobenzene}}$ $-\delta_{\text{benzene}}$ (Table 7.10)
C-1(8)	−31.65	−32.3
C-2(7)	18.6	9.9
C-3(6)	1.4	2.6
C-4(5)	3.45	− 0.4
C-9	− 0.45	—
C-10	2.95	—

49

The following chemical shift differences (pseudocontact shifts) are obtained by comparing Spectra A and B:

Spectrum B [ppm]	Spectrum A [ppm]	$\Delta = \delta_B - \delta_A$ [ppm]	Assignment
203.66	199.35	*4.36*	C-1(CO)
147.72[a]	145.34[a]	*2.38*	C-10[a]
136.05	134.17	*1.88*	C-8
133.67[a]	133.48[a]	*0.19*	C-9[a]
132.11	129.55	*2.56*	C-5
130.10	127.94	*2.16*	*C-7*
128.45	127.40	*1.05*	*C-6*
44.50	39.41	*5.09*	*C-2*
31.22	29.89	*1.33*	*C-4*
25.59	23.44	*2.15*	*C-3*

[a]The bridgehead carbon atoms are recognizable by their low-intensity signals.

On the basis of the pseudocontact shifts (see Section 4.3, p. 36), an assignment of C-2, C-3, C-4, and C-6 and C-7 can be made. Because of the increasing distance from the complexing carbonyl group, the very coarse approximation $\Delta_2 > \Delta_3 > \Delta_4$ and $\Delta_7 > \Delta_6$ is valid. Comparing the chemical shifts of the problem sample with those of tetralin and 1,4-naphthoquinone, one can assign the remaining carbon resonances unequivocally.

Considering the chemical shifts of α-tetralone, the signals of 6-methoxy-1,2,3,4-tetrahydronaphthalene-1-one can be assigned, using the methoxyl group increments (Table 7.10, p. 61) and off-resonance multiplicities (in Spectrum D: S = singlet, D = doublet, T = triplet, and Q = quartet):

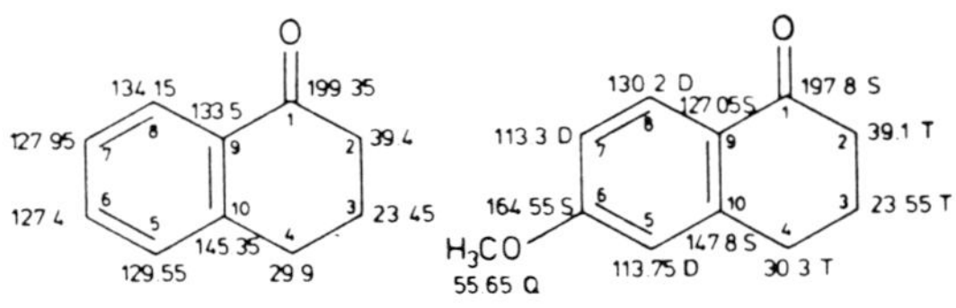

$$\delta_5 = 129.55 - 14.7 = 114.85 \text{ ppm}$$
$$\delta_6 = 127.4 + 30.3 = 157.6 \text{ ppm}$$
$$\delta_7 = 127.95 - 14.7 = 113.25 \text{ ppm}$$
$$\delta_8 = 134.15 + 0.9 = 135.05 \text{ ppm}$$
$$\delta_9 = 133.5 - 8.1 = 125.4 \text{ ppm}$$
$$\delta_{10} = 145.35 + 0.9 = 146.25 \text{ ppm}$$

50

In addition to the *N*-acetyl group signals (CO, 170.4 ppm; CH_3, 24.9 ppm), Spectrum A exhibits a signal at 165.3 ppm in the $\rangle$C=X carbon resonance region, with a marked nuclear Overhauser enhancement. This fits an azomethine carbon atom bonded to hydrogen. According to the molecular formula and the number of signals, the molecule consists of two symmetrical halves. Considering the reaction pathway, the structure should be *bis*-(*o*-acetylaminobenzaldehyde)-*o*-phenylenediimine (I).

The reaction of oxalic acid *o,o'*-diformyldianilide with *o*-phenylenediamine can lead to a polymeric Schiff base or to a macrocycle (II). The molecular mass agrees with the macrocyclic structure, the symmetry of which is also in agreement with ^{13}C NMR Spectrum B.

An assignment of the signals in the spectra of both compounds is possible using the acetylamino increments in Table 11, comparing the chemical shifts with those of 5,14-dihydrodibenzo[b,i]5,9,14,18-tetraaza[14]annulene (III) (see Reference), and considering nuclear Overhauser effects upon the signal intensities (enhanced intensity for C-H carbons, suppressed intensity for quaternary carbons):

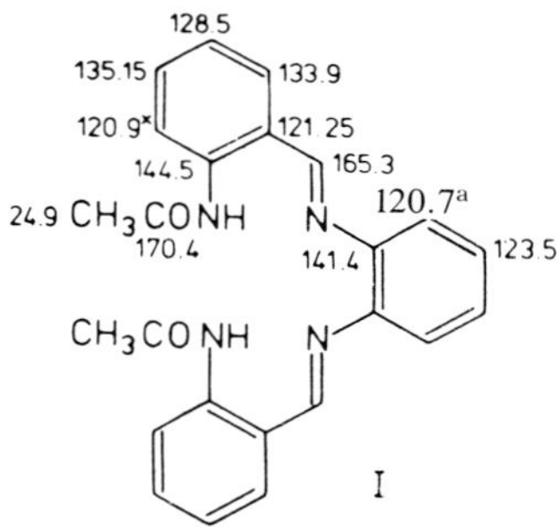

I

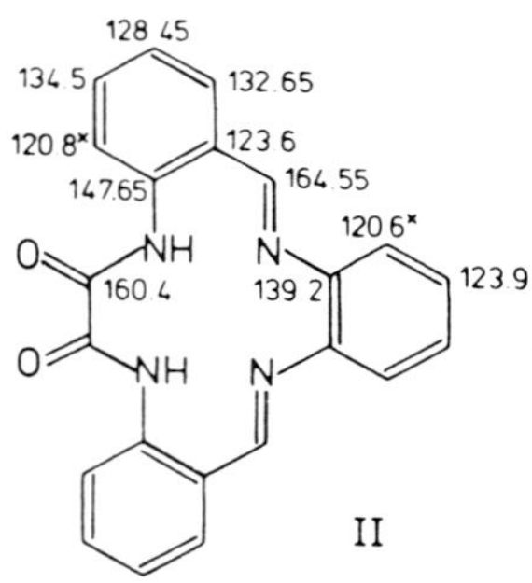

II

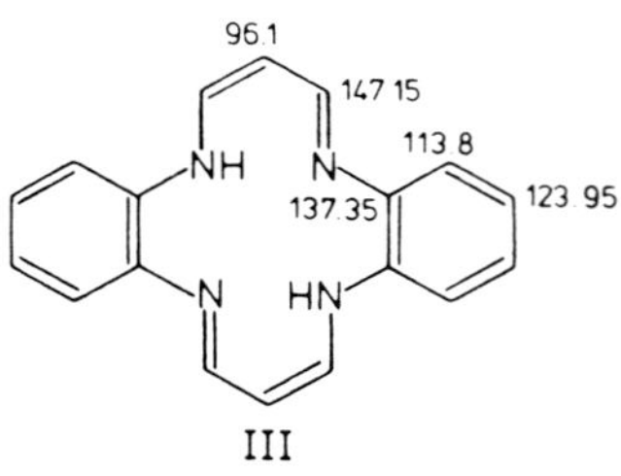

III

[a]Assignment interchangeable.

51

The following assignment results for the ^{1}H-NMR spectrum of 2-acetyl-1-methylpyrrole:

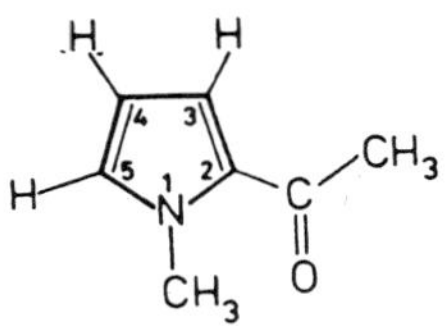

2.31 ppm	singlet	$COCH_3$
3.84 ppm	singlet	NCH_3
6.06 ppm	doublet of doublets	H-4
6.86 ppm	doublet of doublets or triplet	H-3
6.91 ppm	doublet of doublets	H-5

If one decouples one particular kind of proton, the carbon atom coupled with this proton appears as a sharp singlet in the ^{13}C NMR spectrum. All the other carbon atoms give off-resonance residual splittings. The following assignment results from these decoupling experiments for $COCH_3$, NCH_3, and C-4:

Experiment	Perturbed ppm	^{1}H Signal Assignment	Decoupled ppm	^{13}C Signal Assignment
(c)	2.31	$COCH_3$	27.05	$COCH_3$
(d)	3.84	NCH_3	37.55	NCH_3
(e)	6.06	H-4	108.85	C-4
(f)	6.91	H-3	120.65	C-3
			132.2	C-5
			145.85	C-2
			163.8	CO

In experiment F, the signals at 120.65 and 132.2 ppm are decoupled because the proton resonances lie too close together. Despite this, an assignment can be made: Because the ring nitrogen atom deshields more strongly at the α position, the CH signal at 132.2 ppm [off-resonance doublet in experiment E] belongs to C-5. Therefore, the signal at 120.65 ppm must be assigned to C-3. The signal at 163.8 ppm can be assigned without hesitation to the acetyl carbonyl group, and only an extremely weak peak at 132.2 ppm, overlapping with the C-5 signal and only visible in Spectra C, D, and E, is left for assignment to C-2.

52

The singlet at 161.45 ppm fits a carbonyl group carbon, a fact that is also consistent with the molecular formula since $C_5H_4O_3$ – CO_2H = C_4H_3O (furyl). Comparison of the signal multiplicities in the spectrum shows that only the signal at 146.65 ppm does not exhibit any large C-H coupling; therefore, it belongs to the carbon atom bearing the carboxyl group. To decide between furan-2-carboxylic acid and furan-3-carboxylic acid, the known chemical shifts of furan (C-2,5: 142.7 ppm) can be compared with this value. In this way, the signal at 146.65 ppm should be assigned to the quaternary C-2, and the compound is furan-2-carboxylic acid. The assignment of the signals at 119.55 and 113.55 ppm to C-3 and 4 results from comparison of their multiplicities and C-H coupling constants with the values given for furan:

ppm	Multiplicity	Assignment	Coupling Constants
161.45	Singlet	C-6 (CO)	—
148.45	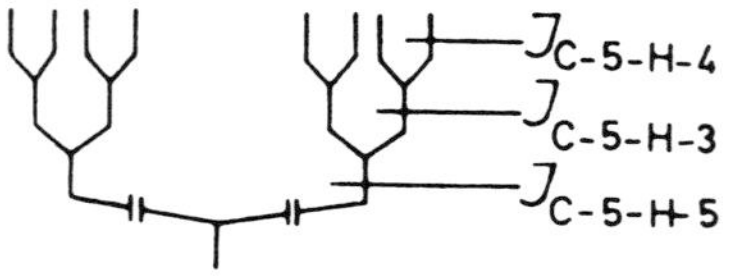	C-5	205.0 Hz with H-5 11.0 Hz with H-3 8.0 Hz with H-4
146.65	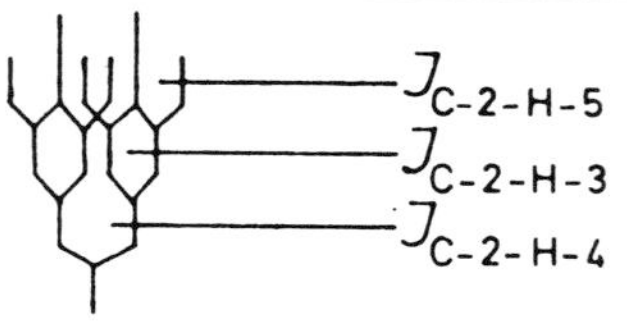	C-2	ca. 12 Hz with H-4 ca. 7 Hz with H-3 ca. 7 Hz with H-5
119.55	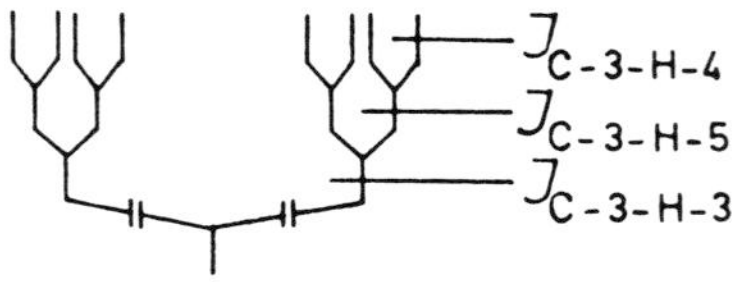	C-3	179.5 Hz with H-3 5.5 Hz with H-5 3.7 Hz with H-4
113.55	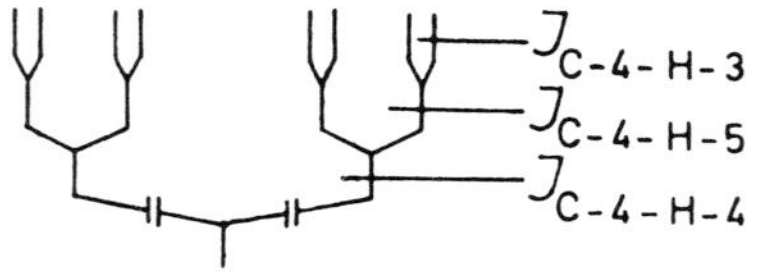	C-4	178.0 Hz with H-4 13.2 Hz with H-5 3.7 Hz with H-3

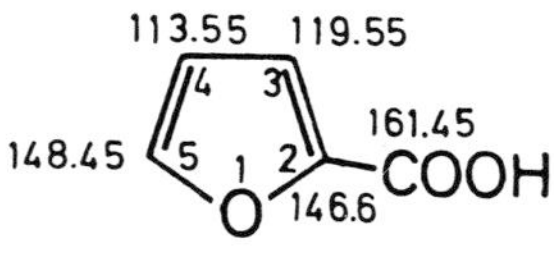

53

Interpretation of the Spectrum

ppm	Off-resonance Multiplicity	Intensity	Assignment
104.75	Doublet	Strong	—O—CH—O—
104.25	Doublet	Medium	—O—CH—O—
63.1	Triplet	Strong	—O—CH_2—
62.95	Triplet	Medium	—O—CH_2—
31.4	Triplet	Strong	—CH_2—
30.55	Triplet	Medium	—CH_2—
15.3	Quartet	Very strong	—CH_3

The similar intensity ratios of the pairs of signals at 104.75/104.25, 63.1/62.95, and 31.4/30.55 ppm, as well as the greater intensity of the signal at 15.3 ppm, suggest that the tetrahydrofuran derivative is present as a *cis-trans* mixture in which one isomer (probably *trans*) predominates.

According to the formula the molecule contains eight carbon atoms, but only four signals or signal pairs appear in the ^{13}C NMR spectrum. Thus the tetrahydrofuran ring is symmetrically substituted with two identical substituents. These are ethoxy groups, as can be seen from the chemical shifts of 63.1/62.95 (triplets) and 15.3 ppm (quartet). The pair of signals at 104.75/104.25 ppm (doublets) fits carbon atoms deshielded by two directly bonded oxygen atoms [acetal grouping, —$CH(OR)_2$]. A mixture of *cis* and *trans*-2,5-diethoxytetrahydrofuran is therefore present:

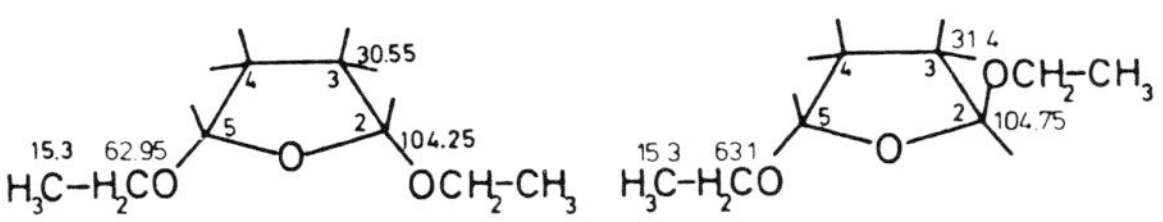

cis *trans*-

-2,5-diethoxytetrahydrofuran

54

The following assignment results from the signal multiplicities and a comparison of the C-H coupling constants with the known values for thiophene:

ppm	Multiplicities	Assignment	Coupling Constants
183.5	Singlet	C-6 (CO)	—
171.8	Long-range quartet of long-range doublets	C-8	J_{CCF} = 36.5 Hz J_{CCH} = 2.2 Hz

ppm	Multiplicities	Assignment	Coupling Constants
139.75	Long-range doublet of doublets	C-2	11.0 Hz with H-4 6.6 Hz with H-3 ca. 5.0 Hz with H-5
135.7	Doublet of long-range doublets of doublets	C-5	186.8 Hz with H-5 11.0 Hz with H-3 6.6 Hz with H-4
133.2	Doublet of long-range doublets of doublets	C-3	169.9 Hz with H-3 9.6 Hz with H-5 5.9 Hz with H-4
129.20	Doublet of long-range triplets (long-range doublet of doublets with identical coupling constants)	C-4	172.1 Hz with H-4 4.1 Hz with H-3 and H-5
118.7	Quartet of long-range doublets	C-9 (CF_3)	280.1 Hz with F_3 2.2 Hz with H-7
93.65	Doublet of long-range quartets	C-7	169.5 Hz with H-7 2.5 Hz with F_3

The C-7 signal at 93.6 ppm is split into a doublet of long-range quartets. The long-range quartet splitting is also visible with proton decoupling and is the result of the coupling with the three fluorine nuclei in the CF_3 group. The C-H coupling constant of 169.5 Hz fits an sp^2 carbon atom; therefore, the compound exists as an enol. Since the signal at 183.5 ppm (carbonyl carbon) exhibits no long-range quartet splitting, whereas the signal at 171.8 ppm (enol carbon) does, even in the proton-decoupled spectrum, the enol carbon is bonded to the CF_3 group. Unlike most other 1,3-diketones, 1-(2-thienylcarbonyl)-3,3,3-trifluoroacetone has only one *cis*-enol form (stabilized by an intramolecular hydrogen bond):

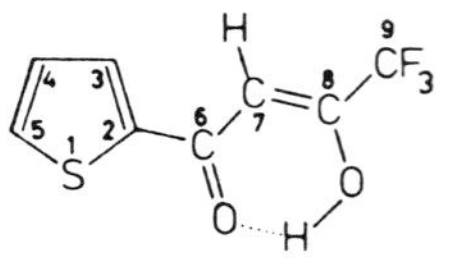

55

According to the structural formula of 3-methyl-5-oxo-1-phenyl-4,5-dihydropyrazole, four signals are expected in the ^{13}C NMR spectrum for the pyrazolone moiety. For reasons of symmetry, another four are expected for the phenyl portion, or eight in all. In addition to eight strong signals, however, the spectrum exhibits seven additional considerably weaker resonances. Therefore, the compound is present as an equilibrium mixture of two tautomers. It is known (see H. Beyer, *Lehrbuch der Organischen Chemie,* S. Hirzel-Verlag, Stuttgart, 18th ed., 1976, p. 648) that 3-methyl-5-oxo-1-phenyl-4,5-dihydropyrazole can exist in three tautomeric forms:

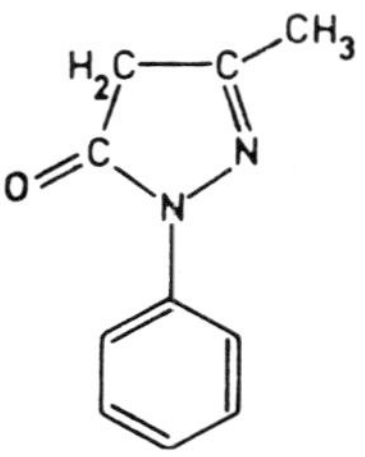

CH form

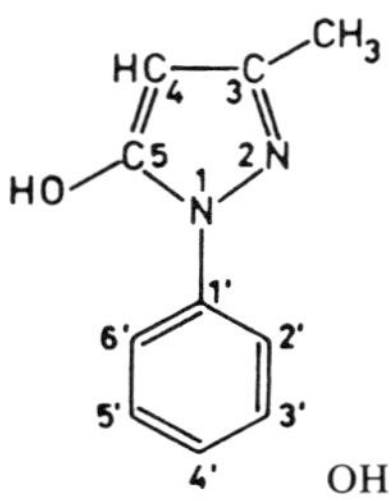

OH form

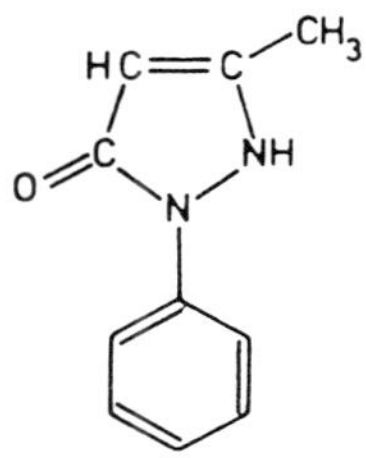

NH form

First, the strong doublet at 89.75 ppm fits C-4 of the OH or NH form. On the contrary, the enol carbon resonance at 156.6 ppm and the azomethine resonance at 149.55 ppm fit C-5 and C-3 of the OH form. The weak signals (e.g., carbonyl at 172.3, azomethine at 159.2, and methylene at 43.25 ppm) agree with the CH form, which, from a comparison of the signal intensities, amounts to about 10% of the sample. Using the numbering system defined above for the OH (or NH) form, the assignment shown on p. 354 results.

Assignment of ^{13}C NMR Signals, Spectrum 55

ppm	Assignment	Tautomer	Multiplicity	Coupling Constants
172.3	C-5	CH	Unresolved	—
159.2	C-3	CH	Unresolved	—
156.6	C-5	OH	Long-range doublet	6.4 Hz with H-4
149.55	C-3	OH	Long-range quintet	5.0 Hz with H-4 and CH_3
139.6	C-1′	OH	Long-range triplet	10.7 Hz with H-3′,5′
134.35	C-1′	CH	Unresolved	—
129.65	C-3′,5′	OH and CH	Doublet of long-range doublets of doublets	162.5 Hz with H-3′,5′; 7.3 Hz with H-5′,3′ and 1.8 Hz with H-4′ or H-2′,6′
126.15	C-4′	OH	Doublet of long-range triplets	161.6 Hz with H-4′; 7.5 Hz with H-2′,6′
125.85	C-4′	CH	Overlap with C-4′ signal of the OH tautomer	—
121.25	C-2′-6′	OH	Doublet of long-range triplets	164.8 Hz with H-2′,6′; 6.2 Hz with H-4′ and H-6′,2′
119.15	C-2′,6′	CH	Doublet of unresolved long-range multiplets	ca. 164 Hz with H-2′,6′
89.75	C-4	OH	Doublet of long-range quartets	176.2 Hz with H-4; 3.2 Hz with CH_3
43.25	C-4	CH	Triplet	ca. 130 Hz
16.75	CH_3	CH	Quartet	ca. 127 Hz
13.95	CH_3	OH	Quartet	127.3 Hz

The OH and NH forms could be distinguished by ^{15}N NMR.

56

Interpretation of the Spectrum

ppm	Off-resonance Multiplicity	Assignment	
185.0	Doublet	—CH=O	(conjugated)
159.35	Doublet	—CH=N—	
156.35	Singlet	>N—C<	(heterocycle)
152.95	Singlet	>N—C≤	(heterocycle)
140.05	Singlet	>N—C≤	(aromatic)
129.45	Doublet	=CH—	(aromatic)
127.45	Doublet	=CH—	(aromatic)
125.0	Doublet	=CH—	(aromatic)
110.45	Singlet	=C<	(heterocycle)
40.6	Quartet	>N—CH_3	
34.55	Quartet	>N—CH_3	
13.55	Quartet	≥C—CH_3	

The doublet at 185.0 ppm corresponds to a formyl carbon atom conjugated with a C-C double bond. A Vilsmeier formylation has therefore taken place on the pyrazole ring. The signals at 159.35 ppm (doublet) and 40.6 and 34.55 ppm (quartets) fit a formamidine resulting from the reaction of dimethylformamide and the aminopyrazole. Since the amidine C-N bond has partial π character, the two CH_3 groups are magnetically nonequivalent. Analogous to the *N*-alkylamides (see G. C. Levy, G. L. Nelson, *J. Am. Chem. Soc.* **94,** 4897, 1972), the N-CH_3 group *trans* to the azomethine nitrogen is more strongly shielded (40.6 ppm). The reaction product thus has the following structure:

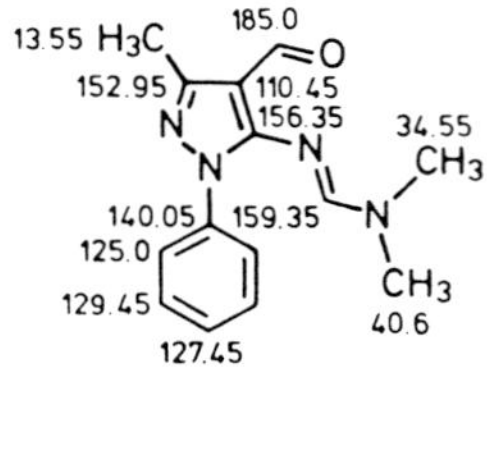

The assignment of the ^{13}C signals of the heterocycle and the phenyl ring results from comparison with the spectrum of 3-methyl-5-oxo-1-phenyl-4,5-dihydropyrazole (Spectrum and Exercise 55).

57

The molecular formulae of the starting compounds trifluoroacetic anhydride (1), $C_4F_6O_3$, and 2-amino-2-methylpropanoic acid (2), $C_4H_9O_2N$, add up to $C_8H_9O_5NF_6$. In comparison, the molecular formula of the reaction product, $C_6H_6O_2NF_3$ has a deficit of $C_2H_3O_3F_3$, corresponding to the loss of trifluoroacetic acid and water, which permits ring closure to a five-membered heterocycle:

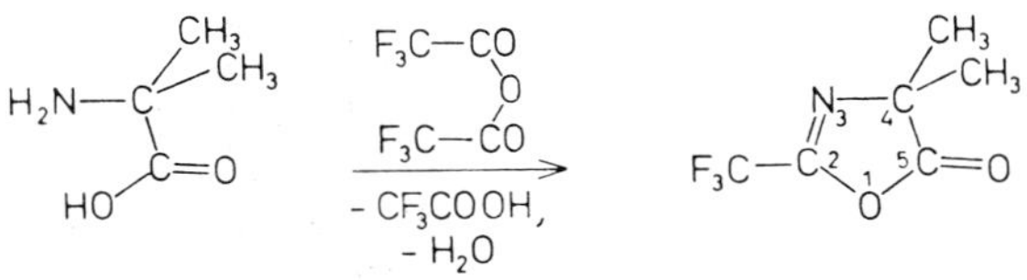

4,4-Dimethyl-5-oxo-2-trifluoromethyl-Δ^2-oxazoline

As the following table shows, ^{13}C NMR Spectrum No. 57 can be assigned to agree with this structural proposal:

ppm	Relative Intensity	^{13}C-^{19}F Coupling	Assignment	Origin
178.2	Medium	—	—COO (C-5)	2
152.4	Weak	Quartet, 43.2 Hz	F_3C—C=N (C-2)	1
116.55	Medium	Quartet, 273.6 Hz	—CF_3 on C-2	1
67.15	Strong	—	CR_4 (C-4)	2
24.15	Extremely strong	—	2 —CH_3 on C-4	2

58

Adding the molecular formulae of the starting compounds, C_6H_7NS and $C_{16}H_{14}O$, $C_{22}H_{21}NO_4S$ results. The reaction product, $C_{22}H_{19}NO_3S$, in comparison, has a deficit of H_2O. The elimination of water can be thought to result from the formation of products 1-3 (Ar = 4-methoxyphenyl):

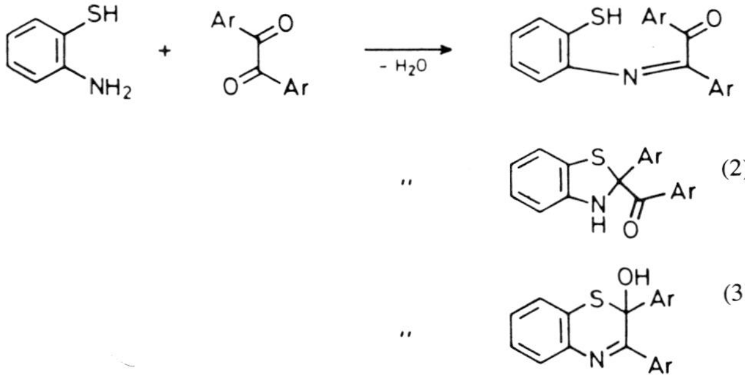

(1) Schiff Base

(2) 2-(4-methoxybenzoyl)-2-(4-methoxyphenyl)-benzothiazoline

(3) 2,3-di-(4-methoxyphenyl)-3-hydroxy-2H-benzo-1,4-thiazine

In the first place, two signals favor 2-(4-methoxybenzoyl)-2-(4-methoxyphenyl)-benzothiazoline (2): the chemical shift of 194.2 ppm is typical of a carbonyl carbon conjugated with a benzene ring and agrees well with the carbonyl resonance of 4,4′-dimethoxybenzil. The signal at 83.9 ppm belongs to a quaternary sp^3 carbon, flanked by two electron-withdrawing heteroatoms (S and N).

On the other hand, a carbonyl resonance and an azomethine resonance are expected near 190 and 150 ppm for the Schiff base (1). If the benzo-1, 4-thiazine were present, an azomethine resonance near 150 pm as well as an sp^3 carbon signal around 90-100 ppm should also appear, but a carbonyl resonance should not.

The following signal assignment is based on a comparison with the ^{13}C chemical shifts of 4,4′-dimethoxybenzil (see Exercise 44), the clearly different signal intensities of all the CH carbons, and the use of chemical shift increments (NH_2, OCH_3, and CO increments according to Table 7.10, p. 61).

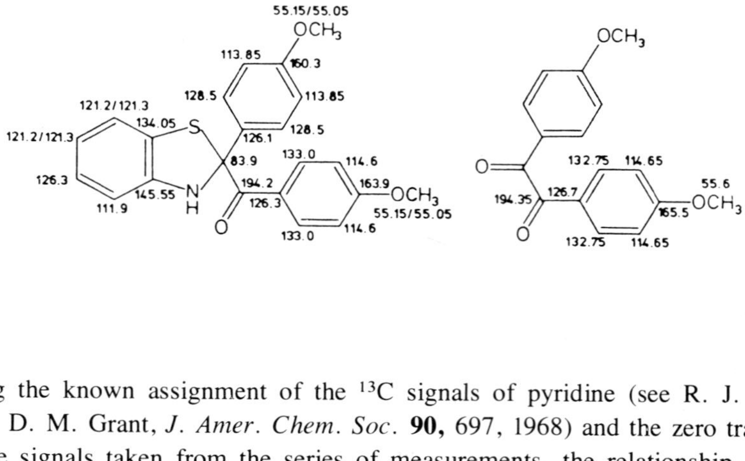

59

Using the known assignment of the ^{13}C signals of pyridine (see R. J. Pugmire, D. M. Grant, *J. Amer. Chem. Soc.* **90,** 697, 1968) and the zero transits of the signals taken from the series of measurements, the relationship

$$T_1 = \frac{\tau_0}{\ln 2}$$

gives the following T_1 values for the carbon atoms of pyridine:

ppm	C–	Zero Transit τ_0 [s]	T_1 [s]
149.9	2,6	5.0–5.5	7.2–8.0
136.2	4	3.5	5.0
124.3	3,5	5.0–5.5	7.2–8.0

Therefore, C-4 relaxes more rapidly than C-2,6 and C-3,5, showing that molecular motion of pyridine in D_2O solution is anisotropic. Because of the (^2H—) hydrogen bond between pyridine and deuterium oxide, the motion is analogous to that of a monosubstituted benzene: rotation around the axis through C-4 and the associated pyridine nitrogen is preferred (see Exercise 32, p. 295, nitrobenzene, and Section 9.5.2, p. 98).

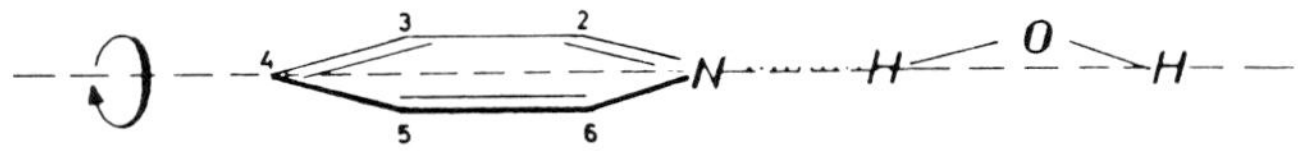

The series of measurements for 2,2′-bipyridyl (see cover illustration) shows, after evaluating the zero transits, that of all the C-H carbons, C-5,5′ relax more rapidly ($T_1 \approx 5.8$ s) than do C-3,3′, C-4,4′, and C-6,6′ ($T_1 \approx 5.8$ s). This also can be understood as a consequence of preferred rotation of the molecule around the axis traversing C-2,2′ and C-5,5′. Note the much slower relaxation ($T_1 \approx 51$ s) of carbons C-2,2′, not bonded to hydrogen.

60

The heterocycle contains a C-methyl group, as the quartet at 20.8 ppm in Spectrum 60A shows. If one substitutes H for CH_3 in the given molecular formula, the formula of pyridine, C_5H_5N, results. Since only three signals appear for the ring carbon atoms instead of five, the pyridine is symmetrically substituted: i.e., at C-4. Therefore, the compound is 4-methylpyridine (γ-picoline). The signal assignment follows from the off-resonance multiplicities, the increased intensity of the signals at 149.9 and 124.75 ppm, and the known values for pyridine (see Exercise 59):

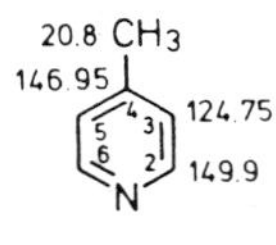

4-Hydroxymethylpyridine and 4-methylpyridine N-oxide are the two derivatives of 4-methylpyridine with the formula C_6H_7NO that should give three ring carbon signals as in Spectrum 60B. 2-Hydroxymethylpyridine is not present, however, since no ^{13}C resonance for a CH_2OH carbon appears (>60 ppm). The spectrum agrees better with 4-methylpyridine N-oxide (γ-picoline N-oxide), for which the following assignment results from the off-resonance multiplicities and signal intensities.

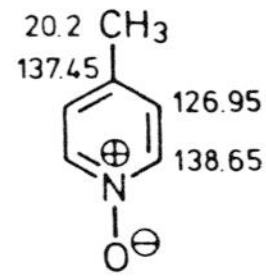

Carbon atoms C-2,6 and C-4 are thus more strongly shielded in the N-oxide than in γ-picoline. This means that the electron density is larger at these positions, as the resonance formulas of 4-methylpyridine N-oxide (γ-picoline N-oxide) illustrate:

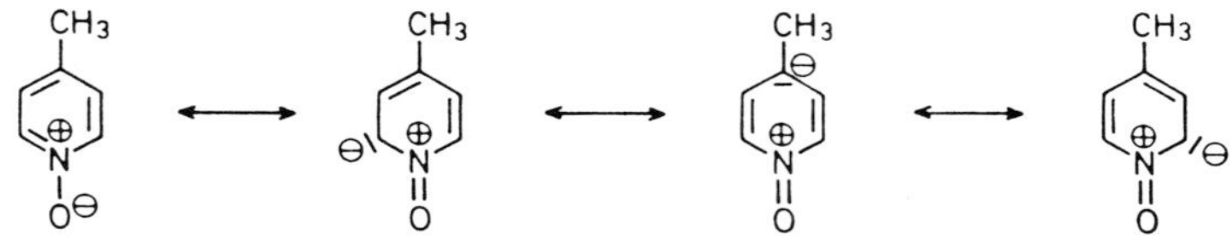

61

If one subtracts the partial formula C_5H_3N of a disubstituted pyridine from the given molecular formula $C_7H_5NO_4$, $C_2H_2O_4$ is left. This residue and the strong signal at 166.4 ppm agree with two carboxyl groups. Despite the magnetic equivalence of both carboxyl groups, the pyridine is not symmetrically dicarboxylated because five (not three) ring carbon signals appear in the spectrum. Since the signals at 151.9 and 129.8 ppm exhibit only long-range couplings, the corresponding carbon atoms are bonded to the carboxyl groups. Thus 2,3- and 2,5-pyridinedicarboxylic acids must be considered because the α substituent effect of COOH is small (see Table 7.10, p. 61), and chemical shifts of $\delta \approx 150$ and $\delta < 130$ ppm are typical of the pyridine ring carbons C-2,6 and C-3,5, respectively (see Spectrum No. 59). The triplet splitting of the signal at 129.8 ppm—a consequence of two identical long-range couplings of 6.6 Hz—is another clue. $J_{C\text{-}5\text{-}H\text{-}6} = J_{C\text{-}5\text{-}H\text{-}3} = 6.6$ Hz agrees with pyridine-2,5-dicarboxylic acid. Values of 7 and 6 Hz, respectively, were measured for pyridine itself. In the case of pyridine-2,3-dicarboxylic acid, on the other hand, $J_{C\text{-}3\text{-}H\text{-}5} \approx 6$ Hz, and $J_{C\text{-}3\text{-}H\text{-}4} \approx 2$ Hz (the latter possibly unresolved) would have been measured. The following scheme gives the assignment of the pyridine resonances and couplings for the 2,5-dicarboxylic acid:

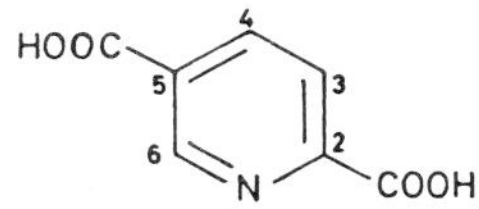

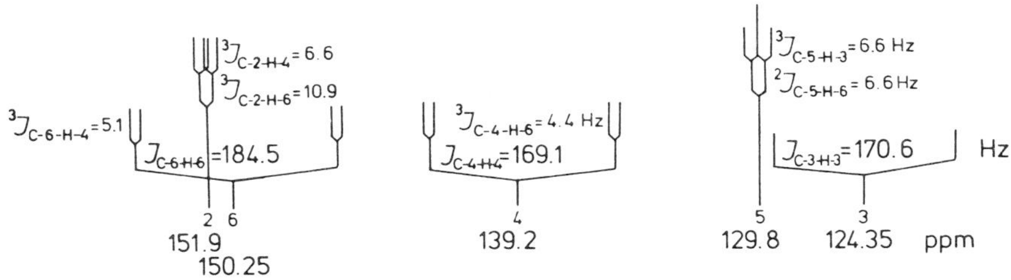

62

Interpretation of the Spectrum

ppm	Off-resonance Multiplicity	Intensity	Assignment
141.95	Singlet	Medium	
139.9	Doublet	Strong	
135.65	Doublet	Strong	
131.2	Singlet	Medium	
108.95	Doublet	Strong	

The signal intensities show that the pyridine ring is disubstituted. If one subtracts C_5H_3N from the molecular formula, two amino groups are left as substituents. Since the spectrum exhibits five signals, the symmetrically substituted diaminopyridines can be ruled out. For example, using the ^{13}C chemical shifts of pyridine and the chemical shift increments for NH_2 (Table 7.12, p. 64), the chemical shift of C-2 can be estimated in 3,4-diaminopyridine:

$$\begin{aligned}\delta_2 = \quad \delta_{2,6} &= 149.7 \text{ ppm (Pyridine)}\\ + Z_{32}(NH_2) &= -11.9 \text{ ppm}\\ + Z_{42}(NH_2) &= +0.9 \text{ ppm}\\ \hline \delta_2 &= 138.7 \text{ ppm}\end{aligned}$$

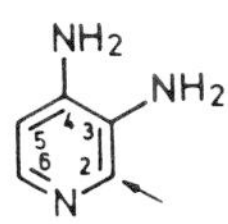

The carbon chemical shifts of 2,3-, 2,4-, 2,5-, and 3,4-diaminopyridine can be calculated in this way (see table below). The results for 3,4-diaminopyridine agree best with Spectrum 62.

-Diamino-pyridine	δ_2	δ_3	δ_4	δ_5	δ_6
2,3-	149.1	131.0	124.3	114.5	138.0
2,4-	161.9	95.7	158.1	99.8	149.7
2,5-	150.4	110.4	124.7	135.1	136.9
3,4-	138.7	131.9	141.6	111.3	139.8
Measured values	*135.65*	*131.2*	*141.95*	*108.95*	*139.9*

63

Both spectra exhibit a doublet each. Therefore, one of the four carbon atoms of each pyrimidine is bonded to hydrogen, and both pyrimidines are thus trisubstituted. When the partial formula of a trisubstituted pyrimidine, C_4HN_2, is subtracted from the given molecular formula, $C_4H_7N_5$, N_3H_6 is left. This corresponds to three amino groups. Since both spectra show three signals, not four, the problem compounds are symmetrically substituted: i.e., in the 2,4,6- and 4,5,6-positions and not in the 2,4,5-positions.

From the known coupling behavior in pyrimidine ($J_{C\text{-}2\text{-}H\text{-}2} > J_{C\text{-}4,6\text{-}H\text{-}4,6} > J_{C\text{-}5\text{-}H\text{-}5}$), it follows that the spectrum (A) with the smaller C-H doublet at 75.5 ppm fits 2,4,6-triaminopyrimidine and that the spectrum (B) with the larger doublet at 144.3 ppm fits 4,5,6-triaminopyrimidine. The complete assignment results from the signal intensities.

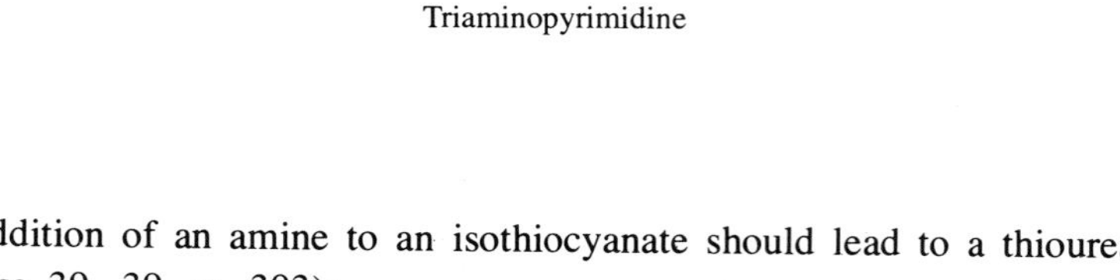

Triaminopyrimidine

64

The addition of an amine to an isothiocyanate should lead to a thiourea (see Exercise 39, 39, p. 303):

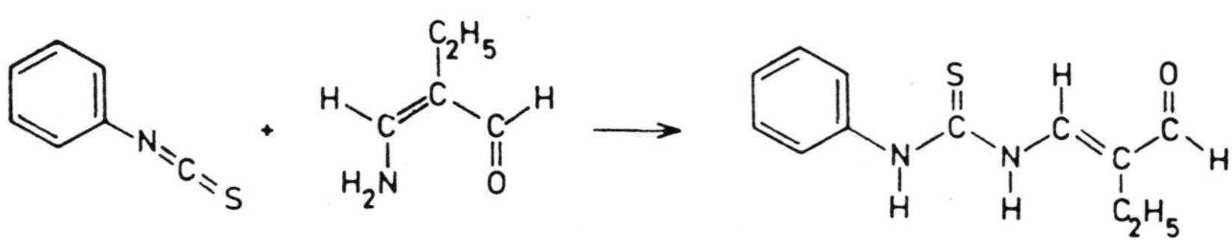

The molecular formula of the reaction product, $C_{12}H_{12}N_2S$, however, contains one water molecule less than the expected thiourea, $C_{12}H_{14}N_2OS$. The elimination of water from the tautomer of thiourea with ring closure to afford a pyrimidine (1) or a 1,3-thiazine-2-one-imine (2) can be postulated:

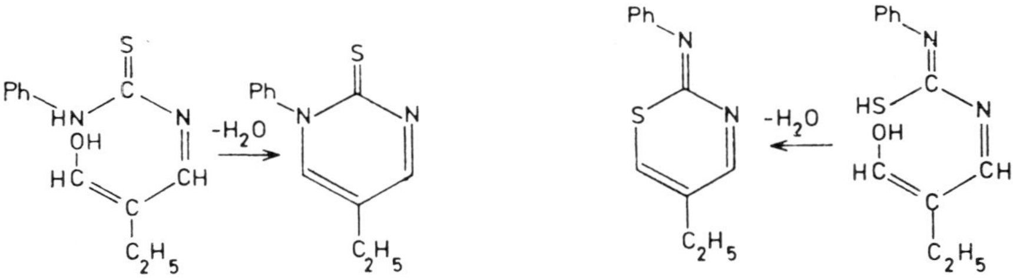

1 2

The signal at 182.7 ppm agrees better with a thiourea function in (1) than with an isothiourea group in (2). The two C-H doublets at 161.9 and 146.45 ppm with C-H coupling constants of 180 and 182.5 Hz also agree well with C-4 and C-6 of a pyrimidine ring (see, for example, J. B. Stothers, *Carbon-13 NMR Spectroscopy*, Academic Press, New York, 1972, p. 343). Owing to its enamine character (namely, the ^{13}C chemical shift of the enamine carbon in 3-amino-2-ethyl-acrolein, 159.55 ppm), C-6 appears here downfield from the azomethine carbon C-4. The multiplicities of the signals in the top spectrum, the double intensity of the signals at 129.95 and 126.4 ppm in the proton-decoupled spectrum, and the fact that the *meta* carbons of the phenyl ring resemble benzene carbons most closely ($\delta \approx 128.5$ ppm, Table 7.10, p. 61) lead to the following assignment:

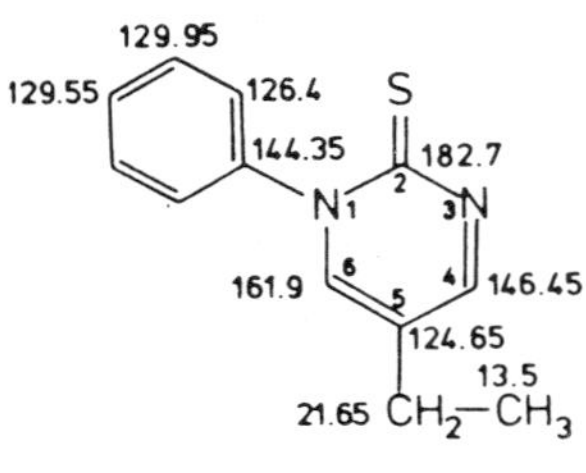

The long-range "quartet" splitting of the C-4 and C-6 signals is a consequence of coupling with the γ ring proton (C-4 with H-6 and vice versa) and the two ethyl group CH_2 protons. The J_{CCCH} coupling constants for the ring and the CH_2 protons are obviously the same; therefore, the long-range doublet of triplets is reduced to a long-range quartet. The coupling with the CH_2 protons can be demonstrated if these are specifically decoupled with low transmitter power: The long-range quartets are further simplified under off-resonance conditions to long-range doublets (Spectrum 64B).

65

In this reaction, the product can have either a linear or an angular ring system:

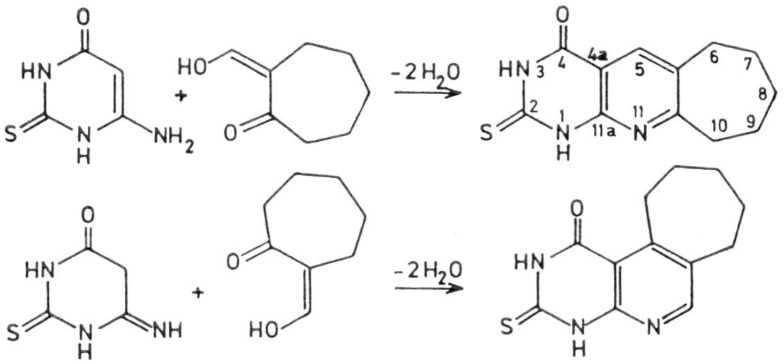

Linear condensation

Angular condensation

If the rings are linearly fused, the γ pyridine carbon is bonded to a hydrogen atom, and one can expect a C-H doublet with $J_{CH} \approx 163$ Hz around 136 ppm from the data for 6-methyl-1,2,3,4-tetrahydropyrido [2,3-d] pyrimidine-2,4-dione. The spectrum shows a doublet (J_{CH} = 165.2 Hz) of long-range triplets (J_{CCCH} = 5.9 Hz) at 135.75 ppm, favoring a linear ring system.

The long-range triplet splitting results from coupling with the protons of the nearest α CH_2 group of the seven-membered ring. With angular condensation, the C-H doublet is expected near 150 ppm with a considerably larger coupling constant ($J_{CH} > 170$ Hz), which corresponds to a pyridine α carbon.

In addition to other long-range couplings, the coupled spectrum shows that the signal at 135.5 ppm belongs to two nonequivalent carbon atoms. The following assignment results from the long-range couplings and the chemical shifts of the reference compounds:

Assignment of the Spectra

ppm	Multiplicity	Coupling Constants	Assignment
176.25	Singlet	—	C-2
170.75	Broad long-range multiplet	Maximum: 5.2 Hz	C-8a (coupling with H-5)
160.5	Long-range doublet	4.1 Hz	C-4 (coupling with H-5)
149.95	Long-range doublet	7.3 Hz	C-7 (coupling with H-5)
	Broad long-range multiplet	—	C-4a
135.75	Doublet of long-range triplets	165.2 Hz 5.9 Hz	C-5 (coupling with H-5 and H_2-13)
109.9	Broad singlet	—	C-6

Assignment of the Spectra

ppm	Multiplicity	Coupling Constants	Assignment
38.65	Triplet	¢a. 130.0 Hz	C-13
33.35	Triplet	125-130 Hz	C-9
31.5	Triplet	125-130 Hz	C-10, C-11, C-12
27.6	Triplet	125-130 Hz	(assignment uncertain)
25.8	Triplet	125-130 Hz	

66

Starting from the *cis*-1,2-dicyanoethene-1,2-dithiolate anion (B), the 1,2,5-trithiepine (I) and the bicyclic structures (IIa and b) can be considered for the product $C_8N_4S_3$:

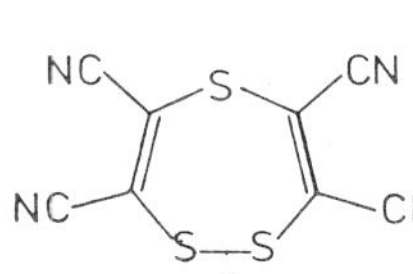

I

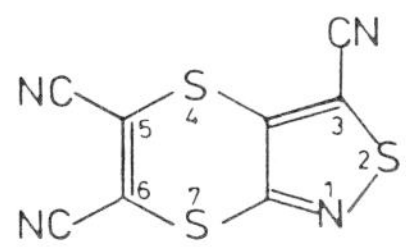

IIa

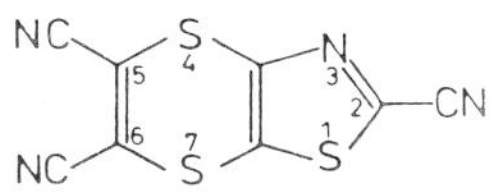

IIb

^{13}C NMR Spectrum 66 exhibits eight signals. Therefore, the trithiepine can be ruled out because this symmetrical ring should only give four signals. In the ^{13}C NMR spectrum of heterocycle (IIb), three resonances should appear with $\delta > 140$ ppm, for C-2 (azomethine carbon), C-8 and C-9 (bridge in the bicyclic structure). Instead, the spectrum shows only one signal in this region (156.2 ppm), which fits C-9 in heterocycle (IIa). The nitrile and ring carbon atoms of IIa can also be assigned. Thus the two neighboring pairs of signals at δ = 122.5/123.8 ppm and δ = 112.8/112.9 ppm belong to the chemically similar C-5 and C-6 and to the nitrile groups bonded to them. Because of the extremely small differences in chemical shifts, the assignments of the pairs of signals at 112.8/112.9, 122.5/123.8, and 130.65/133.3 ppm are interchangeable.

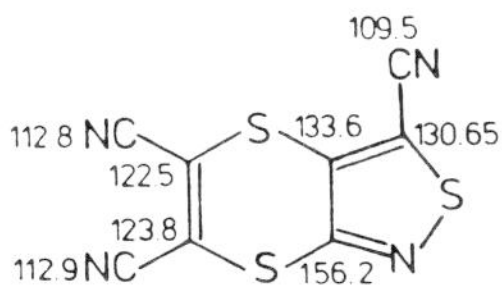

IIa: 3,5,6-tricyano-4,7-dithiino[c]isothiazole

67

The low-intensity signals at 177.6, 59.65, 32.9, 18.75 (obscured by the strong signal of the unknown amino acid) and 18.25 ppm belong to valine. Because of the neighboring asymmetric α carbon, the two methyl groups are *diastereotopic* and exhibit one ^{13}C resonance each (18.75 and 18.25 ppm). The splitting of the signal at 59.65 ppm is a consequence of the ^{13}C-^{15}N coupling (J_{CN} = 4.5 Hz).

The strong signals at 178.35, 53.1, and 18.75 ppm remain for the carboxyl, the α and the β carbons of the problem amino acid, which therefore contains three carbon atoms. The values agree with those published for alanine, as shown by a comparison with the data in W. Voelter, G. Jung, E. Breitmaier, E. Bayer, *Z. Naturforsch.* **26b**, 213, (1971) and by an increment calculation starting from the chemical shift of ethane (5.7 ppm according to J. B. Stothers, *Carbon-13 NMR Spectroscopy*, Academic Press, New York, 1972, p. 56) and the α and β increments of the substituents —COOH or —COO$^-$ and —NH$_2$ or —NH$_3$$^+$ (see Table 7.2, p. 52):

Increment calculation for alanine:

$\overset{\beta}{H_3C}$—$\overset{\alpha}{CH}$(—NH_2)—COOH

C-α:
	=	5.7 ppm
$+Z_{\alpha COOH}$	=	28.3 ppm
$+Z_{\alpha NH_2}$	=	20.1 ppm
		54.1 ppm

C-β:
	=	5.7 ppm
$+Z_{\beta COOH}$	=	2.0 ppm
$+Z_{\beta NH_2}$	=	11.3 ppm
		19.0 ppm

$\overset{\beta}{H_3C}$—$\overset{\alpha}{CH}$(—NH_3^+)—COO^-

C-α:
	=	5.7 ppm
$+Z_{\alpha COO^-}$	=	24.5 ppm
$+Z_{\alpha NH_3^+}$	=	26.0 ppm
		56.2 ppm

C-β:
	=	5.7 ppm
$+Z_{\beta COO^-}$	=	3.5 ppm
$+Z_{\beta NH_3^+}$	=	7.5 ppm
		16.7 ppm

ppm	Amino Acid	Carbon	C-N Coupling	$J_{C\text{-}N}$ Hz
178.35	Alanine	COO	—	—
177.6	Valine	COO	—	—
59.65	Valine	C-α	Doublet	4.5
53.1	Alanine	C-α	Doublet	4.1
32.9	Valine	C-β	—	—
18.75	Alanine	C-β	—	—
18.75	Valine	C-γ	—	—
18.25	Valine	C-γ	—	—

68

Menthol and isomenthol differ in their configuration at C-1:

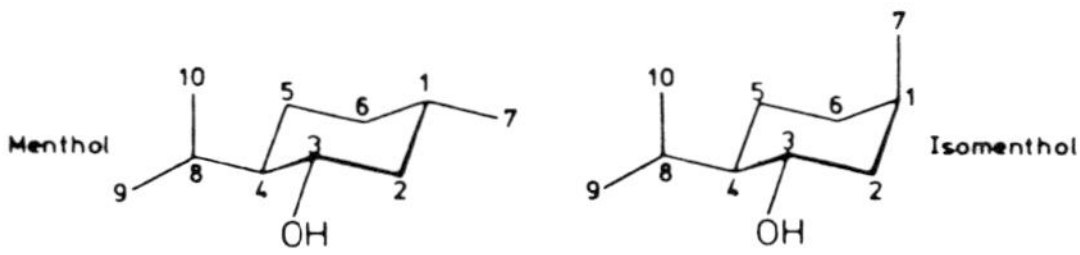

The relationship $\delta_e > \delta_a$ was found for C-1, C-3, and C-5 in 1-methylcyclohexanes, whereas $\delta_e \approx \delta_a$for C-4 (see Table 7.6, p. 56; D. K. Dalling, D. M. Grant, *J. Am. Chem. Soc.* **89**, 6612, 1967; and **94**, 5318, 1972). Therefore, carbon atoms C-1, C-2, C-3, C-5, and C-6, as well as C-1-CH_3 (= C-7) are more strongly shielded in isomenthol than in menthol. and C-4 and the isopropyl carbon atoms are less affected by the configurational change at C-1. Consequently, Spectrum 68B, with the larger ^{13}C chemical shifts, belongs to menthol, and Spectrum 68A belongs to isomenthol. The following assignment results from the chemical shifts and off-resonance multiplicities:

Spectrum	Compound	ppm	Off-Resonance Multiplicity	Assignment	Comparison δ (b)-δ (a) [Hz]
68A	Isomenthol	67.85	Doublet	C-3	
		49.65	Doublet	C-4	
		40.15	Triplet	C-2	
		30.65	Triplet	C-6	
		27.65	Doublet	C-1	
		26.1	Doublet	C-8	
		21.15	Quartet	C-9	
		20.05	Quartet	C-7	
		19.65	Triplet	C-5	
		18.15	Quartet	C-10	
68B	Menthol	71.1	Doublet	C-3	3.25
		50.1	Doublet	C-4	0.45
		45.25	Triplet	C-2	5.10
		34.9	Triplet	C-6	4.25
		31.95	Doublet	C-1	4.30
		25.55	Doublet	C-8	−0.55
		23.55	Triplet	C-5	3.90
		22.35	Quartet	C-7	2.30
		21.2	Quartet	C-9	0.05
		16.05	Quartet	C-10	−2.10

The assignment of the ^{13}C NMR spectrum of menthol has been published [see, for example, M. Jautelat, J. B. Grutzner, J. D. Roberts, *Proc. Nat. Acad. Sci.* **65**, 288, 1970. Because of asymmetrical carbon atom C-4, C-9 and C-10 are *diastereotopic* and give separate signals despite the free rotation of the isopropyl group.

69

D-xylose mutarotates in aqueous solution, and an equilibrium is reached between α- and β-D-xylopyranose. The β anomer predominates in the equilibrium mixture and gives the stronger signals:

Measuring the integrals, the following equilibrium concentration is found for C-1 or α-D-xylopyranose:

$$\% \alpha = \frac{34634}{64156 + 34634} \cdot 100 = 35\%$$

Similarly, the concentrations of the anomeric xylopyranoses can be calculated for all pairs of carbon atoms:

C-1: $\alpha:\beta$ = 35:65
C-2: $\alpha:\beta$ = 36:64
C-3: $\alpha:\beta$ = 36:64
C-4: $\alpha:\beta$ = 38:62
C-5: $\alpha:\beta$ = 34:66

The result for C-4 is less reliable because the signals of both anomers overlap.

70

The following assignment can be made on the basis of the off-resonance multiplicities in Spectrum B (S = singlet, D = doublet, T = triplet) and using the given chemical shifts of adenine and β-D-ribofuranose (see Fig. 4.2, p. 33):

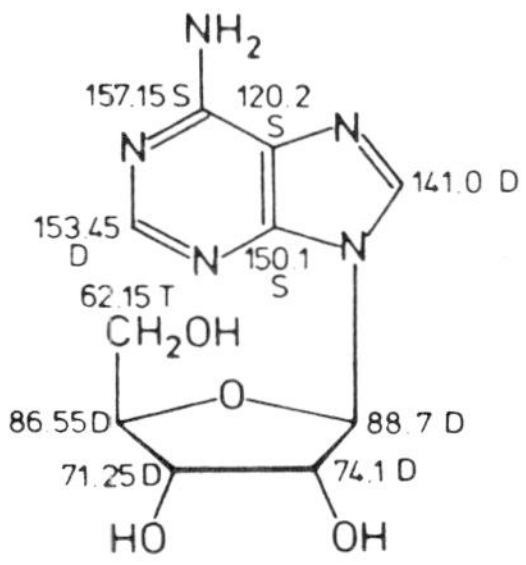

Adenosine

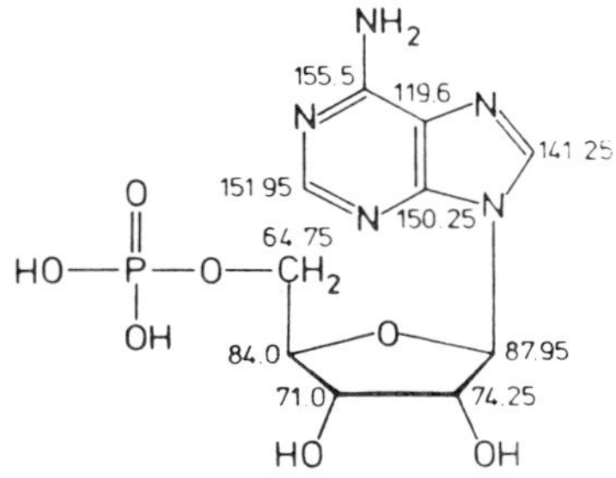

Adenosine-5'-monophosphate

This assignment is also supported by the couplings of carbon atoms C-5' and C-4' with phosphorus (4.3 and 7.3 Hz) in Spectrum C of the nucleotide (see Table 8.4, p. 82).

71

Spectrum 71A exhibits six signals with $\delta > 100$ ppm. Three signals are split into doublets with C-H coupling constants between 155.3 and 158.5 Hz (partial Spectrum E), which corresponds to carbon atoms with trigonal planar bond geometry. Therefore, the compound contains a trisubstituted benzene ring. The signal at 156.75 ppm (no directly bonded hydrogen) then agrees well with a phenolic carbon atom (see Table 7.10, p. 61). The two other signals with no direct C-H coupling, according to their ^{13}C chemical shifts of 139.6 and 133.3 ppm, belong to alkylated benzene carbons, with the more strongly shielded carbon atom (133.3 ppm) *para* with respect to the OH group (see Table 7.10). Since the steroid exhibits no more than 20 signals in ^{13}C NMR Spectrum A, but the steroid skeleton without the side chain already contains 17 carbon atoms, the benzene ring is part of the steroid nucleus. Therefore, the compound is an estrane derivative.

In spectra A, B, and C, a methyl group can be recognized as a side chain (off-resonance quartet at 13.45 ppm); similarly, in spectra A, D, and E, an ethynyl group can be distinguished (doublet with J_{CH} = 235.3 Hz at 75.15 ppm, and long-range doublet with $^{2}J_{CH}$ = 43.0 Hz at 89.35 ppm).

The singlets at 80.85 and 48.45 ppm agree best with C-17 and C-13 of an estra-3,17-diol, where C-17 is bonded to the ethynyl group; otherwise, C-17 (at 80.85 ppm) would not be quaternary.

The C-13-CH_3 resonance in 3,17 β-estradiol appears at 11.47 ppm, whereas in the α isomer it lies at 17.53 ppm (see reference in Exercise 71). The cause of this difference in chemical shift is the steric compression of the C-17-β-OH and the C-13-CH_3 groups which, like the γ effect (see Section 7.2, p. 52), leads to shielding of the methyl carbon. Because of the ethynyl group's linear geometry, a similar steric interaction cannot be considered for this function, so in 17-ethynyl-3,17-estradiol, the C-13-methyl chemical shift is also influenced mainly by its interaction with the OH group.

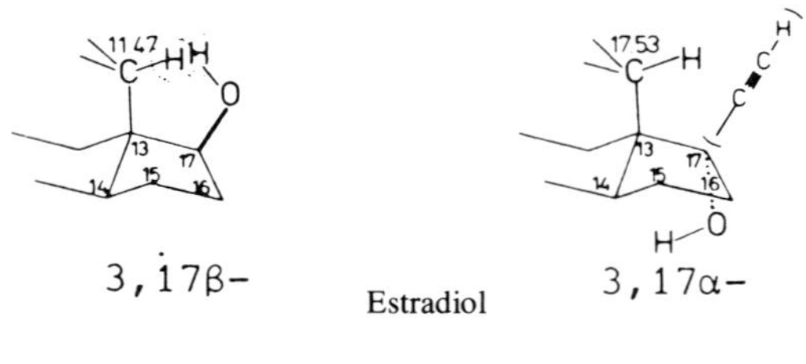

3,17β- Estradiol 3,17α-

The methyl resonance in Spectrum 71A lies closer to the CH_3 chemical shift of 3,17β-estradiol. This favors the 17α-ethynyl-3,17β-estradiol structure.

The assignment of the remaining ^{13}C signals is based on the off-resonance multiplicities (S = singlet, D = doublet, T = triplet, Q = quartet) and on the comparison of the chemical shifts with those of 3,17β-estradiol (see reference in Exercise 71).

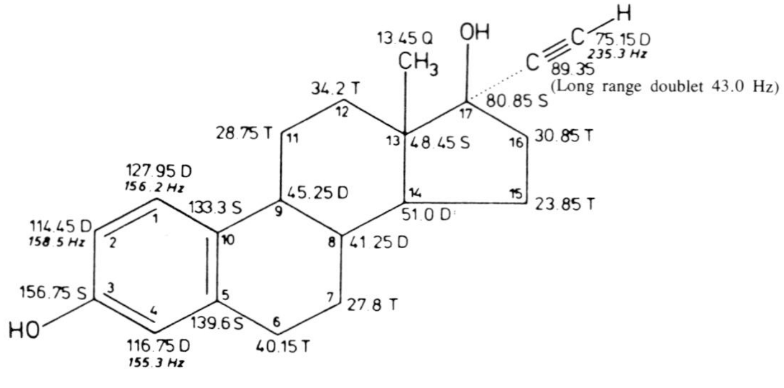

72

Using low-power noise decoupling, the quaternary carbon signals appear sharp in Spectrum C, whereas the carbon atoms bonded to hydrogen give broadened, mostly weaker signals. The assignment, based on the decoupling experiments and comparison of chemical shifts, is summarized in the following table.

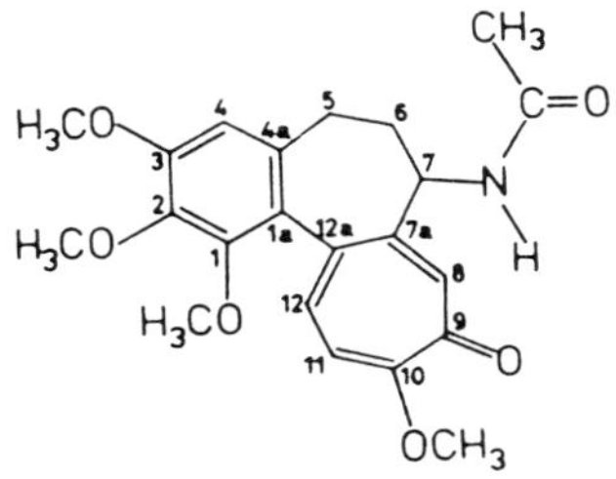

Colchicine

Assignment of Spectrum 72

^{1}H-Spectrum		^{13}C-Spectrum			
ppm	Assignment	ppm Without Direct C-H bond	ppm With Direct C-H bond	Assignment	Assignment Criteria
	—	180.1		C-9	Conjugated carbonyl carbon
	—	170.1		CO (-NHCOCH$_3$	N-acetyl CO
	— —	164.6		C-7a	C-β in α, β-unsaturated Ketones: ca. 150 ppm; alkylation effect: + ca. 5-10 ppm; β effect of the NHCOCH$_3$ group: + ca. 5-10 ppm.
		154.0		C-10	Remains after assigning C-1, 2, and 3
	—	153.2		C-3	Chemical shift similar to that of C-3 in 3,4,5-trimethoxybenzoic acid (153.1 ppm)
	—	151.6		C-1	*o*-alkenyl- + *o*-methoxy effect, thus more strongly shielded than C-10
	—	142.0		C-2	Chemical shift similar to that of C-4 in 3,4,5-trimethoxybenzoic acid; (143.6 ppm) + *p*-alkyl effect (−2 to −3 ppm)

Assignment of Spectrum 72

^{1}H-Spectrum		^{13}C-Spectrum			
ppm	Assignment	ppm Without Direct C-H bond	ppm With Direct C-H bond	Assignment	Assignment Criteria
		137.4		C-12a	Remains after assigning C-1a, 4a, and 7a
7.41	H-11		135.8	C-11	Strong singlet in decoupling experiment M
		134.6		C-4a	α-alkyl, *o*-alkenyl, *p*-methoxy effects
7.70	H-8		130.8	C-8	Strong singlet in decoupling experiment N
		125.9		C-1a	α-alkyl, *o*-alkyl, *o*- and *p*-methoxy effects
6.97	H-12		113.2	C-12	Strong singlet in decoupling experiment L
6.58	H-4		107.7	C-4	Strong singlet in decoupling experiment K
3.68	10-OCH_3		61.3	10-OCH_3	Strong singlet in decoupling experiment F
3.95	OCH_3		61.1	OCH_3	Strong singlet in decoupling experiment H
4.04	OCH_3		56.2	OCH_3	Strong singlet in decoupling experiment I
3.92	OCH_3		56.0	OCH_3	Strong singlet in decoupling experiment G
4.70	H-7		52.6	C-7	Strong singlet in decoupling experiment J
2.62	H-6		36.0	C-6	β effect of the $NHCOCH_3$ groups (decoupling experiment E unclear because of excessive line width at 29.6 and 36.0 ppm)
2.62	H-5		29.6	C-5	Remains after assigning C-6
1.97	-$NHCOCH_3$		22.2	-$NHCOCH_3$	Strong singlet in decoupling experiment D

73

By comparing the ^{13}C chemical shifts measured for LSD with the assigned data for methyl lysergate using the best fit criterion, the following assignment can be made:

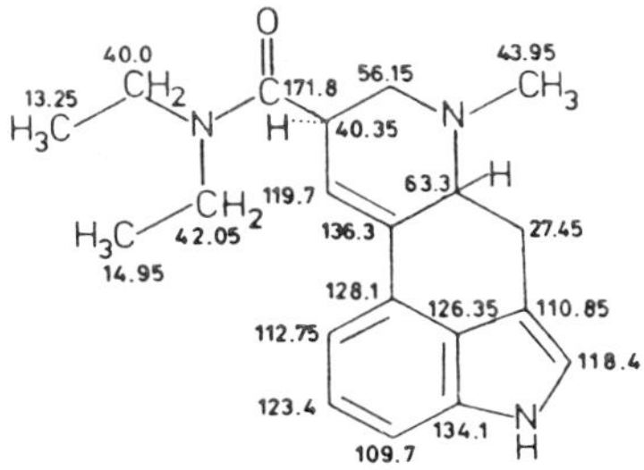

The distinctly reduced intensity of the quaternary carbon signals, owing to the lack of nuclear Overhauser effects, is helpful in the assignment. Because of the partial π character of the amide C-N bond, the two ethyl groups are nonequivalent and four separate resonances appear. Similar to the assignment discussed by G. C. Levy and G. L. Nelson for *N,N*-dialkylamides (*J. Am. Chem. Soc.* **94**, 4897, 1972), the ethyl carbons *cis* to the carbonyl group are identified as the more strongly shielded pair.

74

The following ^{13}C chemical shift assignment has been published for scopolamine (see reference to Wenkert et al, Exercise 74).

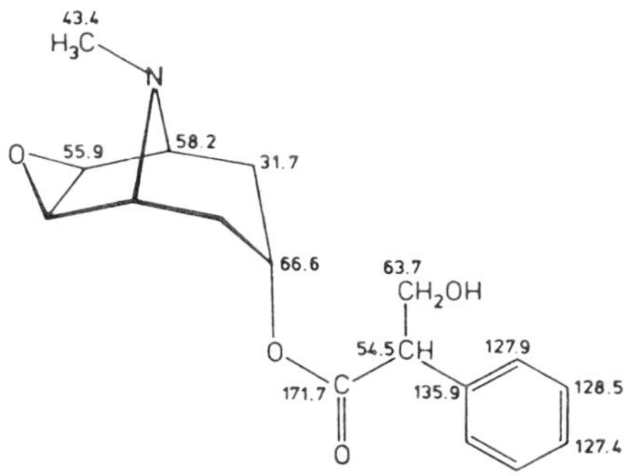

The oxiran ring carbons appear here at 55.9 ppm. The spectrum of scopoline exhibits no signals in this region, so the molecule no longer contains an oxiran ring. On the other hand, scopoline cannot be a tropan-3,6,7-triol $C_8H_{15}O_3N$, which formally would contain an extra water molecule as compared with the given molecular formula $C_8H_{13}O_2N$. Consequently, C-7 and C-3 are bridged by an oxygen atom in scopoline:

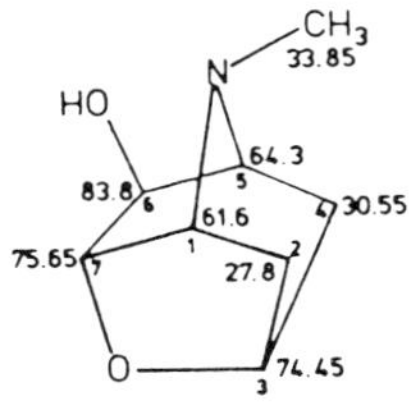

The assignment found is based on the off-resonance multiplicities of the ^{13}C signals and the somewhat stronger shielding of carbon atoms in five-membered rings in comparison with larger rings (compare, for example, J. B. Stothers, *Carbon-13 NMR Spectroscopy*, Academic Press, New York, 1972, pp. 60 and 270).

According to its chemical shift of 33.85 ppm, the N-CH_3 group has the *syn* configuration, which follows from low-temperature measurements on other tropane conformers (see reference to Schneider and Sturm, Exercise 74: *syn*-NCH_3, 32-33 ppm; *anti*=NCH_3, 41-42 ppm).

75

Spectrum 75B of the aglycone exhibits ten signals: four in the sp^2 carbon region and six in the sp^3 carbon region. Because there are no signals with $\delta > 160$ ppm, no carbonyl or carboxyl groups are present. Therefore, the two oxygen atoms belong to alcohol or ether functions. As the analysis of the off-resonance spectrum shows (see table below), only 16 of the 18 hydrogen atoms are bonded to the carbon skeleton. The molecule thus incorporates the oxygen as two OH groups. According to the molecular formula $C_{10}H_{18}O_2$, the compound is also doubly unsaturated ($C_{10}H_{18}O_2$ − 2 OH + 2 H = $C_{10}H_{18}$ = C_nH_{2n-2}) and is not cyclic but open-chained since four ^{13}C resonances appear in the alkene carbon region.

Analysis of the Spectrum of the Aglycone

ppm	Off-Resonance Multiplicity	Assignment
145.10	Doublet	⋛CH
135.15	Singlet	⋛C—
126.05	Doublet	⋛CH
111.85	Triplet	$=CH_2$
75.45	Singlet	⋛C—O⁄
68.90	Triplet	$-CH_2$—O⁄
41.95	Triplet	$-CH_2$—
27.90	Quartet	$-CH_3$
22.40	Triplet	$-CH_2$-
13.65	Quartet	$-CH_3$
	Carbon framework	$C_{10}H_{16}$

A terpene structure can be suspected from the molecular formula $C_{10}H_{18}O_2$, more specifically a linalool derivative, because this terpenol contains a tertiary hydroxyl group like that found in the problem aglycone. The corresponding off-resonance singlet appears at 75.45 ppm for the aglycone (deuteriochloroform) and at 72.7 ppm for linalool (hexadeuterioacetone; compare E. Breitmaier, W. Voelter, *^{13}C NMR Spectroscopy: Methods and Applications*, "Monographs in Modern Chemistry," Vol. 5, H. E. Ebel, ed., Verlag Chemie Gmbh, Weinheim, 2nd ed., 1978, p. 222). In this case, the chemical shift difference is largely a consequence of the different solvents. The chemical shifts of C-4, C-5, C-7, C-8, and C-10 of linalool appear close to the values for the aglycone. On the other hand, no chemical shifts can be found in the aglycone spectrum that resemble the values for C-1, C-2, and C-9. It follows that either C-1 or C-9 is the primary alcohol carbon with $\delta = 68.9$ ppm. The decision in favor of 9-hydroxylinalool results from adding the α chemical shift increment of an OH group (42.9 ppm) in the allylic position to the C-1 ($17.15 + 42.9 = 60.05$ ppm) and C-9 ($25.25 + 42.9 = 68.15$ ppm) methyl chemical shifts of linalool. Isobutene and 2-methylallyl alcohol are appropriate model compounds to calculate the allyl OH increment (compare J. B. Stothers, *Carbon-13 NMR Spectroscopy*, Academic Press, New York, 1972, pp. 71, 82, 188).

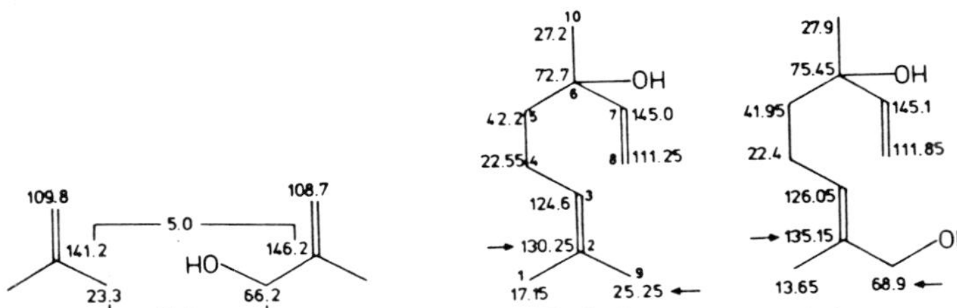

Linalool Aglycone

The change from the aglycone to the glucoside is reflected less through the inductive effect than through steric effects, particularly at the γ positions to the glucosidic oxygen. Since the chemical shifts of the γ carbons with respect to the C-6 OH group are nearly unchanged in the glucoside tetraacetate, whereas the carbon γ to the C-9 OH group is distinctly deshielded (129.9 ppm) in comparison with the aglycone (126.05 ppm), the compound is the C-9 glucoside.

α- and β-D-glucose, their anomeric pentaacetates and methyl glucopyranosides are model compounds for the decision between the α- and β-glycosidic linkages (compare Breitmaier and Voelter, p. 244). The C-1 chemical shifts of methyl α- and β-glucopyranoside tetraacetates can be estimated as follows from the C-1 chemical shifts of the model compounds:

Methyl α-D-glucopyranoside:	99.6 ppm	Methyl β-D-glucopyranoside:	103.4 ppm
α-D-glucopyranose:	92.1 ppm	β-D-glucopyranose:	95.9 ppm
$\delta_{1\text{-}OCH_3} - \delta_{1\text{-}OH}$:	7.5 ppm	$\delta_{1\text{-}OCH_3} - \delta_{1\text{-}OH}$:	7.5 ppm
Pentaacetyl-α-D-glucopyranose:	88.6 ppm	Pentaacetyl-β-D-glucopyranose:	91.3 ppm
Methyl α-D-glucopyranoside tetraacetate:	96.1 ppm	Methyl β-D-glucopyranoside tetraacetate:	98.8 ppm

In ^{13}C NMR Spectrum 75A, a glycosidic carbon resonance is found at 98.85 ppm. This value agrees better with a β-glycosidic linkage. The proposed structure is thus 9-hydroxylinaloyl-β-D-glucopyranoside tetraacetate. The following assignment of Spectrum A is based on a comparison of chemical shifts with β-D-glucopyranose. Interchangeable (i.e., ambiguous) assignments are marked with an X.

Acetyl CO : 170.8, 170.4, 169.55, 169.4
Acetyl CH_3: 20.65 (one signal)

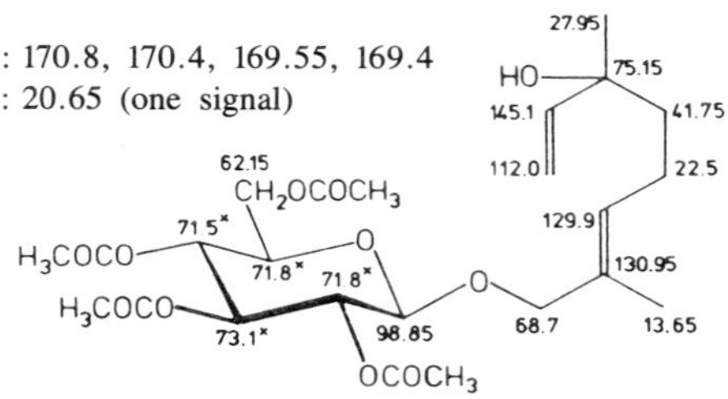

76

The molecular mass agrees with the elementary composition summarized in the molecular formula. To determine the number of double-bond equivalents, one can formally replace the nitrogens in this molecular formula by CH, neglect the oxygen, and thus obtain $C_{14}H_{20}$. This substitute formula lacks ten hydrogens, as compared with $C_{14}H_{30}$. Therefore, the number of double-bond equivalents is five.

In addition to a vinyl group (doublet and triplet at 142.6 and 116.2 ppm), a carbonyl resonance at 179.9 ppm and an azomethine, amidine, or guanidine carbon at 158.6 ppm appear in the sp^2 carbon chemical shift region. Therefore, on the basis of the ^{13}C chemical shifts, only three double bonds are recognizable. The two remaining double-bond equivalents point to a bicyclic compound.

Assuming that the singlets at 179.9 and 158.6 ppm belong to an amide and a guanidyl carbon, a partial formula results containing all the compound's heteroatoms:

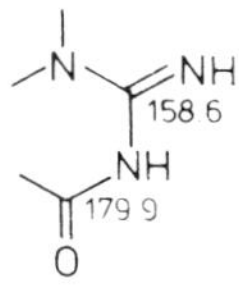

This formula also agrees with two more signals, the singlet at 65.2 ppm and the doublet at 63.2 ppm, which, according to their chemical shifts, are bonded to one of the guanidyl nitrogens. A bond to oxygen can be ruled out because this atom has already been recognized as part of the amide carbonyl group, and the molecule contains only one oxygen:

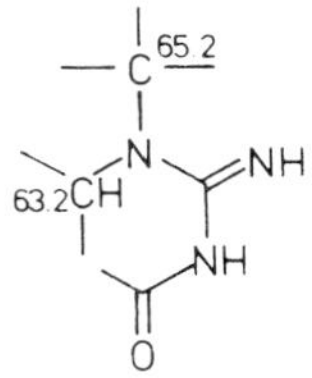

In addition to the vinyl group mentioned above, two nonequivalent methyl groups (quartets at 20.7 and 11.6 ppm) can be recognized as skeletal substituents:

142.6		
$-CH=CH_2$	$-CH_3$	$-CH_3$
116.2	20.7	11.6

The methyl group with the larger ^{13}C chemical shift lies in the β position with respect to nitrogen (β effect: compare Table 7.2, p. 52) and therefore is bonded to the quaternary carbon resonating at 65.2 ppm. The vinyl group is also bonded to this carbon, which explains the nonequivalence of both C-methyl groups (N-methyl groups appear near 40 ppm). Finally, another doublet appears at 42.4 ppm. The corresponding carbon bears the second methyl group and closes either a pyrrolidine ring (Proposal I) with one of the remaining methylene carbons (triplets at 40.8 and 30.0 ppm) or a perhydropyrimidine ring (Proposal II) in the heterobicycle:

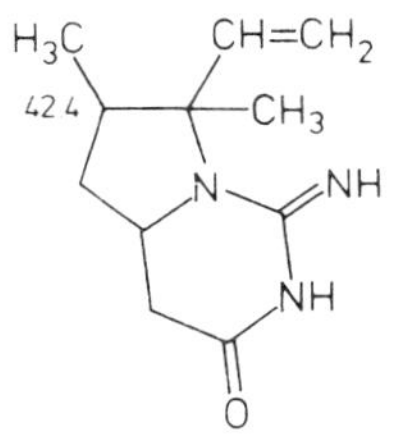

I

oder

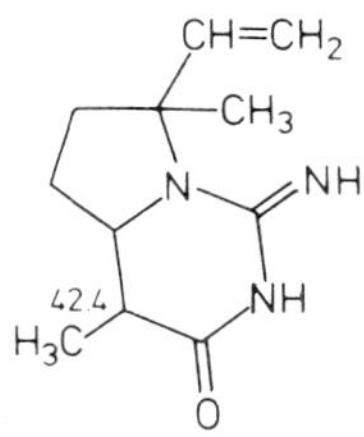

II

Proposal II agrees better with the experimental methylene chemical shift of 30.0 ppm because the deshielding β effect of the ring nitrogen is compensated for by the shielding γ effect of the methyl group in this structure only. Proposal II is also in consonance with the proton resonance findings (compare A. Rabaron et al., *J. Am. Chem. Soc.* **93**, 6270, 1971). The analysis of the ^{13}C NMR spectrum can be summarized as follows:

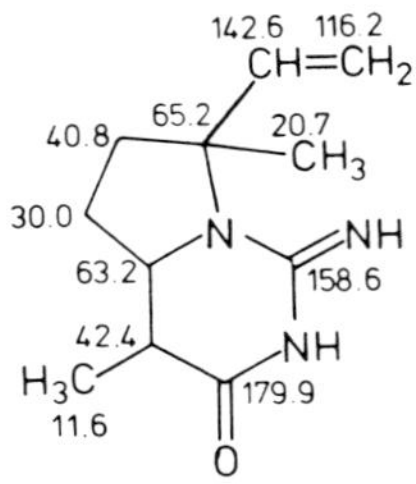

Bibliography

Breitmaier, E., Spohn, K.-H., Berger, S. *Angew. Chem.* **87,** 152 (1975).

Breitmaier, E., Voelter, W. *^{13}C NMR Spectroscopy: Methods and Applications,* "Monographs in Modern Chemistry," Vol. 5, H. E. Ebel, ed., Verlag Chemie GmbH, Weinheim, 2nd ed., 1978.

Clerc, J. T., Pretsch, E., Sternhell, S. *^{13}C-Kernresonanzspektroskopie,* Akademische Verlagsgesellschaft, Frankfurt am Main, 1973.

Dalling, D. K., Grant, D. M. *J. Am. Chem. Soc.* **89,** 6612 (1967).

Dalling, D. K., Grant, D. M. *J. Am. Chem. Soc.* **94,** 5318 (1972).

Dorman, D. E., Jautelat, M., Roberts, J. D., *J. Org. Chem.* **36,** 2757 (1971).

Farrar, T. C., Becker, E. S. *Pulse and Fourier Transform NMR,* Academic Press, New York-London, 1971.

Friebolin, H., et al. *NMR-Spektroskopie, Eine Einführung mit Übungen,* Verlag Chemie GmbH, Weinheim, 1974, Taschentext Bd. 15.

Grant, D. M., Paul, E. G., *J. Am. Chem. Soc.* **86,** 2984, (1964).

Günther, H. *NMR-Spektroskopie, Eine Einführung,* Georg-Thieme-Verlag, Stuttgart, 1973.

Hagaman, E. W. *Org. Magn. Res.* **8,** 389 (1976).

Levy, G. C. *Accounts Chem. Res.* **6,** 161 (1973).

Levy, G. C., Nelson, G. L. *Carbon-13 Nuclear Magnetic Resonance for Organic Chemists,* Wiley-Interscience, New York-London, 1972.

McInnes, A. G., Smith, D. G., Vinning, L. C., Johnson, L. *Chem. Commun.,* 325 (1971).

Noggle, J. H., Schirmer, R. E. *The Nuclear Overhauser Effect: Chemical Applications,* Academic Press, New York-London, 1971.

Oláh, G. A. *Angew, Chem.* **85,** 183 (1973).

Paul, E. G., Grant, D. M. *J. Am. Chem. Soc.* **85,** 1701 (1963).

Schneider, H. J., Price, R., Keller, T. *Angew. Chem.* **83,** 759 (1971).

Stothers, J. B. *Carbon-13 NMR Spectroscopy,* Academic Press, New York-London, 1972.

Suhr, H. *Anwendungen der Kernmagnetischen Resonanz in der Organischen Chemie,* Springer Verlag, Berlin-Heidelberg-New York, 1965.

v. Ammon, R., Fischer, R. D. *Angew. Chem.* **84,** 737 (1972).

Wehrli, F. W., Wirthlin, T. *Interpretation of Carbon-13 Nuclear Magnetic Resonance Spectra,* Heyden & Son Ltd., London, 1976.

Woessner, D. E. *J. Chem. Phys.* **37,** 647 (1962).

Appendices

1. List of Chemical Substances

Page numbers for compounds for which ^{13}C data are listed completely in the text are designated by ordinary numbers. Those for compounds for which spectra are reproduced in the text are designated by numbers in *italics*. Page numbers for compounds for which spectra are reproduced in the Exercises are designated by numbers in *underlined italics.*

2. List of Compounds Discussed in the Exercises

Exercise No.	
1.	2,2,4-Trimethylpentane (isooctane)
2	Butylmercaptan
3	Pentylamine
4	2-Octanol
5	2-Methylbutyl nitrite, 3-methylbutyl nitrite
6	Ethyl cyanoacetate
7	Methyl trifluoroacetate

8 Hydroxyacetylsuccinic acid lactone
9 N,N-Dimethylacetamide
10 N,N-Dimethylformamide dimethyl acetal
11 2-Isopropylmalonic dialdehyde dimethyl acetal
12 Methoxyacetone
13 2-Methyl-2-(3-nitropropyl)-1,3-dioxolane
14 Ethyl acetoacetate
15 Ethyltrimethylsilylcarbodiimide, ethyltrimethylsilylcyanamide
16 *trans*-Polyisoprene (gutta percha), *cis*-Polyisoprene (natural rubber)
17 Phosphoenolpyruvic acid cyclohexylammonium salt
18 *cis*-Ethyl propenyl ether, *trans*-ethyl propenyl ether
19 3-Acetylamino-2-methylacrolein
20 Diketene
21 Dicyclopentadiene
22 1-Acetylaminoadamantane
23 *cis*-1,2-Dimethylcyclohexane
24 *cis*-4-*t*-Butylcyclohexanol, *trans*-4-*t*-butylcyclohexanol
25 2-Cyclohexenone
26 Cyclooctatetraene dimer, Diels-Alder adduct with tetracyanoethylene
27 *p*-Cyclophane
28 Polystyrene, atactic and isotactic
29 Mesitylacetylene
30 Benzocyclopropene
31 Ethylphenylmalonic acid diethyl ester
32 Nitrobenzene
33 Picric acid
34 2,4,6-Trinitrophenylacetone
35 2,4,6-Trinitrophenylacetone potassium Meisenheimer salt
36 Carvacrol, thymol
37 2,4-Dimethylaniline
38 1,3-Bis(2-aminophenylthio)propane
39 N-Phenylthiourea
40 3-(N-Phenylcarbamoyl)amino-2-methylacrolein
41 *p*-Ethoxyacetanilide (phenacetin)
42 4-*t*-Butyl-2-nitroacetanilide
43 Vanillin (4-hydroxy-3-methoxybenzaldehyde)
44 4,4′-Dimethoxybenzil
45 Ethyl 3-dimethylamino-3-phenylpropanoate, ethyl cinnamate
46 3-(4-Phenylazophenyl) propanoic acid
47 3,6-Dihydroxy-1,8-diphenyl-*trans*-4-octen-1,7-diine,
3,6-Dioxo-1,8-diphenyl-*trans*-4-octen-1,7-diine
48 1,8-Diiodonaphthalene
49 1,2,3,4-Tetrahydronaphthalene-1-one,
6-methoxy-1,2,3,4-tetrahydronaphthalene-1-one

50 Bis-(*o*-acetylaminobenzaldehyde)-*o*-phenylenediimine, 1,4,8,11-tetraazatribenzo[b.f.j]tetradeca-6,11-dien-2,3-dione
51 2-Acetyl-1-methylpyrrole
52 Furan-2-carboxylic acid
53 *cis*-2,5-Diethoxytetrahydrofuran, *trans*-2,5-diethoxytetrahydrofuran
54 1-(2-Thienylcarbonyl)-3,3,3-trifluoroacetone
55 3-Methyl-5-oxo-1-phenyl-4,5-dihydropyrazole
56 5-(Dimethylaminomethyleneimino)-4-formyl-3-methyl-1-phenyl-pyrazole
57 4,4-Dimethyl-5-oxo-2-trifluoromethyl-Δ^2-oxazoline
58 2-*p*-Methoxybenzoyl-2-*p*-methoxyphenylbenzothiazoline
59 Pyridine, 2,2′-bipyridyl
60 4-Methylpyridine, 4-methylpyridine N-oxide
61 Pyridine-2,5-dicarboxylic acid
62 3,4-Diaminopyridine
63 2,4,6-Triaminopyrimidine, 4,5,6-triaminopyrimidine
64 5-Ethyl-1-phenyl-1,2-dihydropyrimidine-2-thione
65 4-Oxo-2-thioxo-1,2,3,4,7,8,9,10-octahydro-6H-cyclohepta[5,6]-pyrido[2,3-d]pyrimidine
66 3,5,6-Tricyano-4,7-dithiino[c]isothiazole
67 Alanine, lysine
68 Isomenthol, menthol
69 α-D-xylopyranose, β-D-xylopyranose
70 Adenosine, adenosine 5′-monophosphate
71 17α-Ethynyl-3,17β-estradiol
72 Colchicine
73 Lysergic acid diethylamide
74 Scopoline
75 Betulalbuside tetraacetate, 9-hydroxylinalool
76 Arenaine

Subject Index